2024 IEEE 24th Topical Meeting on Silicon Monolithic Integrated Circuits in RF Systems (SiRF 2024)

San Antonio, Texas, USA
21-24 January 2024

IEEE Catalog Number:	CFP24SMI-POD
ISBN:	979-8-3503-4331-1

Copyright © 2024 by the Institute of Electrical and Electronics Engineers, Inc.
All Rights Reserved

Copyright and Reprint Permissions: Abstracting is permitted with credit to the source. Libraries are permitted to photocopy beyond the limit of U.S. copyright law for private use of patrons those articles in this volume that carry a code at the bottom of the first page, provided the per-copy fee indicated in the code is paid through Copyright Clearance Center, 222 Rosewood Drive, Danvers, MA 01923.

For other copying, reprint or republication permission, write to IEEE Copyrights Manager, IEEE Service Center, 445 Hoes Lane, Piscataway, NJ 08854. All rights reserved.

*** This is a print representation of what appears in the IEEE Digital Library. Some format issues inherent in the e-media version may also appear in this print version.

IEEE Catalog Number: CFP24SMI-POD
ISBN (Print-On-Demand):   979-8-3503-4331-1
ISBN (Online):   979-8-3503-4330-4
ISSN: 2475-2983

Additional Copies of This Publication Are Available From:

Curran Associates, Inc
57 Morehouse Lane
Red Hook, NY 12571 USA
Phone: (845) 758-0400
Fax: (845) 758-2633
E-mail: curran@proceedings.com
Web: www.proceedings.com

TABLE OF CONTENTS

Session Mo1C: Power Amplifiers

Mo1C-1
Toward High-Power and Multi-Way Silicon-Based mmWave Doherty Power Amplifiers .... (NA)
Taiyun Chi

Mo1C-2
A Ku-Band Power Amplifier in 22nm FDSOI 1
Alexander Haag, Ahmet Çağrı Ulusoy

Mo1C-3
A W-Band Amplifier in FinFET Technology 5
Yuen-Sum Ng, Yunshan Wang, Huei Wang

Mo1C-4
A D-Band 28nm CMOS-Bulk Power Amplifier with 12.8dBm Output Power and 31.3GHz 3dB Bandwidth 9
Pascal Stadler, Hakan Papurcu, Justin Romstadt, Nils Pohl

Session Mo2C: Phase Shifters and Tunable Components

Mo2C-1
The Role of Varactor, the Nonlinear Semiconductor, for Next Generation of Intelligent and Reconfigurable Radio Nodes 13
Najme Ebrahimi

Mo2C-2
A Compact, High Tuning Accuracy and Enhanced Linearity 37–43GHz Digitally-Controlled Vector Sum Phase Shifter 17
Mehran Hazer Sahlabadi, Hang Yu, Jingjing Xia, Slim Boumaiza

Mo2C-3
A Novel High Q-Factor Structure of Digitally Tunable Capacitor for High RF Power Handling Applications 21
Wonwoo Seo, Sunghyuk Kim, Byunghun Ko, Jehwan Lee, Yongbae Choi, Taejoo Sim, Junghyun Kim

Mo2C-4
Compact D-Band Passive Phase Shifters with Fine and Coarse Control Steps in BiCMOS-55nm 25
Lorenzo Piotto, Guglielmo De Filippi, Mahmoud M. Pirbazari, Andrea Mazzanti

Session Mo3C: Millimeter-Wave Signal Generation

Mo3C-1
A 4.5dBm SiGe Doubler-Amplifier Chain Covering the Entire D-Band 29
Matthias Möck, İbrahim Kağan Aksoyak, Ahmet Çağrı Ulusoy

Mo3C-2
A SiGe-Based Quadrature D-Band Up-Converter with High Output Power 33
İbrahim Kağan Aksoyak, Matthias Möck, Ahmet Çağrı Ulusoy

Mo3C-3
A 300GHz ×9 Multiplier Chain with 9.6dBm Output Power in 0.13-μm SiGe Technology 37
Arjith Chandra Prabhu, Janusz Grzyb, Philipp Hillger, Thomas Bücher, Holger Rücker, Ullrich Pfeiffer

Mo3C-4
230GHz Signal Generator for High-Bandwidth Data Links in 130nm SiGe BiCMOS 41
Christian Hoyer, Luca Steinweg, Florian Protze, Franz Alwin Dürrwald, Tilo Meister, Frank Ellinger

Session Mo4C: Devices, Technology, and Integration

Mo4C-1
2.5D/3D Heterogeneous Integration .... (NA)
Ankush Mohan, John Carlson, Tina Seeholzer, Clayton Tu, Avantika Sodhi, Hasan Sharifi

Mo4C-2
The Chip-Level In-Plane Stress Distribution Over BiCMOS Wafers 45
Zhibo Cao, Thomas Voss, Matthias Wietstruck, Corrado Carta, Mehmet Kaynak

Mo4C-3
Characterization of Silicon Substrates for Sub-THz Electronics, Benefit of the Beatty Resonator Test-Structure 48
Luca Lucci, Olivier Valorge, Alexandre Oliviera, Herve Boutry, Christophe Dubarry, Fred Gaillard, Blandine Duriez

Mo4C-4
f_T Extraction of HEMT Transistors at mm-Waves through EM-Simulated De-Embedding Devices 52
Mohammed Medbouhi, Jose Lugo-Alvarez, Philippe Ferrari, Erwan Morvan

Session Tu1B: Amplifier Design

Tu1B-1
CMOS LNA and VGA for 5G NR Using Gain-Linearity-Boosting and Body Floating Techniques 56
Jin-Fa Chang, Yo-Sheng Lin

Tu1B-2
Parametric-Oscillation-Free Efficient SiGe:C Power Amplifier Design for Ku-/Ka-Band SATCOM 60
Tsung-Ching Tsai, Václav Valenta, Ahmet Çağrı Ulusoy

Tu1B-3
A 94GHz Bandwidth Transimpedance Amplifier in 55nm SiGe BiCMOS for High Speed Optical Receivers 63
Lachlan Cuskelly, Christopher Falt, Peter Schvan

Tu1B-4
A 200–325GHz Gain-Boosted J-Band Low-Noise Amplifier in a 130nm SiGe BiCMOS Technology 67
Manuel Koch, Sascha Breun, Robert Weigel

Session Tu1C: Radar and Sensor Circuits and Architectures

Tu1C-1
Advances in mmWave Radar Architectures .... (NA)
Brian Ginsburg

Tu1C-2
Analysis of a SiGe BiCMOS Detector for a Broadband mmW-Integrated EPR Spectrometer 71
Selina Eckel, Ahmet Çağrı Ulusoy

Tu1C-3
27Gb/s PRBS Generator with In-Operation Programmable Taps for PMCW Radar 75
Florian Probst, Andre Engelmann, Robert Weigel

Session Tu3C: Voltage-Controlled Oscillators

Tu3C-1
A 23–30GHz Low-Phase-Noise 5-Bit Voltage-Controlled Oscillator in 90-nm CMOS Process 79
Po-Yuan Chen, Jun-Liang Chen, Hong-Yeh Chang

Tu3C-2
Low Phase Noise 104GHz Oscillator Using Self-Aligned On-Chip Voltage-Tunable Spherical Dielectric Resonator in 130-nm SiGe BiCMOS 83
Yu Zhu, Georg Sterzl, Jan Hesselbarth, Tilo Meister, Frank Ellinger

Tu3C-3
A 34GHz CMOS VCO with Transformer Tail-Node Filter and TSPC Frequency Divider in 22nm FDSOI 87
Andre Engelmann, Florian Probst, Philip Hetterle, Robert Weigel

Tu3C-4
D-Band VCO with Uniformly Low Phase Noise versus Frequency and Temperature 91
Isabel Kraus, Herbert Knapp, Nils Pohl

Tu3C-5
Voltage-Controlled-Oscillator Using an 8-Shaped Transformer-Coupled Transmission Line 95
Sheng-Lyang Jang, Yi-Ping Hsieh, Wen-Cheng Lai

Session IF1: Interactive Forum Poster Session

IF01-2
Single-Voltage-Supply pHEMT/mHEMT 2.4 and 5.8GHz LNAs Using Power Constrained Design 98
Chinchun Meng, Chung-Yo Lin, Guo-Wei Huang

IF01-23
A Comprehensive Approach to Extracting Coupling Matrix from Filtenna Measurements 102
Sara Javadi, Behrooz Rezaee, Manfred Stadler, Michael Leitner, Wolfgang Bösch

IF01-25
Design of a Six-Stage W-Band Low-Noise Amplifier Using a 90-nm CMOS Technology 106
Rou-Yin Huang, Yu-Chia Su, Hong-Yeh Chang

IF01-30
A 1.28mW K-Band Modified Gilbert-Cell Mixer Design in 22nm FDSOI CMOS 110
Adilet Dossanov, Vadim Issakov

2024 IEEE Radio and Wireless Week

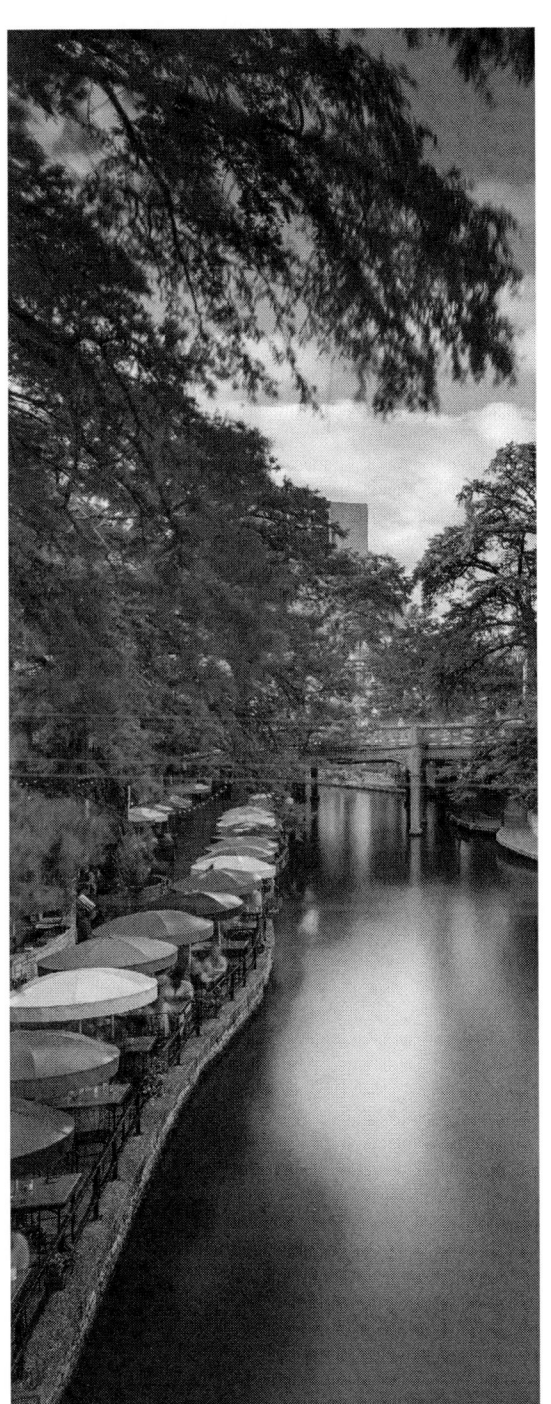

2024 IEEE 24th Topical Meeting on Silicon Monolithic Integrated Circuits in RF Systems

21–24 January 2024
San Antonio, Texas, USA

www.radiowirelessweek.org

RWW 2024 sponsored by:

Image Credit: Shutterstock.com/Sean Pavone

Welcome Messages

GREETINGS FROM THE GENERAL CHAIR OF RADIO & WIRELESS WEEK 2024

General Chair
Changzhi Li

Welcome to the 19th IEEE Radio & Wireless Week (RWW) in San Antonio, TX, USA! The city is located in the heart of Texas and is known for an eclectic mix of Mexican, German, French, and Old West cultures that combine to create an "only in San Antonio" experience. The conference venue Grand Hyatt San Antonio River Walk celebrates the history and charm of the Alamo City with a welcoming spirit and an elegant style. The hotel is located right on the River Walk in downtown San Antonio, steps from top attractions like the Alamodome, AT&T Center and The Alamo.

Over the years, RWW has become an icon of "January family reunion" for the wireless and microwave community from all over the world to discuss the latest trends based on five co-located conferences: the IEEE Topical Conference on Power Amplifiers for Wireless and Radio Applications (PAWR), the Topical Meeting on Silicon Monolithic Integrated Circuits in RF Systems (SiRF), IEEE Topical Conference on Wireless Sensors and Sensor Networks (WiSNet), IEEE Space Hardware and Radio Conference (SHaRC), and Radio and Wireless Symposium (RWS). Each topic-conference has its focus days: PAWR on Monday, SiRF on Monday and Tuesday, WiSNet and SHaRC on Wednesday, all of which take place along with RWS sessions Monday through Wednesday. This core technical program is strengthened by several workshops, technical panels, and short courses to address the latest trends in radio and wireless.

The 2024 class of MTT-S Distinguished Microwave Lecturers will present their talks during a dedicated track on Monday morning, which is a perfect opportunity to get a comprehensive overview on some topic areas. Professional panel sessions of female and young professional role models will be again organized by the Women in Microwave and Young Professional teams on Sunday night. Two world renowned keynote speakers will give their insights during the plenary session on Tuesday morning, where the winners of the student paper competition will also be awarded. Attendees are also welcome to a PAWR/ARFTG joint panel and a Space Night on Monday and Tuesday nights, respectively.

Besides four exciting technical workshops on Sunday afternoon, an entrepreneurial workshop "From Lab to Market: Empowering Researchers for Entrepreneurial Success" will be open to all conference attendees on Monday afternoon. For the first time, MTT-S is organizing an information session to "Demystify the IEEE Fellow Nomination" on Monday afternoon. RWW is also very proud to announce the collaboration with the editorial offices of the IEEE Transactions on Microwave Theory and Techniques (T-MTT), IEEE Microwave and Wireless Technology Letters (MWTL), and IEEE Journal of Microwaves (JMW), which offer journal/letter authors of the past year opportunities to present at RWW and enjoy professional interaction with conference attendees.

The ARFTG is again co-locating their conference with RWW. Besides the ARFTG conference with technical papers and short courses, a joint ARFTG & RWW exhibition will be hosted to showcase the latest products and solutions from industry. The exhibition floor will also highlight an industry reception open to all conference attendees, as well as a demo track. ARFTG and PAWR are also hosting a joint panel session on Monday night. Another highlight is the co-located Internet of Things (IoT) summit, one of the very successful collaborations of RWW with other communities.

At this moment, we would like to warmly welcome you and thank you for coming to RWW 2024. We wish you enjoy the week in San Antonio!

RWW2024 General Chair,
Changzhi Li, Texas Tech University

Technical Program Chair's Welcome Message

TPC Chair
Václav Valenta

Dear colleagues, RWW and ARFTG participants, it is a great pleasure to welcome you on the behalf of the technical program committee to another edition of the famous radio wireless week! Our committee put together a very attractive program to ensure fruitful exchange and establish collaboration links between scientific community, industry and academia, young professionals, and radio enthusiasts. More than 180 papers from academia and industry from more than 20 countries worldwide were thoroughly reviewed to assure high standards of scientific publishing and over 30 technical sessions are open this year.

Given the success from past years, specific panel sessions, workshops or hands-on activities and summits are held again in addition to the focused technical sessions, bringing the main players together to share the latest results from wide spectrum of RF fields in an interactive way.

Beside the technical program, please check the timing for social events, networking and light receptions that are the unique opportunities for technical exchange, dissemination of your research outcomes and promotion of radio-wireless technologies and techniques worldwide!

RWW is also an important platform for the technical and professional development of young professionals and students. Based on the initial evaluation by the TPC, a group of finalists were selected to participate in the student paper competition, which will feature both elevator pitch presentations and interactive forum discussions. Don't miss out on the opportunity to meet our bright young researchers! The best student papers will be recognized during the plenary session.

Let me finish this welcome message with the famous quote of Edward Everett Hale, suitable not only to RWW newcomers – "Coming together is a beginning, keeping together is progress, working together is success!". Enjoy your time at RWW and in San Antonio!

RWW2024 Technical Program Chair,
Václav Valenta, European Space Agency

Welcome Messages

RWW 2024 STEERING COMMITTEE

General Co-Chair
Holger Maune

Finance Chair
Roberto Gomez-Garcia

General Chair:
Changzhi Li, Texas Tech University

General Co-Chair:
Holger Maune, Otto-von-Guericke-Universität Magdeburg

Technical Program Chair:
Václav Valenta, European Space Agency

Finance Chair:
Roberto Gomez-Garcia, University of Alcala

Topical Conferences

PAWR Co-Chairs:
Vittorio Camarchia, Politecnico di Torino
John Dooley, Maynooth University

WiSNet Co-Chairs:
Paolo Mezzanotte, University of Perugia
Fabian Lurz, Otto-von-Guericke-Universität Magdeburg

SHaRC Co-Chairs:
Marie Piasecki, NASA Glenn Research Center
Jan Budroweit, German Aerospace Center

SiRF Chair:
Robert Schmid, Johns Hopkins Applied Physics Lab

Distinguished Microwave Lecturers Chair:
Markus Gardill, Brandenburg University of Technology

Workshops Chairs:
Jan Budroweit, German Space Agency
Pushkar Bajirao Kulkarni, Qualcomm

Technical Lectures:
Juan A. Becerra, Universidad de Sevilla

IoT Summit Liaison:
Charlie Jackson, Northrop Grumman
Jasmin Grosinger, Graz University of Technology

Women in Engineering Chair:
Jasmin Grossinger, Graz University of Technology

Student Paper Contest Co-Chairs:
Ken Kolodziej, MIT Lincoln Laboratory
Ifana Mahbub, UT Dallas

Student Initiative Chair
Michael Chung-Tse Wu, Rutgers University

University Demo Chair:
Mario Pauli, Karlsruhe Institute of Technology

Young Professionals Chair:
Davi Rodrigues, Texas Tech University

Publications Chair:
Markus Gardill, Brandenburg University of Technology
Glauco Fontgalland, Universidade Federal de Campina Grande

Publicity Co-Chairs:
Venkata Vanukuru, GlobalFoundries
Eduardo Rojas-Nastrucci, Embry-Riddle Aeronautical University

Microwave Magazine Special Issue Editor:
Chia-Chan Chang, National Chung-Cheng University

MTT Transactions Mini Special Issue Editors:
Václav Valenta, European Space Agency

Exhibition/Sponsorships Chair:
Elsie Vega, IEEE MCE
Susie Horn, SMH Consulting

RWW Executive Committee Chair:
Robert Caverly, Villanova University

Conference Management:
Elsie Vega, IEEE MCE
Cassandra Carollo, IEEE MCE

Visa Letters:
Cassandra Carollo, IEEE MCE

Webmasters:
Min Hua, Raysilica
Joel Arzola, Raytheon Technologies

At Large (Advisors):
Alexander Koelpin, Hamburg University of Technology
Kevin Chuang, Analog Devices
Nuno Borges Carvalho, Universidade de Aveiro
Rashaunda Henderson, University of Texas at Dallas

Conference Information

REGISTRATION HOURS:

Registration will be open during the following times in the Texas Ballroom Foyer:

- Sunday, 21 January 2024 7:00AM - 6:00PM
- Monday, 22 January 2024 7:00AM – 6:00PM
- Tuesday, 23 January 2024 7:00AM - 6:00PM
- Wednesday, 24 January 2024 7:00AM - 12:00PM

EXHIBIT HOURS:

The joint RWW/ARFTG Exhibition area will be open during the following times:

- Monday, 22 January 2024 1:00PM – 7:00PM
- Tuesday, 23 January 2024 9:00AM – 5:00PM

Please refer to the conference website at http://www.radiowirelessweek.org/exhibits for the latest information and details on how to become a sponsor and exhibit at RWW.

SOCIAL EVENTS, NETWORKING, AND LITE RECEPTIONS:

- Sunday 21 January 2024 before the WiM event in Bowie
- Joint RWW/ARFTG Welcome Reception Monday, 22 January 2024 5:30PM – 6:30PM Location: Exhibit Hall – Texas Ballroom
- Tuesday 23 January 2024 at 5:30 PM before the MTT-S Space Night event in the Texas B

EXHIBITORS & SPONSORS

Gold Sponsors

Exhibitors

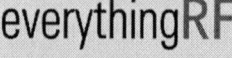

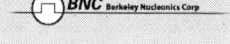

Delegate Bag Sponsor

Media Partner

RWW Topical Conferences

RADIO AND WIRELESS SYMPOSIUM

RWS2024 Chair
Changzhi Li, Texas Tech University

RWS2024 Co-Chair
Holger Maune, Otto-von-Guericke-Universität Magdeburg

RWS2024 Technical Program Committee

High-speed and Broadband Wireless Technologies:
Upkar Dhaliwal, Jennifer Kitchen, Masaaki Kojima, Jing Wang, Muh-Dey Wei, Dietmar Kissinger, Kevin Chuang

Emerging Wireless Technologies & Novel Engineered Materials:
Hyun Kyu Chung, Alessandro Cidronali, Ahmad Hoorfar, Sangkil Kim, Syed Abdullah Nauroze, Spyridon Pavlidis, Junyu Shen, Hjalti Sigmarsson

Wireless System Architecture and Propagation Channel Modeling:
Juan Antonio Becerra, Ugo Dias, Aly Fathy, Paulo Ferreira, Maria J. Madero-Ayora, Chenming Zhou, Pravin Premakanthan

Wireless Digital Signal Processing and Artificial Intelligence:
Nuno Carvalho, Markus Gardill, Rui Ma, Eiji Okamoto, Arnaldo Oliveira, Ken Kolodziej, Pushkar Kulkarni

Applications to Bio-Medical, Environmental, and Internet of Things:
Chia-Chan Chang, Robert Caverly, Syed Islam, Mohammad-Reza Tofighi, Chau Yuen, Changzhan Gu, Daniel Rodriguez, Jenshan Lin

Antenna Technologies, MIMO and Multi-Antenna Communications:
Wasif Khan, Dariush Mirshekar, Jiang Zhu, You Zou, Rashaunda Henderson, Jeremy Muldavin, Edward Niehenke

Passive Components & Packaging:
Roberto Gomez-Garcia, T.-S. Jason Horng, Dimitra Psychogiou, Yu-Chen Wu, Li Yang, Jong Gwan Yook, Bayaner Arigong, Sai-Wa Wong

MM-Wave to THz Systems & Applications:
Shanthi Bhagavatheeswaran, Yi-Jan (Emery) Chen, David Delrio, Nathalie Deltimple, Glauco Fontgalland, Minoru Fujishima, Renato Negra, Hiroshi Okazaki, Sergio Pacheco, Xin Wang, Xinwei Wang, Yu Ye

POWER AMPLIFIERS FOR RADIO AND WIRELESS APPLICATIONS (PAWR)

Power amplifiers are often the most critical component of RF/microwave communications systems and consequently the focus of intense research to achieve increased linearity and power efficiency. New forms of power amplification are being developed to meet the needs of the wireless communication equipment industry and the world's demand for greater information transmission. PAWR2024 will feature innovative work in (but not limited to) the following areas of RF/microwave power amplifier technology:

- High Power/Wideband Active Devices
- Power Amplifiers for Mobile, Avionics and Space
- Modeling and Characterization
- Advanced Circuit Design and Topologies
- Green Power Amplifier Technology
- Integration Technology
- Packaging and Reliability
- Linearization and Efficiency Enhancement Techniques
- Applications, Novel Architectures and System Analysis

PAWR2024 Chair
Vittorio Camarchia, Politecnico di Torino

PAWR2024 Co-Chair
John Dooley, Maynooth University

PAWR2024 Technical Program Committee

Modeling and Characterization:
Ehsan Azad, Filipe Barradas, Vittorio Camarchia, Stephen Maas, Jose Pedro, Zoya Popovic, Patrick Roblin, David Runton, Kefei Wu

Advanced Circuit Design and Topologies:
Paolo Colantonio, Nathalie Deltimple, Paolo de Falco, Jose A. Garcia, William Hallberg, Wolfgang Heinrich, Bumman Kim, Chao Lu, Anna Piacibello, Francesc Purroy, Frederick Raab

Packaging and Reliability:
Florinel Balteanu, Robert Caverly, Murat Eron, Ming Ji, Chang-Ho Lee, Don Lie

Linearization and Efficiency Enhancement Techniques: Taylor Barton, Juan A. Becerra, Wenhua Chen, Kevin Chuang, Armando Cova, Christian Fager, Pere Gilabert, Allen Katz, Morten Olavsbråten, Anding Zhu

WIRELESS SENSORS AND SENSOR NETWORKS (WiSNet)

Wireless sensors and wireless sensor networks are crucial components for manufacturing, structural health, security monitoring, environmental monitoring, smart agriculture, transportation, commercial applications, localization, tracking systems and other important and emerging applications. WiSNet2024 is intended to stimulate discussion and foster innovation on these components and applications.

WiSNet2024 Chair
Paolo Mezzanotte, University of Perugia

WiSNet2024 Co-Chair
Fabian Lurz, Otto-von-Guericke-Universität Magdeburg

WiSNet2024 Technical Program Committee

Wireless Sensors for IoT Communication and Applications:
Georg Fischer, Tuami Lasri, Federico Alimenti, Reinhard Feger, Davi Valerio de Queiroz Rodrigues

Wireless Sensors for Radar, Positioning, Tracking, and Imaging:
Alexander Koelpin, Paolo Mezzanotte, Changzhi Li, Zahir Alsulaimawi, Arne Jacob, Mario Pauli, Hendrik Rogier, Valentina Palazzi, Spyridon Daskalakis

Wireless Sensors Circuits& System Technologies:
Alessandra Costanzo, Diego Masotti, Wang Wang, J-C Chiao, Serioja Tatu, Fabian Lurz, Guoan Wang

WSN Hardware-Software CoDesign:
Amr Fahim, Manos Tentzeris, Jennifer Williams, Kamal Samanta, Nils Pohl, Emanuele Cardillo

Innovations in Wireless Sensor Networks:
Marco Dionigi, Rahul Khanna, Luciano Tarricone, Maurizio Bozzi, Xianming Qing, Kai-Ten Feng, Xuyu Wang

RWW Topical Conferences

SPACE HARDWARE AND RADIO CONFERENCE (SHaRC)

The IEEE Space Hardware and Radio Conference (IEEE SHaRC) addresses new concepts, novel implementations, as well as emerging applications for space-based hardware for communications, earth observation, and other novel disruptive services. To meet recent needs, there has been a renaissance of interest and investment in space- and suborbital- based systems especially for high-data-rate communications networks. These new global satellite networks are disruptive, rely on new system and subsystem design paradigms, and are an enabler for many novel applications. The IEEE Space Hardware and Radio Conference provides a forum for discussions on this new frontier.

SHARC2024 Chair
Marie Piasecki, NASA Glenn Research Center

SHARC2024 Co-Chair
Jan Budroweit, German Aerospace Center

SHARC2024 Technical Program Committee

Systems, Hardware, and Electronics for Space:
Thomas Ussmueller, Nuno Carvalho, Jasmin Grosinger, Ramesh Gupta, James McSpadden, Steven Reising, Steven Rosenau, Rick Sturdivant, Vaclav Valenta, Robert Weigel, Markus Gardill, Federico Clazzer

Mission Concepts, Operations, Regulation, and Standardization:
Jan Budroweit, Goutam Chattopadhyay, Rudy Emrick, Dale Force, Charles Jackson, Holger Maune, Thomas Royster, Klaus Schilling, Zizung Yoon, Sachidananda Babu, Dustin Schroeder, Marwan Younis

SILICON MONOLITHIC INTEGRATED CIRCUITS IN RF SYSTEMS (SIRF)

SiRF2024 will mark the 24th topical meeting on SiRF, with a renewed emphasis on promoting a dialogue between IC designers and researchers promoting non-standard technologies, exploiting the maturity of Silicon processes, but addressing the challenges of tomorrow. SiRF2024 will chronicle recent advances in our dynamic field, and provide the platform for developing new ideas, and candid exchange. Invited speakers will stimulate our discussions, with an emphasis on emerging technologies.

SIRF2024 Chair
Robert Schmid, Johns Hopkins APL

SIRF2024 TPC Chair
Mehmet Kaynak, Texas Instruments

SIRF2024 TPC Co-Chair
Ickhyun Song, Hanyang University

SIRF2024 Publicity Chair
Austin Chen, Peraso, Inc.

SIRF2024 Executive Committee
Yi-Jan Emery Chen, Julio Costa, Vadim Issakov, Mehmet Kaynak, Chien-Nan Kuo, Donald Lie, Venkata Koushik Malladi, Monte Miller, Sergio Pacheco, Nils Pohl, Hasan Sharifi, Ahmet Cagri Ulusoy, Václav Valenta, Roee Ben-Yishay, Saeed Zeinolabedinzadeh

SIRF2024 Technical Program Committee

RF, Millimeter-wave and THz Integrated Circuit Front Ends:
Amit Jha, Michael Oakley, Ickhyun Song, Cagri Ulusoy, Robert Schmid, Roee Ben-Yishay, Rahul Kodkani, Austin Chen, Christopher Coen

Wireline Communication Circuits and Silicon-Photonics Integrated Circuits:
Saeed Zeinolabedinzadeh, Juergen Hasch, Vadim Issakov, Aleksey Dyskin, Ankur Guha Roy

High Speed Data Converters & Mixed Signal Circuits:
Wei-Min (Lance) Kuo, Hsieh-Hung Hsieh, Monte Miller, Chien-Nan Kuo, Arindam Sanyal

Devices, Materials, Modeling, and Measurement:
Mehmet Kaynak, Ming-Ta Yang, Katsuyoshi Washio, Julio Costa, Jean-Pierre Raskin, Pierre Blondy, Venkata Malladi, Vikas Shilimkar, Florian Herrault, Xun Gong

IEEE INTERNET OF THINGS (IoT) SUMMIT

The Seventh IEEE Internet of Things (IoT) Vertical and Topical Summit at Radio & Wireless Week (RWW) 2024 is devoted to biomedical IoT applications. Technologies for sensing, signal processing and computing, data storage, and communications are important ingredients in architecting biomedical solutions that are important to human health and human well-being. There are at least five aspects to biomedical IoT applications that are important for the general population. They include the following:

- The practice of medicine
- Public health
- Personal lifestyle
- Environmental exposure
- Biomedical research

Many of the advancements in the field have significantly benefited from the pace of innovation in the capabilities and performance improvements in electronics and the exploitation of methods and techniques for measuring and generating electromagnetic signals across the spectrum - from light to microwave frequencies. They have also benefited from the ability to create new materials, miniaturize electronic components, embed sophisticated signal processing and computation in devices, and lastly, harness new sensing and actuation methods, such as the uses of microelectromechanical systems and microfluidic devices, and new data interpretation techniques, such as the breakthroughs in artificial intelligence (AI) and machine learning (ML) algorithms. The applications range from personal wearables, widely available as consumer goods, to specialized medical instrumentation focused on specific diseases or conditions. They cover the detection of pathogens in our environment, the direct monitoring and management of chronic disease symptoms, and ways of dealing with human impairments through devices that restore or enhance body functions. The science and engineering involved are multidisciplinary and require collaboration between technical and professional disciplines and communities of practice.

Event Overview • Sunday, 21 January 2024

Room: Crockett AB	Room: Crockett CD	Room: Republic A	Room: Republic BC	Room: Seguin AB	Room: Bowie

8:00

ARFTG/NIST Short Course (Day 1)

9:40

10:10

11:50

13:30

Workshop
RF and Millimeter-Wave Communication Systems: Design and Analysis

Workshop
Power Amplifier Design and Linearization Techniques

Workshop
Ground Stations and Advanced Ground Station Networks

15:10

15:40

17:30

19:00

Women in Microwaves

20:30

Young Professionals Panel

21:30

Event Overview • Monday, 22 January 2024

Room: Texas A	Room: Texas B	Room: Texas C	Room: Crockett AB	Room: Republic	Room: Seguin AB	Room: Texas D-F

8:00

| MTT-S Distinguished Microwave Lecturers' Talks Part 1 | PAWR Mo1B Advanced Circuit Design and Topologies 1 | SIRF Mo1C Power Amplifiers | Journal Paper Session 1 Metasurfaces, Transmission Media, and Circuit Modeling | ARFTG NVNA User's Forum | ARFTG/NIST Short Course (Day 2) | |

9:40

Coffee Break

10:10

| MTT-S Distinguished Microwave Lecturers' Talks Part 2 | PAWR Mo2B Modeling and Characterization | SIRF Mo2C Phase Shifters and Tunable Components | WISNET Panel RFID as a Sustainable Route to Digital Twins | ARFTG On-Wafer User's Forum | | |

11:50

Lunch Break

Exhibition

13:30

| RWS Mo3A High-speed and Broadband Wireless Technologies | PAWR Mo3B Advanced Circuit Design and Topologies 2 | SIRF Mo3C Millimeter-Wave Signal Generation | Student Paper Contest Elevator Pitches | ARFTG Session A Advances in Measurements I | Journal Paper Session 2 Electronic Devices and Active Circuits | |

15:10

Coffee Break

15:40

| Workshop From Lab to Market: Empowering Researchers for Entrepreneurial... | PAWR Mo4B Linearization and Efficiency Enhancement Techniques | SIRF Mo4C Devices, Technology, and Integration | Demystify IEEE Fellow Nomination | ARFTG Session B On-Wafer Measurements and Calibration | Journal Paper Session 3 Transmitters and Receivers | Student Paper Contest Interactive Forum |

17:20
17:30

RWW/ARFTG Joint Reception

18:30

| PAWR/ARFTG Joint Panel | | | | | | |

19:30

Event Overview · Tuesday, 23 January 2024

Room: Texas A	Room: Texas B	Room: Texas C	Room: Republic	Room: Lone Star A-C	Room: Seguin A-B	Room: Texas D-F

8:00

RWS Tu1A Antenna and Beamforming Technologies	SiRF Tu1B Amplifier Design	SIRF Tu1C Radar and Sensor Circuits and Architectures	ARFTG Session C Advances in Measurements II		Journal Paper Session 4 Radar, Localization, and Sensory Systems	Exhibition

9:40

Coffee Break

10:10

Joint RWW/ARFTG Plenary Session

11:50

Lunch Break

13:30

			ARFTG Session D On-Wafer and EVM Measurements			Poster Session RWW & ARFTG Including demo track presentations.

15:10

Coffee Break

15:40

RWS Tu3A Bio-Medical Applications		SIRF Tu3C Voltage-Controlled Oscillators	ARFTG Session E Generalized Network Analysis and Load-Pull		Journal Paper Session 5 Power Amplifiers	

17:20
17:30

MTT-S Space Night

19:30

Event Overview • Wednesday, 24 January 2024

Room: Texas A	Room: Texas B	Room: Texas C	Room: Republic A	Room: Republic B	Room: Republic C

8:00

| RWS We1A — Wireless Digital Signal Processing and Artificial Intelligence | WisNet We1B — Wireless Sensing and Localization Concepts | SHaRC We1C — Microwave Subsystems and Antennas for Space | | ARFTG Workshop — Traditional vs. Data and Artificial Intelligence Driven Modeling: Battle of the Ages | |

9:40

Coffee Break

10:10

| RWS We2A — Passive Components and Filters | WisNet We2B — Recent Developments of Smart Radar Sensors | SHaRC We2C — Space Communication Systems | | | |

11:50

Lunch Break

13:30

| RWS We3A — Emerging Wireless Technologies | WisNet We3B — Advanced Signal Processing and Machine Learning Concepts in Radar Sensing | Journal Paper Session 6 — Phase Shifters, Switches, and Resonators | | Short Course — Characterization of MMIC DPAs: from wafer screening to system level | IoT Summit |

15:10

Coffee Break

15:40

| RWS-WiSNet We4A — Advancements in wireless sensing and communication | WisNet We4B — Emerging Concepts for Wireless Sensors | | | Short Course — High Efficiency CMOS Power Amplifiers: Design Challenges and Outlook | |

17:20

Sunday, 21 January 2024 • Workshops

Workshop	Workshop	Workshop
RF and Millimeter-Wave Communication Systems: Design and Analysis	**Power Amplifier Design and Linearization Techniques**	**Ground Stations and Advanced Ground Station Networks**
Organizer: Mathworks	Organizers: John Dooley, Maynooth University and Pushkar Kulkarni, Qualcomm	Organizer: Jan Budroweit, DLR
Room: Crockett AB	Room: Crockett CD	Room: Republic BC

13:30

RF and Millimeter-Wave Communication Systems: Design and Analysis

Abstract:

The current trend for wireless systems to operate at millimeter wave (mmWave) frequencies and over wide bandwidth drives challenging requirements for RF front ends. In this webinar, you will learn how MATLAB and Simulink can be used for modeling RF and mmWave transceivers, performing RF budget analysis, and simulating wideband adaptive architectures.

We will first address typical RF data analysis tasks, such as reading and writing Touchstone files, transforming, visualizing, and fitting S-parameters for distributed elements. As a second step, we will discuss how to model and simulate amplifiers, matching networks, and antenna arrays operating at mmWave frequencies. Using virtual prototypes, we will simulate wideband transmitters and receivers including co-existence and interference scenarios, beam-squinting and antenna coupling, and dynamic EVM measurements for different communications standards such as 5G FR2.

With practical examples, we will demonstrate how to optimize baseband signal processing algorithms and control logic together with RF transceivers to compensate for RF impairments, to increase resilience to interfering signals, and to support multiple communication standards.

Power Amplifier Design and Linearization Techniques

Abstract:

PA design paradigm changes from mm-wave to sub-mm-wave frequencies. We will look at the device technology (Si and III-V) capabilities and the amplifier architectures, derive compact transistor modeling and its translation to system figures of merit. The device level reliability mechanisms, its circuit-level impacts and the architecture-level mitigation techniques, including the case for PMOS based PAs will be discussed. We will survey the sub-mm-wave PA designs which tend to be simpler architectures with power combining and more stages, with class-A bias and lesser harmonic control. These exacerbate the reliability, stability and thermal concerns.

Program:

1. General overview of Wireless Systems; Prof. John Dooley, Maynooth University

2. Current pinch-points for mmWave system design; Rahul Mushini, Maynooth University/NXP

3. Reliable mm-Wave and Sub-THz PA Design; Dr. Jefy Jayamon, Qualcomm

4. Agile Transmitters: Efficiency Enhancements using Digital Predistortion and Envelope Tracking Power Amplifiers; Dr. Paul Draxler, MaXentric Technologies LLC

5. Neural Networks for DPD; Andrius Vaicaitis, Maynooth University/Analog Devices

Ground Stations and Advanced Ground Station Networks

Abstract:

Ground stations are the backbone for any kind of satellite communication. The evolution of those terrestrial satellite communication infrastructures has become extremely diverse from classic ground stations as everyone knows with big dishes for deep space communication, small ground stations on the rooftop of university buildings as well as new technologies such as inter satellite link communication for 24/7 access to low Earth orbit satellites. In this workshop we will provide a great set of speakers with insights from different perspectives, incl. space agencies, academia and commercial services.

13:30

14:30

15:30

16:30

17:30

Sunday, 21 January 2024 • Women in Microwaves Event

Distinguished Women in Microwaves Event

Organizer: Jasmin Grosinger, Graz University of Technology, Austria

Room: Bowie

=== 19:00 ===

Distinguished Women in Microwaves Event

The Women in Microwaves (WiM) event at the upcoming IEEE Radio & Wireless Week (RWW) 2024 will spotlight distinguished women who have advanced the field of microwave theory and technology considerably. Three outstanding women in microwaves will talk about their respective research fields and careers. A light reception will accompany the event, allowing us to network and connect. Prof. Lei Guo from the Dalian University of Technology, China, will talk about Wireless Power Harvesters: A Charging Solution for IoT Applications. Dr. Kiki Ikossi, a 2020-2022 ASEE Science and Technology Policy Fellow at the National Science Foundation, Alexandria, VA, USA, will give details on Semiconductors for GHz to THz Devices. And Sara Barros from Thales Nederland B.V., The Netherlands, will dive deep into The Evolution of Surface Radar and Naval Platform Integration.

Wireless Power Harvesters: A Charging Solution for IoT Applications

Speaker: Lei Guo, Dalian University of Technology, China

Abstract: The realization of Internet-of-Things (IoT) relies on a large amount of geographically distributed wireless sensor nodes (WSNs). Traditional power supply cords prevent the large-scale utilization and mobility of the WSNs, while the batteries as substitutes for supply cords are not optimal solutions due to the limited lifetime, high cost, and undesired ecological effects. In this scenario, wireless power harvesting technologies provide a new approach to remotely charging or powering WSNs. This talk will discuss analytical methods for evaluating and designing radio frequency (RF) power harvesters by considering the wide adaptability to frequencies, input power, and load conditions. Based on the analysis, high-efficiency multi-band or wide-power-range RF power harvester systems will be designed and discussed. The feasibility of the proposed RF power harvesters will also be demonstrated in a real low-power wireless sensor platform. The proposed wireless power harvesting techniques have the potentials to be implemented in IoT applications where powering issues are critical.

Semiconductors for GHz to THz Devices

Speaker: Kiki Ikossi, 2020-2022 ASEE Science and Technology Policy Fellow at the National Science Foundation, Alexandria, VA, USA

Abstract: High-performance devices, integrated circuits for future energy-efficient high-speed communication networks, and IoT sensors demand high-performance semiconductor devices that can be integrated into advanced systems. This talk will examine the properties sought for heterostructure devices for GHz to THz applications. Challenges faced by the key high-speed device technologies that set in motion the communication revolution of our times will be discussed, along with some of the efforts underway for fulfilling the demands of future applications. We will see how the presence of defects and carrier traps in semiconductors hinder device performance and affect efficiency and how future technology can exploit these traps.

The Evolution of Surface Radar and Naval Platform Integration

Speaker: Sara Pena Barros, Thales Nederland B.V., The Netherlands

Abstract: Naval forces are exposed to several different threats with elaborate behaviors, especially uncrewed air and surface vehicles, in a complex above-water environment. This translates into a need for superior air and surface detection, tracking, and classification performance. In a high-intensity Above Water Warfare context, the integration of all sensors provides a higher quality and faster tactical picture, giving a clear, rapid full situational awareness (air, surface) and increased defense capabilities of the whole task group. This talk will focus on the hand-in-hand evolution of surface radar with an evergrowing-in-complexity threat scenario and the challenges of naval platform integration of such (multi-) sensor systems.

=== 20:15 ===

Sunday, 21 January 2024 • Young Professionals Panel

Young Professionals Panel

Organizer: Davi Rodrigues, The University of Texas at El Paso

Room: Bowie

20:30

Young Professionals Panel

This interactive session will offer a rare opportunity to tap into the minds of accomplished professionals who have navigated the complexities of either industry or academia. The panelists will provide guidance to young professionals and answer questions from the audience. Join us at the Young Professionals Panel for a chance to connect, learn, and ignite your career!

Panelists:

Dr. Tejinder Singh

Tejinder Singh received the Ph.D. degree (with highest academic honor) in electrical and computer engineering from the University of Waterloo, Waterloo, ON, Canada, in 2020. He is currently a Principal Member of Technical Staff, Office of the CTO at Dell Technologies, ON, Canada and an Adjunct Assistant Professor at the University of Waterloo, ON, Canada.

Dr. Taiyun Chi

Taiyun Chi received the B.S. degree (with highest Hons.) from the University of Science and Technology of China (USTC), Hefei, China, in 2012, and the Ph.D. degree from the Georgia Institute of Technology, Atlanta, GA, USA, in 2017. He is currently an Assistant Professor with the Department of Electrical and Computer Engineering, Rice University, Houston, TX, USA.

Dr. Valentina Palazzi

Valentina Palazzi received the M.Sc. degree (magna cum laude) in electronic engineering and the Ph.D. degree in industrial and information engineering from the University of Perugia, Italy, in 2014 and 2018, respectively. Since 2019, she has been a researcher with the High Frequency Electronics Laboratory, Department of Engineering, University of Perugia, Italy.

Sara Pena Barros

Sara Pena Barros received her MSc in Electronics Engineering and Telecommunications in 2013 from the University of Aveiro, Portugal. She is an Advanced Development Engineer at Thales and involved in low TRL research activities related to the development of advanced waveform concepts and novel radar technology; she also leads Integrated Topside Design studies for Navies worldwide.

Dr. Ricardo Figueiredo

Ricardo Figueiredo received the M.Sc. degree and the Ph.D. degree in electronics and telecommunications engineering from the Universidade de Aveiro, Aveiro, Portugal, in 2018 and in 2023, respectively. He is currently a Research Associate at the Instituto de Telecomunicações, Aveiro.

Dr. Kiki Ikossi

Kiki Ikossi has Ph.D. and M.S. degrees in Electrical and Computer Engineering from the University of Cincinnati with an emphasis in Solid State Electronics and a B.S.E.E. from the National Technical University of Athens (EMP), Greece. Dr. Ikossi has held several academia, major research labs, and government positions. Currently, she is an independent Research and Development Science and Technology consultant.

Monday, 22 January 2024 • Early Morning Sessions

DML Part 1	PAWR Session Mo1B
DML Special Session	**Advanced Circuit Design and Topologies 1**
Chair: Markus Gardill	Chair: Frederick Raab, Green Mountain Radio Research LLC. Co-Chair: Vittorio Camarchia, Politecnico di Torino
Room: Texas A	Room: Texas B

8:00

Maximum Power Transfer Efficiency of MIMO-WPT System

Speaker: Qiaowei Yuan

Abstract: This lecture introduces a universal approach for calculating the power transfer efficiency (PTE) and maximum power transfer efficiency (MPTE) of a Multiple Input Multiple Output Wireless Power Transfer (MIMO-WPT) system. The method is applicable to various wireless power coupling techniques and can accommodate any number of transmitters and receivers. The approach, known as E-MIMO, utilizes the Rayleigh quotient and is based on an equivalent M+N ports S-parameters circuit network of the MIMO-WPT system. The process of computing PTE and MPTE for a MIMO-WPT system will be thoroughly explained in a step-by-step manner, accompanied by numerous practical applications, demonstrating the versatility of the approach. Additionally, the lecture will introduce an essential potential application for array beamforming.

Additive Manufacturing: Emerging Opportunities for Microwave Components

Speaker: Cristiano Tomassoni

Abstract: The Additive Manufacturing (AM) technology, also known as 3D-printing technology, offers several interesting and attractive features, including fast prototyping, geometry flexibility, easily customizable products, and low cost (in some cases). However, using such technologies for microwave devices is not straightforward as AM has not been specifically developed for microwave components, and in most cases, some adaptation and post-processing is necessary. Furthermore, there are many AM technologies available, and it is important to understand their characteristics before selecting one.

In the presentation, an overview of the different AM technologies available will be provided. Additionally, an analysis of some of the most common AM technologies used for the manufacturing of microwave components will be conducted in more detail, with the help of several examples. Several microwave components manufactured with some of the most popular AM technologies will be shown, along with a detailed description of the manufacturing process, post-processing, and all actions necessary to make the component perform well. Furthermore, it will be shown how the flexibility of this technology allows the development of new classes of components with non-conventional geometries that can be exploited to obtain high-performing components in terms of compactness, weight, losses, etc.

Mo1B-1: Load mismatch and isolator removal: an old new rabbit hole for power amplifier design

Authors: Roberto Quaglia, Cardiff University

Mo1B-2: A Concurrent 2.45-5.8 GHz Power Amplifier with an Optimal Dual-band Matching Method

Authors: Sunwoo Lee, Hanyang University; Byeongcheol Yoon, Hanyang University; Jooyoung Jeon, Gangneung-Wonju National University; Junghyun Kim, Hanyang University

Mo1B-3: High-Efficiency GaN Doherty Power Amplifier based on Inverse Class-F Operation

Authors: Anna Piacibello, Politecnico di Torino; Vittorio Camarchia, Politecnico di Torino

9:00

Mo1B-4: A 2.6GHz broadband LDMOS Doherty Power Amplifier for Small-Cell Applications

Authors: Alexis Courty, Ampleon; Christophe Quindroit, Ampleon; Valentin Favard, Ampleon; Mariano Ercoli, Ampleon

Mo1B-5: A compact 27 dBm triple-stack power amplifier for 13 GHz operation in CMOS-SOI

Authors: Sravya Alluri, University of California, San Diego; Vincent W Leung, Baylor University; Peter Asbeck, Ucsd

Monday, 22 January 2024 • Early Morning Sessions

SiRF Session Mo1C

Power Amplifiers

Chair: Robert Schmid, Johns Hopkins Applied Physics Laboratory
Co-Chair: Kiki Ikossi, George Mason University

Room: Texas C

Journal Paper Session JP1

Metasurfaces, Transmission Media and Circuit Modeling

Chair: Kamal Samanta, Sony Europe

Room: Crocket AB

8:00

Mo1C-1: Toward High-Power and Multi-Way Silicon-Based mmWave Doherty Power Amplifiers

Authors: Taiyun Chi, Rice Univ.

JP1-1: Dispersion Analysis of Metasurfaces with Hexagonal Lattices with Higher Symmetries

Authors: Oscar Quevedo-Teruel; S. Yang; O Zetterstrom; F. Mesa; O. Quevedo-Teruel

8:20

JP1-2: Maximizing the Gap Height in Gap-Waveguides with Helical Wires Operating at the Vicinity of Resonance

Authors: Walid Dyab; Mourad Ibrahim; Ahmed Sakr; Ke Wu

8:40

Mo1C-2: A Ku-Band Power Amplifier in 22nm FDSOI

Authors: Alexander Haag, Karlsruhe Institute of Technology; Ahmet Cagri Ulusoy, Karlsruhe Institute of Technology

JP1-3: Ultraprecise Printing of D-Band Transmission Lines

Authors: Martin Roemhild; Georg Gramlich; Holger Baur; Thomas Zwick; Norbert Fruehauf

9:00

Mo1C-3: A W-Band Amplifier in FinFET Technology

Authors: Yuen-Sum Ng, National Taiwan University; Yunshan Wang, National Taiwan University; Huei Wang, National Taiwan University

JP1-4: Electromagnetic Parametric Modeling Using Combined Neural Networks and RLGC-Based Eigenfunctions for Two-Port Microstrip Structures

Authors: Wei Liu; Feng Feng; Yan Zhuo; Jianan Zhang; Qian Lin; Kaixue Ma; Qi-Jun Zhang

9:20

Mo1C-4: A D-Band 28 nm CMOS-Bulk Power Amplifier with 12.8 dBm Output Power and 31.3 GHz 3 dB Bandwidth

Authors: Pascal Stadler, Ruhr University Bochum; Hakan Papurcu, Ruhr University Bochum; Justin Romstadt, Ruhr University Bochum; Nils Pohl, Ruhr University Bochum

JP1-5: Signal-Flow-Graph Analysis of Weakly Nonlinear Microwave Circuits Around a Large-Signal Operating Point

Authors: Shuhei Amakawa; Ryotaro Sugimoto; Korkut Kaan Tokgoz; Sangyeop Lee; Hiroyuki Ito; Ryoko Kishikawa

9:40

Monday, 22 January 2024 · Late Morning Sessions

DML Part 2	PAWR Session Mo2B	SiRF Session Mo2C
DML Special Session	**Modeling and Characterization**	**Phase Shifters and Tunable Components**
Chair: Markus Gardill	Chair: Pere Gilabert, University Politècnica de Catalunya Co-Chair: Roberto Quaglia, Cardiff University	Chair: Cagri Ulusoy, Karlsruhe Institute of Technology Co-Chair: Vaclav Valenta, European Space Agency
Room: Texas A	Room: Texas B	Room: Texas C

10:10

100-300 GHz Wireless Communications

Speaker: Mark Rodwell

Abstract: 100-300 GHz wireless systems can provide very high data rates per signal beam, and, given the short wavelengths, even compact arrays can contain many elements, permitting massive spatial multiplexing for further increased capacity. We will describe the underlying transistor technology, plus IC, antenna, array module, and systems design of 140 GHz massive MIMO wireless hubs and 210 GHz and 280 GHz MIMO backhaul links.

Mo2B-1: Microwave Transistor Nonlinear Modeling for Power Amplifier Designers: The Revealed Truth

Authors: Antonio Raffo, Università di Ferrara

Mo2C-1: The role of Varactor, the Nonlinear Semiconductor, for Next Generation of Intelligent and Reconfigurable Radio Nodes

Authors: Najme Ebrahimi, University of Florida

Mo2B-2: Application of the Cardiff Model for Orthogonal LMBA Response Prediction

Authors: Mengyue Tian, Cardiff University; Jean-Baptiste Urvoy, Cradiff University; Roberto Quaglia, Cardiff University; James Bell, Cardiff University; Paul Tasker, Cardiff University; Steve Cripps, Cardiff University; Jeff Powell, Skyarna Ltd.

Mo2B-3: Large-signal characterization and behavioral modeling of mm-wave GaN HEMT switches tailored for advanced power amplifier architectures

Authors: Seyed Urman Ghozati, Cardiff University; Alexander Baddeley, Cardiff University; Roberto Quaglia, Cardiff University

Mo2C-2: A Compact, High Tuning Accuracy and Enhanced Linearity 37-43 GHz Digitally-Controlled Vector Sum Phase Shifter

Authors: Mehran Hazer Sahlabadi, University of Waterloo; Hang Yu, University of Waterloo; Jingjing Xia, Synopsys Inc; Slim Boumazia, University of Waterloo

11:10

Mo2B-4: Electromagnetic Coupling Between Passive Circuits and Non-Uniform Transistor Operation in High-Power Microwave Packaged Devices

Authors: Harutoshi Tsuji, Sumitomo Electric Industries, Ltd.; Ken Kikuchi, Sumitomo Electric Industries, Ltd.; Ayumu Honda, Sumitomo Electric Industries, Ltd.; Hiroshi Yamamoto, Sumitomo Electric Industries, Ltd.

Mo2C-3: A Novel High Q-factor Structure of Digitally Tunable Capacitor for High RF Power Handling Applications

Authors: Wonwoo Seo, Hanyang University; Sunghyuk Kim, Hanyang University; Byunghun Ko, Hanyang University; Jehwan Lee, Hanyang University; Youngbae Choi, Hanyang University; Taejoo Sim, Hanyang University; Junghyun Kim, Hanyang University

Mo2B-5: Impact of Matching Networks on the Impedance Settling Time in High Frequency Power Amplifiers

Authors: Roberto Quaglia, Cardiff University; Steve Cripps, Cardiff University; Jeff Powell, Skyarna Ltd

Mo2C-4: Compact D-Band Passive Phase Shifters with Fine and Coarse Control Steps in BiCMOS-55nm

Authors: Lorenzo Piotto, University of Pavia; Guglielmo De Filippi, University of Pavia; Mahmoud M Pirbazari, University of Pavia; Andrea Mazzanti, University of Pavia

11:50

Monday, 22 January 2024 · WiSNet Panel

WiSNet Panel "RFID as a Sustainable Route to Digital Twins"

Organizers: Valentina Palazzi, University of Perugia
Mahmoud Wagih, University of Glasgow

Room: Crockett AB

10:10

RFID as a Sustainable Route to Digital Twins

Panelists:

- Valentina Palazzi, University of Perugia, Moderator & Organizer

- Mahmoud Wagih, University of Glasgow, Moderator & Organizer

- CJ Reddy, Altair

- Nuno Borges Carvalho, University of Aveiro

- Jasmin Grosinger, TU Graz

- Eduardo Rojas, Embri-Riddle Aeronautical University

- John McVay, Sandia National Laboratories

- Mohammad Zarifi, University of British Columbia

11:50

Monday, 22 January 2024 • Early Afternoon Sessions

RWS Session Mo3A

High-speed and Broadband Wireless Technologies

Chair: Changzhan Gu, Shanghai Jiao Tong University
Co-Chair: Michael Brown, Texas Tech University

Room: Texas A

PAWR Session Mo3B

Advanced Circuit Design and Topologies 2

Chair: Peter Asbeck, University of California, San Diego
Co-Chair: Anna Piacibello, Politecnico di Torino

Room: Texas B

13:30

Mo3A-1: 25.9-Gb-s 259-GHz Phased-Array CMOS Receiver Module with 28° Steering Range

Authors: Shinsuke Hara, National Institute of Information and Communications Technology; Mohamed H Mubarak, National Institute of Information and Communications Technology; Akifumi Kasamatsu, National Institute of Information and Communications Technology; Yoshiki Sugimoto, Nagoya Institute of Technology; Kunio Sakakibara, Nagoya Institute of Technology; Kyoya Takano, Tokyo University of Science; Takeshi Yoshida, Hiroshima University; Shuhei Amakawa, Hiroshima University; Minoru Fujishima, Hiroshima University

Mo3B-1: A Ku-band High Gain 40 W GaN HPA MMIC for Satellite Systems in a 0.25-um GaN Technology

Authors: Taejoo Sim, Hanyang University; Seungju Lee, Hanyang University; Dongmin Lee, Hanyang University; Wonseok Choe, MMII Laboratory Co., Ltd.; Minchul Kim, MMII Laboratory Co., Ltd.; Sangmo Kim, RFHIC Corporation; Youngwan Lee, RFHIC Corporation; Kyoungil Na, Agency for Defense Development; Junghyun Kim, Hanyang University

---

13:50

Mo3A-2: Low-Additive Phase Noise Low-Power Static Frequency Dividers

Authors: Samin Hanifi, University of Virginia; Steven M Bowers, University of Virginia

Mo3B-2: Harmonic-Injection Doherty Power Amplifier: Benefits and Limitations

Authors: Moïse Safari Mugisho, Fraunhofer Institute for Applied Solid State Physi; Christian Friesicke, Fraunhofer Institute for Applied Solid State Physics; Mohammed Ayad, United Monolithic Semiconductors; Thomas Maier, Fraunhofer Institute for Applied Solid State Physics; Ruediger Quay, Fraunhofer Institute for Applied Solid State Physics

---

14:10

Mo3A-3: An RoE-based Real-Time Radio Spectral Probe

Authors: Guilherme Lourenço, Universidade de Aveiro; Francisco F Serôdio, Instituto De Telecomunicacoes; Luís F Almeida, University of Aveiro; Hugerles S Silva, Instituto De Telecomunicacoes; Arnaldo R Oliveira, Univ. de Aveiro - Inst. de Telecom.

Mo3B-3: An All-Analog Sampled-Line VSWR Sensor

Authors: Grace Gomez, University of Colorado; Devon Donahue, University of Colorado; Robert Macfarland, University of Colorado; Taylor Barton, University of Colorado

---

14:30

Mo3A-4: Track, Hold, and Reset Network for Eliminating Transient Distortion in Direct Sampling Front-Ends

Authors: Daniel Vorobiev, Texas A&M University; Sakshi Vastrad, Texas A&M University; Linda Katehi, Texas A&M University

Mo3B-4: Ultra-Fast Operating Point Switching for Watt-Level 3.6 GHz Power Amplifiers

Authors: Maximilian G Becker, Technische Universitaet Dresden; Andres Seidel, Technische Universitaet Dresden; Marco Gunia, Technische Universitaet Dresden; Frank Ellinger, Technische Universitaet Dresden

---

14:50

Mo3A-5: A Wideband 2.18-13.51 GHz Ultra-Low Additive Phase Noise Power Amplifier in InP 250nm HBT

Authors: Pedram Shirmohammadi, University of Virginia; Steven M Bowers, University of Virginia

Mo3B-5: A 22FDX® 2 Stack Power Amplifier for 5G Applications with 19dBm Psat and 49% Peak PAE

Authors: Zaid Al-Husseini, GLOBALFOUNDRIES; Paolo Valerio Testa, GLOBALFOUNDRIES; Shafi Syed, GLOBALFOUNDRIES; Mayuri Padmakar Wanve, GLOBALFOUNDRIES; Chris Boyer, GLOBALFOUNDRIES; Tianbing Chen, GLOBALFOUNDRIES

15:10

Monday, 22 January 2024 • Early Afternoon Sessions

SiRF Session Mo3C

Millimeter-Wave Signal Generation

Chair: Taiyun Chi, Rice University
Co-Chair: Austin Chen, Peraso, Inc.

Room: Texas C

Journal Paper Session JP2

Electronic Devices and Active Circuits

Chair: Almudena Suarez Rodriguez, IEEE T-MTT

Room: Seguin AB

13:30

Mo3C-1: A 4.5 dBm SiGe Doubler-Amplifier Chain Covering the Entire D-Band

Authors: Matthias Moeck, Karlsruhe Institute of Technology; Ibrahim Kagan Aksoyak, Karlsruhe Institute of Technology; Cagri Ulusoy, Karlsruhe Institute of Technology

JP2-1: W-Band Graded-Channel GaN HEMTs With Record 45% Power-Added-Efficiency at 94 GHz

Authors: Jeong-Sun Moon; Bob Grabar; Joel Wong; Chuong Dao; Erdem Arkun; Haw Tai; Dave Fanning; Nicholas C. Miller; Michael Elliott; Ryan Gilbert

---

13:50

Mo3C-2: A SiGe-Based Quadrature D-Band Up-Converter with High Output Power

Authors: Ibrahim Kagan Aksoyak, Karlsruhe Institute of Technology; Matthias Moeck, Karlsruhe Institute of Technology; Cagri Ulusoy, Karlsruhe Institute of Technology

JP2-2: Fully Integrated Avalanche Noise Sources: Reproducibility and Stability Assessment

Authors: Guendalina Simoncini; Federico Alimenti; Valentina Palazzi; Giulia Orecchini

---

14:10

Mo3C-3: A 300 GHz x9 Multiplier Chain With 9.6 dBm Output Power in SiGe Technology

Authors: Arjith Chandra Prabhu, University of Wuppertal; Janusz Grzyb, University of Wuppertal; Philipp Hillger, University of Wuppertal; Thomas Buecher, University of Wuppertal; Holger Ruecker, IHP Microelectronics; Ullrich Pfeiffer, University of Wuppertal

JP2-3: A Fractional-N Synthesizer Based on Programmable Frequency Multiplier for 5G+ Communication System

Authors: Nam-Pyo Hong; Kyu-Hyun Nam; Jun-Seok Park

14:30

Mo3C-4: 230 GHz Signal Generator for High-Bandwidth Data Links in 130 nm SiGe BiCMOS

Authors: Christian Hoyer, Technische Universitaet Dresden; Luca Steinweg, Technische Universität Dresden; Florian Protze, Technische Universität Dresden; Franz Alwin Dürrwald, Technische Universitaet Dresden; Tilo Meister, Technische Universität Dresden; Frank Ellinger, Technische Universität Dresden

JP2-4: A Dual-Band Rectifier Using Half-Wave Transmission Line Matching for 5G and Wi-Fi Bands RFEH/MPT Applications

Authors: Md. Ahsan Halimi; Taimoor Khan; Shiban K. Koul; Sembiam R. Rengarajan

---

15:10

Monday, 22 January 2024 • Late Afternoon Sessions

Workshop

From Lab to Market:
Empowering Researchers for Entrepreneurial Success

Organizer:
Weston Waldo, Director NSF I-Corps Southwest Region Hub

Room: Texas A

PAWR Session Mo4B

Linearization and Efficiency Enhancement Techniques

Chair: Antonio Raffo, University di Ferrara
Co-Chair: John Dooley, Maynooth University

Room: Texas B

15:40

Abstract:

Are you a researcher with groundbreaking technology seeking to take your innovations from the lab to the market? Join us for an inspiring and informative workshop as part of the prestigious IEEE Radio & Wireless Week! This workshop is dedicated to nurturing entrepreneurship among researchers like you, who are determined to bridge the gap between cutting-edge research and real-world impact.

Led by Weston Waldo, an expert in entrepreneurship and a key member of the newly funded National Science Foundation $15M Southwest Hub I-Corps grant, this workshop offers a unique opportunity to learn essential strategies for commercializing your research findings. The Southwest Hub I-Corps grant's mission is to empower and guide early-stage researchers, providing them with the tools and knowledge needed to navigate the challenging journey of technology commercialization.

During this interactive workshop, you will gain invaluable insights into:

- Identifying Market Opportunities: Understand how to identify and assess potential markets for your technology, ensuring your innovations meet real-world demands and needs.
- Developing a Business Model: Learn how to create a solid business model that aligns with your research, effectively translating your ideas into a commercially viable product or service.
- Market Validation: Discover the importance of market validation and how to engage with potential customers and stakeholders to refine your offerings.
- Funding and Resources: Explore various funding options and support resources available to early-stage researchers, including the opportunities offered by the Southwest Hub I-Corps grant.
- Entrepreneurial Mindset: Cultivate the entrepreneurial mindset required to navigate the challenges of entrepreneurship, including risk-taking, adaptability, and resilience.

Whether you're just starting to explore the world of entrepreneurship or have already taken your first steps, this workshop is designed to equip you with the knowledge, skills, and confidence to successfully bring your technology to the market. Be prepared to engage in stimulating discussions, practical exercises, and networking opportunities with like-minded researchers and seasoned entrepreneurs.

Don't miss this chance to unlock your potential as a technology entrepreneur! Join us at the IEEE Radio & Wireless Week and embark on a transformative journey from lab to market.

Mo4B-1: Artificial neural networks for digital predistortion linearization: more than an academic solution?

Authors: Pere L. L Gilabert, Univ. Politecnica de Catalunya; Wantao Li, University Politècnica de Catalunya; David López-Bueno, Centre Tecnològic de Telecomunicacions de Cataluny; Gabriel Montoro, Universitat Politècnica de Catalunya

---

Mo4B-2: DPD Algorithm with Long-term Memory Effects Compensation for AlGaN-GaN HEMTs

Authors: Zhijian Yu, Ampleon; Yi Zhu, Ampleon; Radjindrepersad Gajadharsing, Ampleon

---

Mo4B-3: High Accuracy DPD Approach for Hybrid Beamformer using Novel Training Symbol Mapping

Authors: Rahul Mushini, Maynooth University; Ciara Mcdonald, Maynooth University; Peter Rashev , NXP Semiconductors; Ronan Farrell, Maynooth University; John Dooley, Maynooth University

---

16:40

Mo4B-4: Investigating Feeding Techniques for High power and High efficiency E band Power Amplifiers

Authors: Bharath Kumar Cimbili, Fraunhofer Institute for Applied Solid State Physi; Christian Friesicke, Fraunhofer IAF; Mingquan Bao, Ericsson; Sandrine Wagner, Fraunhofer IAF; Moise Safari Mugisho, Fraunhofer Institute for Applied Solid State Physi; Ruediger Quay, Fraunhofer Institute for Applied Solid State Physi

---

Mo4B-5: A Linearized Calibration Technique Using Modulated Signals for Wideband Dual-Input Doherty Characterization

Authors: Andreas Illmer, Friedrich-Alexander-Universität Erlangen-Nürnberg; Alexander Deublein, Friedrich-Alexander-Universität Erlangen-Nürnberg; Robert Weigel, Friedrich-Alexander-Universität Erlangen-Nürnberg; Thomas Ackermann, Friedrich-Alexander-Universität Erlangen-Nürnberg

Monday, 22 January 2024 • Late Afternoon Sessions

SiRF Session Mo4C

Devices, Technology, and Integration

Chair: Kamel Haddadi, University of Lille
Co-Chair: Mehmet Kaynak, Texas Instruments

Room: Texas C

Journal Paper Session JP3

Transmitters and Receivers

Chair: Roberto Gómez García, IEEE MWTL

Room: Seguin AB

Demystify IEEE Fellow Nomination

Organizer: Jenshan Lin

Room: Crocket AB

15:40

Mo4C-1: 2.5D-3D Heterogeneous Integration

Authors: Ankush Mohan, HRL Laboratories, LLC; John Carlson, HRL Laboratories - Advanced Packaging Group; Tina Seeholzer, HRL Laboratories - Advanced Packaging Group; Clayton Tu, HRL Laboratories - Advanced Packaging Group; Avantika Sodhi, HRL Laboratories - Advanced Packaging Group; Hasan Sharifi, HRL Laboratories - Advanced Packaging Group

JP3-1: In Situ RF Current Assessment for Array Transmission and Optimization

Authors: Adam C. Goad; Charles Baylis; Trevor Van Hoosier; Austin Egbert; Robert J. Marks

JP3-2: Scalable Metamaterial Integrated Digital Coding Transmitting Bit Array for Concurrent Multistream Direct Antenna Modulation

Authors: Shuping Li; Chung-Tse Michael Wu

IEEE Fellow Nomination Info Session

This IEEE Fellow Nomination Information Session at the upcoming IEEE Radio & Wireless Week (RWW) 2024 will offer an opportunity to learn about the IEEE Fellow nomination and elevation process as well as the statistics in the past. In addition to providing information and encouraging nominations, this session aims to provide last-minute help to nominators before the upcoming Fellow nomination deadline on 07 February 2024 (note the much earlier new deadline). IEEE Fellow is a distinction reserved for select IEEE members whose extraordinary accomplishments in any of the IEEE fields of interest are deemed fitting of this prestigious grade elevation. The total number of Fellow recommendations in any one year must not exceed one-tenth of one percent of the IEEE voting membership on record as of 31 December of the year preceding (IEEE Bylaw I-305.9).

Organizer Bio
Dr. Jenshan Lin is a Program Director at the U.S. National Science Foundation (NSF), and a Professor Emeritus at the University of Florida (UF). In his early career prior to joining UF, he was with AT&T/Lucent Bell Laboratories and its spinoff Agere Systems from 1994 to 2003. He has published more than 300 papers and produced 24 patents. Prior to his retirement from UF in January 2022, he graduated 29 Ph.D. students. Dr. Lin was elevated to IEEE Fellow in 2010. He was as an elected IEEE Microwave Theory and Techniques Society (MTT-S) Administrative Committee (AdCom) member from 2006 to 2011. He chaired the MTT-S Fellow Evaluating Committee from 2021 to 2022, after serving as a member of the committee from 2014 to 2017 and the vice chair of the committee in 2018.

Mo4C-2: The Chip-Level in-Plane Stress Distribution over BiCMOS Wafers

Authors: Zhibo Cao, IHP Microelectronics; Thomas Voss, IHP – Leibniz-Institut für innovative Mikroelektro; Matthias Wietstruck, IHP – Leibniz-Institut für innovative Mikroelektro; Corrado Carta, IHP – Leibniz-Institut für innovative Mikroelektro; Mehmet Kaynak, Texas Instruments

JP3-3: A mmW Receiver Exploiting Complementary Current Reuse and Power Efficient Bias Point

Authors: Jesse Moody; Stefan Lepkowski; Travis Forbes

16:40

Mo4C-3: Characterization of Silicon Substrates for Sub-THz Electronics, Benefit of The Beatty Resonator Test-Structure

Authors: Luca Lucci, CEA-LETI; Olivier Valorge, CEA-LETI; Alexandre Oliveira, CEA-LETI; Herve Boutry, CEA-LETI; Christophe Dubarry, CEA-LETI; Fred Gaillard, CEA-LETI; Blandine Duriez, CEA-LETI

JP3-4: Phase Offset Calibration in Multi-Channel Radio-Frequency Transceivers

Authors: Guoyi Xu; Edwin Kan

Mo4C-4: ft extraction of HEMT transistors at mm-waves through EM-simulated de-embedding devices

Authors: Mohammed Medbouhi, CEA-LETI; José Lugo-Alvarez, CEA-LETI; Philippe Ferrari, TIMA Laboratory, CNRS-Grenoble INP-UJF; Erwan Morvan, CEA-LETI

17:20

Monday, 22 January 2024 • Joint RWW Student Paper Contest

The RWW Student Paper Contest

The purpose of the Student Paper Contest is to reward students for exceptional work and consider group projects as well as individual projects. The RWW Student Paper Contest provides students with the opportunity to share their work and discuss their results with experts from industry and academia. It is open to all students attending the RWW and presenting a paper at one of the topical conferences (RWS, PAWR, WiSNet, SiRF, and SHaRC). Starting from 2017 the Steering Committee established a new format of the Student Paper Contest, which is now a single event for the whole RWW.

The following rules apply for participating at the Student Paper Contest:

- First author must be a student (a full time or a part time). A letter is required from major advisor (Professor) stating that the first author is a registered full-time student or part-time student and has done a substantial portion of the work. Failure to provide this letter will result in disqualification.

- Number of authors on the paper: No limit, including outside authors. Outside authors are defined as co-authors from industry or from other institutions (government labs, other universities, etc.). The outside authors are included to encourage group-project submissions. The review committee will consider the number of authors vs. the level of work presented in the paper in order not to penalize the individual project submissions.

- All students wishing to participate in the contest are required to follow the regular Symposium submission process for papers including registration. Please check the checkbox during the submission process for being considered in the contest.

Student Paper Contest submissions are first evaluated by the Technical Program Committee (TPC), along with all other manuscript submissions, and receive no special consideration when being considered for acceptance to the symposium. Those papers that are accepted for oral presentation, identified as Student Paper submissions, and that meet the criteria (Relevance, Novelty, Quality, and Content) become eligible for the Student Paper Contest. The TPC and the Student Paper Contest Chairs will separately evaluate these papers again to select Student Paper Finalists.

Each Student Paper Finalist is required to prepare a short elevator pitch and a poster. Judges from all topical conferences will select the first and second place winners among the Student Paper Finalists based on the quality of the final paper, the poster presentation, and the oral presentation. The winners will be announced at the Plenary Session on Tuesday. Like last year, the first place winner is invited to write a paper for the IEEE MTT-S Microwave Magazine.

Ken Kolodziej, MIT Lincoln Laboratory
Ifana Mahbub, UT Dallas
RWW Student Paper Contest Chairs

Process & Rules of the Student Paper Contest

Each Student Paper Finalist is required to prepare a short elevator pitch and a poster.

The elevator pitches take place on Monday 22 January from 1:30PM to 3:10PM in room Crockett AB.

The poster presentations take place on Monday 22 January from 3:40PM to 5:20PM during the coffee break in room Texas DF.

For the elevator pitches finalists will be in the role of the entrepreneur and present their idea to potential investors (the judges). There will only be a flip chart for supporting the presentation. Maximum two persons per paper can bring whatever they can carry for a presentation of four minutes.

The winners will be announced in the Plenary Session on Tuesday 23rd January.

Finalists 2024

- **Low Phase Noise 104 GHz Oscillator Using Self-Aligned On-Chip Voltage-Tunable Spherical Dielectric Resonator in 130-nm SiGe BiCMOS**, Yu Zhu, Technische Universitaet Dresden

- **Investigating Feeding Techniques for High power and High efficiency E band Power Amplifiers**, Bharath Kumar Cimbili, Fraunhofer Institute for Applied Solid State Physics

- **A Planar Monopulse Comparator Network Design from Port-Transformation Rat-Race Coupler**, Hanxiang Zhang, Florida State University

- **A 4.5 dBm SiGe Doubler-Amplifier Chain Covering the Entire D-Band**, Matthias Moeck, Karlsruhe Institute of Technology

- **Distributed Radar Network with Polymer Microwave Fiber (PMF) Based Synchronization**, Andawattage Samarasekera, Johannes Kepler University Linz

- **A Modular 61 GHz Vital Sign Sensing Radar System for Long-term Clinical Studies**, Marvin Wenzel, Hamburg University of Technology

- **Device-Free Occupant Counting Using Ambient RFID and Deep Learning**, Guoyi Xu, Cornell University

- **Investigation of a Simple and Versatile Concept for OFDM Radar Target Simulator Enhancement**, Christoph Birkenhauer, Friedrich-Alexander-Universität Erlangen-Nürnberg

- **High-Performance Compact Diplexer Based on the Alternative Low-Cost AF-SIW Technology**, Maxime Le Gall, Exens Solutions

- **DDS-based Multiphase Local Oscillator Generator for Fast-Beam-Switching Phased-Array Antennas**, Shuichi Inaguma, Ritsumeikan University

- **High Accuracy DPD Approach for Hybrid Beamformer using Novel Training Symbol Mapping**, Rahul Mushini, Maynooth University

- **A 23-30 GHz Low-phase-noise 5-Bit Voltage-Controlled Oscillator in 90-nm CMOS Process**, Hong-Yeh Chang, National Central University

- **A 34 GHz CMOS VCO with Transformer Tail-Node Filter and TSPC Frequency Divider in 22 nm FDSOI**, Andre Engelmann, Friedrich-Alexander-Universität Erlangen-Nürnberg

- **Concurrent Vibration and Location Detection Using W-band On-chip Super-Regenerative Oscillator-Based Pulsed Radar**, Donglin Gao, Rutgers Univ.

Monday, 22 January 2024 • Joint PAWR/ARFTG Panel

Joint PAWR/ARFTG Panel
"Exploring the Potential of 6G: Building Upon 5G's Lessons."

Organizer: Vittorio Camarchia, Politecnico di Torino

Room: Texas A

18:30

Exploring the Potential of 6G: Building Upon 5G's Lessons.

Abstract:
Get ready for a vivid discussion on the future of wireless technology. Our panel of industry experts will delve into the possibilities of 6G, leveraging the lessons learned from the era of 5G. Don't miss this exciting opportunity to gain valuable insights into the next generation of wireless communication.

Meet our expert panelists featuring experts from HRL, Fraunhofer, Virginia diodes, Keysight, and NI.

Panelists:

- Jeong Moon, HRL

- Gerhard Schoenthal, Virginia Diodes

- Nuno Borges Carvalho, IT Aveiro

- Mark Pierpoint, Keysight Technology

- Fabian Thome, Fraunhofer Research Institute

- Markus Dasilva, National Instruments

- Alessandro Fonte, Siae Microelectronics

19:30

Tuesday, 23 January 2024 • Early Morning Sessions

RWS Session Tu1A

Antenna and Beamforming Technologies

Chair: Vaclav Valenta, European Space Agency
Co-Chair: Ricardo Figueiredo, University of Aveiro

Room: Texas A

SiRF Session Tu1B

Amplifier Design

Chair: Mehmet Kaynak, Texas Instruments
Co-Chair: Robert Schmid, Johns Hopkins Applied Physics Laboratory

Room: Texas B

8:00

Tu1A-1: Beamforming-based Spatial Precoding with Channel Estimation for Massive MIMO-OFDM System

Authors: Chen-Hao Chiu, National Taiwan University; Ju-Hong Lee, National Taiwan University

Tu1B-1: CMOS LNA and VGA for 5G NR Using Gain-Linearity-Boosting and Body Floating Techniques

Authors: Jin-Fa Chang, National Changhua University of Education; Yo-Sheng Lin, National Chi Nan University

---

Tu1A-2: A Pencil Beam Parabolic Reflector Antenna Using LSE-NRD Guide at 140 GHz

Authors: Daiya Miyamoto, National Institute of Technology, Kure College; Futoshi Kuroki, National Institute of Technology, Kure College

Tu1B-2: Parametric-Oscillation-Free Efficient SiGe:C Power Amplifier Design for Ku--Ka-Band SATCOM

Authors: Tsung-Ching Tsai, Karlsruhe Institute of Technology; Vaclav Valenta, European Space Agency; Cagri Ulusoy, Karlsruhe Institute of Technology

---

Tu1A-3: Spatial Processing with High-Fidelity Antenna Models and Quantized Analog Weights

Authors: John Spitzmiller, Parsons

Tu1B-3: A 94 GHz Bandwidth Transimpedance Amplifier in 55nm SiGe BiCMOS for High Speed Optical Receivers

Authors: Lachlan Cuskelly, University of California, Los Angeles; Christopher Falt, Ciena, Corp.; Peter Schvan, Ciena, Corp.

9:00

Tu1A-4: A Compact and Highly Efficient Circularly Polarized UWB Rectenna for Wireless Power Transfer Application

Authors: Nabanita Saha, University of Texas at Dallas; Sunanda Roy, The University of Texas at Dallas; Ifana Mahbub, University of Texas at Dallas

Tu1B-4: A 200 - 325 GHz Gain-Boosted J-Band Low-Noise Amplifier in a 130 nm SiGe BiCMOS Technology

Authors: Manuel Koch, Friedrich-Alexander-Universität Erlangen-Nürnberg; Sascha Breun, Friedrich-Alexander-Universität Erlangen-Nürnberg; Robert Weigel, Friedrich-Alexander-Universität Erlangen-Nürnberg

---

9:40

Tuesday, 23 January 2024 • Early Morning Sessions

SiRF Session Tu1C

Radar and Sensor Circuits and Architectures

Chair: Saeed Zeinolabedinzadeh, Arizona State University
Co-Chair: Davi Rodrigues, University of Texas at El Paso

Room: Texas C

Journal Paper Session JP4

Radar, Localization, and Sensory Systems

Chair: Nuno Borges Carvalho, Universidade de Aveiro

Room: Seguin AB

8:00

Tu1C-1: Advances in mmWave Radar Architectures
Authors: Brian Ginsburg, Texas Instruments

JP4-1: A Miniaturized Millimeter-Wave Radar Sensing Microsystem with High Isolation Full-duplex Microstrip Patch Antenna
Authors: Lina Ma; Zesheng Zhang; Jingyun Lu; Changzhan Gu; Junfa Mao

8:20

JP4-2: A Low-IF Doppler Radar With Asynchronous Bandpass Sampling for Accurate Measurement of Displacement Motions
Authors: Fei Tong; Jingtao Liu; Changzhi Li; Changzhan Gu; Junfa Mao

8:40

Tu1C-2: Analysis of a SiGe BiCMOS Detector for a Broadband mmW-integrated EPR Spectrometer
Authors: Selina Eckel, Karlsruhe Institute of Technology; Ahmet Cagri Ulusoy, Karlsruhe Institute of Technology

JP4-3: A Wi-Fi Frequency Band Passive Biomedical Doppler Radar Sensor
Authors: Dongyang Tang; Victor G. Rizzi Varela; Davi V. Q. Rodrigues; Daniel Rodriguez; Changzhi Li

9:00

Tu1C-3: 27 Gb-s PRBS Generator with In-Operation Programmable Taps for PMCW Radar
Authors: Florian Probst, Friedrich-Alexander-Universität Erlangen-Nürnberg; Andre Engelmann, Friedrich-Alexander-Universität Erlangen-Nürnberg; Robert Weigel, Friedrich-Alexander-Universität Erlangen-Nürnberg

JP4-4: 3-D indoor localization and identification through RSSI-Based angle of arrival estimation with real Wi-Fi signals
Authors: Ho-Chun Yen; Liang-Yu Ou Yang; Zuo-Min Tsai

9:20

JP4-5: Noninvasive Internal Body Thermometry With On-Chip GaAs Dicke Radiometer
Authors: Jooeun Lee; Gabriel Santamaria Botello; Robert Streeter; Zoya Popovic

9:40

Tuesday, 23 January 2024 • Plenary Session

Joint RWW/ARFTG Plenary Session

Chair: Holger Maune, Otto von Guericke University
Co-Chair: David Blackham, Keysight Technologies

Room: Texas A

10:10

Joint RWW/ARFTG Plenary Session

The Future of Heterogeneous Integration for mmWave Systems: Challenges and Opportunities

Speaker: Madhavan Swaminathan, The Pennsylvania State University, USA

Madhavan Swaminathan is the Department Head of Electrical Engineering and is the William E. Leonhard Endowed Chair at Penn State University. He also serves as the Director for the Center for Heterogeneous Integration of Micro Electronic Systems (CHIMES), an SRC JUMP 2.0 Center.Power Amplifier Design and Linearization Techniques Prior to joining Penn State University, he was the John Pippin Chair in Microsystems Packaging & Electromagnetics in the School of Electrical and Computer Engineering (ECE), Professor in ECE with a joint appointment in the School of Materials Science and Engineering (MSE), and Director of the 3D Systems Packaging Research Center (PRC), Georgia Tech (GT). Prior to GT, he was with IBM working on packaging for supercomputers. He received his MS and PhD degrees in Electrical Engineering from Syracuse University in 1989 and 1991, respectively.

Abstract:

Emerging electronic systems require the dense integration of many chiplets in either 2D or 3D form. The metrics for these systems will be dictated by power, performance, form factor, cost, and reliability. The complexity of these systems is expected to be large given the integration of sensing, wireless, computing, and other functionality on a single packaging platform that combines electronics and photonics together. Such systems pose immense integration challenges but also provide opportunities for innovation on several fronts that include architecture, design, thermal, materials, embedded intelligence, and many more? This presentation will provide a discussion of the State of the Art and opportunities for the future.

Redefining ICs Metrics for OTA Characterization

Speaker: Anouk Hubrechsen, ANTENNEX B.V.

Anouk Hubrechsen received the B.Sc. and M.Sc. degrees in Electrical Engineering from the Eindhoven University of Technology, Eindhoven, The Netherlands, in 2017 and 2019, respectively, where she finished her Ph.D. in 2023 on reverberation-chamber measurements of mmWave antennas. She was a Guest Researcher with the National Institute of Standards and Technology at Boulder, Boulder, CO, USA, in 2018 and 2019. There she was involved in reverberation-chamber metrology for Internet-of-Things applications. She is co-founder and CEO of ANTENNEX B.V., a company that develops instrumentation for measuring integrated antenna systems, based on reverberation-chamber technology. Anouk received the Regional and District Zonta Women in Technology Awards in 2019.

Abstract:

With higher frequencies, integration of antennas and RF electronics means that many measurements now need to be performed over the air. In this talk, we explain the challenges of testing RF electronics in phased-array and antenna-on-chip configurations. They require new types of testing methods. We detail the newest over-the-air measurement techniques for metrics such as noise figure, out-of-band emissions, radiated power spectrum, and field distribution in advanced, highly-integrated devices.

12:00

Tuesday, 23 January 2024 • Interactive Forum Poster Session

Interactive Forum IF1 - Room: Texas D-F

13:30

13:30

IF1-1: A Compact 2.45-5.5-GHz Dual-Band LNA Design Using Bridged-T Coils

Authors: Jing-Xuan Chou, National Central University; Yo-Shen Lin, National Central University

IF1-5: Broadband GaN Power Amplifier MMIC with a Nonuniform Transmission Line Output Matching Network

Authors: Paul Flaten, University of Colorado; Zoya Popovic, University of Colorado

IF1-9: A Linear Simulation Technique for a Power Traveling-Wave Amplifier

Authors: Waleed Joudeh, Amcom Communications Inc.; Amin Ezzeddine, AMCOM Communications, Inc.

IF1-13: Multi-Antenna Array for All Space Communications

Authors: Pavlo Molchanov, AMPAC Science

IF1-2: Single-Voltage-Supply pHEMT-mHEMT 2.4 and 5.8 GHz LNAs Using Power Constrained Design

Authors: Chinchun Meng, National Yang Ming Chiao Tung University; Chung-Yo Lin, National Yang Ming Chiao Tung University; Guo-Wei Huang, Natinal Applied Research Institute

IF1-6: Integrated GaN Power Detector for High Power Millimeter-Wave Applications

Authors: Thomas Ufschlag, Institute of Robust Power Semiconductor Systems; Benjamin Schoch, Institute of Robust Power Semiconductor Systems; Dominik Wrana, Institute of Robust Power Semiconductor Systems; Sandrine Wagner, Fraunhofer Institute for Applied Solid State Physics; Dirk Schwantuschke, Fraunhofer Institute for Applied Solid State Physics; Friedbert Van Raay, Fraunhofer Institute for Applied Solid State Physics; Peter Brückner, Fraunhofer Institute for Applied Solid State Physics; Ingmar Kallfass, Institute of Robust Power Semiconductor Systems

IF1-10: Study of AM-PM Deviation on Power Amplifier Linearization Performances for 5G Applications

Authors: Christophe Quindroit, Ampleon; Kaisseh Houssein, Ampleon; Alexis Courty, Ampleon; Stephan Maroldt, Ampleon

IF1-14: Energy-Efficient D-Band Power Amplifier Linearization Adopting Back-Gate Feedforward Technique in 22nm FD-SOI

Authors: Helia Ordouei, Technische Universitat Berlin; Friedel Gerfers, Technische Universität Berlin

IF1-3: Scalable Multi-tap RF Canceller with Arduino Control for STAR Systems

Authors: Pierre-Francois W. Wolfe, MIT Lincoln Laboratory; Kenneth E Kolodziej, Mit Lincoln Laboratory

IF1-7: An X-band Spatial Power Combining Using Rectangular Waveguide with Dielectric Lens

Authors: Takuma Kinoshita, NIT, Kure college; Yuki Shinhama, NIT, Kure college; Masaru Sato, Fujitsu Ltd.; Futoshi Kuroki, NIT, Kure college

IF1-11: A 3.6GHz Highly Efficient Dual-Driver Doherty Power Amplifier

Authors: Ioannis Peppas, Graz University of Technology; Marco Pitton, Infineon Technologies Austria AG; Mustazar Iqbal, Infineon Technologies; Peter Singerl, Infineon Technologies AG; Bhagath Talluri, Infineon Technologies Nijmegen BV; Martin Mataln, Infineon Technologies Austria AG; Helmut Paulitsch, Graz University of Technology; Wolfgang Bösch, Graz University of Technology

IF1-15: Impact on Beam-Forming Processes in the Near Field for 5G Ultra-Wideband Waveform

Authors: Maryna Nesterova, APREL Inc.; Stuart Nicol, APREL; Yuliya Nesterova, Carleton University

IF1-4: Robust Estimation for Digital Predistortion with Non-ideal Equalization

Authors: Richard N Braithwaite, Keysight Technologies

IF1-8: An Ultra-wideband Off-axis Reflector Lens

Authors: Mingyan Zhong, University of Glasgow; Yunan Jiang, University of Glasgow; Yufei Ma, University of Glasgow; Chong Li, University of Glasgow

IF1-12: Cost-Effective Allan Deviation Measurement in SDRs Using Integrated ADC

Authors: Alastair L Wiegelmann, Flinders University; Samuel Drake, Flinders.University; Saeed Rehman, Flinders University; Shengjian Chen, Flinders University

IF1-18: High Gain Metamaterial Superstrate Loaded Antenna For S band Communication

Authors: Prutha P Kulkarni, Vishwakarma Institute of Information Technology; Vivek S Deshpande, Vishwakarma Institute of Information Tehcnology

Tuesday, 23 January 2024 • Interactive Forum Poster Session

IF1-19: Software Configurable Multi-mode Radar Sensor System for Range Tracking and Life Sensing

Authors: John T Crainer, Texas Tech University; Changzhi Li, Texas Tech University

---

IF1-23: A Comprehensive Approach to Extracting Coupling Matrix From Filtenna Measurements

Authors: Sara Javadi, Graz University of Technology; Behrooz Rezaee, Graz University of Technology; Manfred Stadler, Qualcomm Europe, Inc.; Michael Leitner, Qualcomm Europe, Inc.; Wolfgang Bösch, Graz University of Technology

---

IF1-27: Low Power Gesture Sensing System Based on Target Range Using Spiking Neural Networks for Portable Devices

Authors: Muhammad Arsalan, Technische Universität Braunschweig; Avik Santra, Infineon Technologies AG; Vadim Issakov, Technische Universität Braunschweig

---

IF1-31: Over-the-Air LoS Propagation Characteristics of Various Indoor Materials at 28 GHz

Authors: Mohammad Alavirad, Dell Technologies; Tejinder Singh, Dell Technologies; Morris Repeta, Dell Technologies

---

IF1-20: ISAR Imaging of Drones Based on Time Domain Correlation Algorithm Using Millimeter-Wave Fast Chirp Modulation MIMO Radar

Authors: Kenshi Ogawa, National Defense Academy of Japan; Dovchin Tsagaanbayar, National Defense Academy of Japan; Ryohei Nakamura, National Defense Academy of Japan

---

IF1-24: A Wideband Patch Antenna Array with Improved Isolation for Integrated Sensing and Communication

Authors: Lina Ma, Shanghai Jiao Tong University; Changzhan Gu, Shanghai Jiao Tong University; Junfa Mao, Shanghai Jiaotong University

---

IF1-28: Enhancing the Output Power and Efficiency for a Set Noise-Power Ratio of a K-Band Power Amplifier By Means of Analog Pre-Distortion

Authors: Tommaso Cappello, Villanova University; Sarmad Ozan, University of Bristol; Andy Tucker, Filtronic Broadband Ltd.; Peter Krier, Filtronic Broadband Ltd.; Tudor Williams, Filtronic Broadband Ltd.; Kevin Morris, University of Leeds

---

IF1-21: Wireless Network Deployment Survey

Authors: Arash Ahmadi, École de Technologie Supérieure de Montreal; Zahra Sepehri, École de Technologie Supérieure de Montreal; Marouane Indja, École de Technologie Supérieure de Montreal; Vladan Jevremovic, iBwave Solution Inc.; Ali Jemmali, iBwave Solutions Inc.; Marc-Antoine Lamontagne, iBwave Solutions Inc.; Sylvain G Cloutier, École de Technologie Supérieure de Montreal; Ivanka Iordanova, École de Technologie Supérieure; Chahé Nerguizian, Polytechnique Montreal; Ali Motamedi, École de Technologie Supérieure de Montreal

---

IF1-25: Design of a Six-stage W-band Low-Noise Amplifier Using a 90-nm CMOS Technology

Authors: Yu-Chia Su, National Central University; Rou-Yin Huang, National Central University; Hong-Yeh Chang, National Central University

---

IF1-29: Class S Power Amplifier System for Radio Applications in the HF Band

Authors: Alexander Ruderer, University of Innsbruck; Alex Putzer, University of Innsbruck; Thomas Ussmueller, University of Innsbruck

---

IF1-22: A Compact 6-12 GHz MMIC Power Amplifier

Authors: Muhammad Y Mahsud, University of Colorado; Prathamesh Pednekar, University of Colorado Boulder - ECEE; Taylor Barton, University of Colorado

---

IF1-26: Compact Dual-Band Negative Group Delay Circuit

Authors: Nathan B Gurgel, Federal University of Campina Grande; Glauco Fontgalland, Federal University of Campina Grande; Isaac Barros, Federal Rural University of Semiarid; Blaise Ravelo, Nanjing University of Science and Technology

---

IF1-30: A 1.28 mW K-Band Modified Gilbert-Cell Mixer Design in 22nm FDSOI CMOS

Authors: Adilet Dossanov, Technische Universität Braunschweig; Vadim Issakov, Technische Universität Braunschweig

15:10

Tuesday, 23 January 2024 • Demo Track Presentations

Demo Track - Room: Texas D-F

13:30

Bringing EDA-Tool Schematics into LaTeX - An Open-Source Solution

Presenter:
Christof Pfannenmüller

Affiliation:
Friedrich-Alexander-Universität Erlangen-Nürnberg,
Lehrstuhl für Technische Elektronik,
Team Radio & Biomedical Applications

Abstract:
Current EDA tools like PathWave Advanced Design System or Cadence Virtuoso support various formats for data interchange and documentation exports. We propose an open-source solution for directly transferring schematic data to the node-based drawing environment TikZ and its extension CircuiTikZ for electrical schematic drawings. Schematics are exported as XML-based file structures via ADS Board Link (ABL). Using this input, the proposed tool can recalculate components and their placements and convert them to TikZ-based source code. With these, the schematic can be used in LaTeX documents.

Demo Overview:
This demo will contain a live view of several Keysight ADS schematics and their conversion with the proposed open-source JavaScript-based tool. Afterward, the results are shown in a LaTeX document for comparison. A computer with the necessary installed software will be provided. No additional equipment will be required for the demo.

Nature-Inspired All-Space Multi-Antenna Array Architecture

Presenter:
Dr. Pavlo Molchanov

Affiliation:
IPD Scientific LLC

Abstract:
The nature-inspired all-space multi-antenna array architecture combines various techniques to achieve wide-area multi-orbit observation, fast simultaneous signal processing, high directional accuracy, enhanced reliability, and communication quality. The key elements and features of the architecture: Holographic wide area of observation with staring antenna array; Monopulse method of fast simultaneous signals processing; Direct Digitizing Signals on Multi-Axis Overlap Directional Antennas; Integration of Antennas with Signal Conditioning Circuits and SDR; Distributed Placement for System Protection; Transformation and Processing in Multiple Domains. Moreover, the application of wide numerical aperture overlap antennas allows the detection and recognition of different objects in high-scattering mediums by measurement of the Fresnel diffraction patterns and angular spectrum of scattered and diffraction components. Fourier transform and digital multi-domain digitizing can provide reliable recognition of objects by spectrum signatures and separation of transferring medium and objects.

Demo Overview:
The demonstration setup will consist of a low-power transmitter module and a receiving module connected to two separate directional antennas by flexible coaxial cables. The receiving module can be connected to a phase detector or Software Defined Radio (SDR). This setup allows for a demonstration of key features of the array architecture, as well as the recognition of different objects using their spectrum signatures. Remote detection and recognition of concealed objects will be also demonstrated.

mmW-OAI: The Easiest Way to Establish a 5G FR2 End-to-End Test Network

Presenter:
Ethan Lin

Affiliation:
Tmytek

Abstract:
Do you happen to have one or two spare SDRs on hand? Have you ever thought about establishing a complete 5G FR2 end-to-end network with your SDRs?

TMYTEK will introduce the mmW-OAIBOX, an FR2-enabled OAI testbed that TMYTEK worked on with Allbesmart. It is worth noting that this solution has already been delivered to Japan. Incorporating the best of millimeter-wave and OpenAirInterface (OAI), we provide a comprehensive test environment from UE to the core network. The mmW-OAIBOX offers 5G beamformers to mimic gNB and UE array antennas, a frequency converter, a powerful PC installed with the latest OAI stack, including OAI gNB, CN5G, a dashboard, and more.

We will show you how to use the APIs to control TMYTEK FR2 devices, including a 24-44 GHz up/down converter (UD Box) and a 28 GHz mmWave beamformer (BBox), with your SDR development environment. This will include an API introduction, control calls, DLL imports, and more.

There are many topics that need to be addressed in wireless research. We have built the most advanced tools to unleash your creativity, so you are able to develop innovative solutions for the next generation of wireless technology.

15:10

Tuesday, 23 January 2024 • Late Afternoon Sessions

RWS Session Tu3A	SiRF Session Tu3C	Journal Paper Session JP5
Bio-Medical Applications	**Voltage-Controlled Oscillators**	**Power Amplifiers**
Chair: Robert Caverly, Villanova University Co-Chair: Ifana Mahbub, University of Texas at Dallas	Chair: Austin Chen, Peraso, Inc. Co-Chair: Chung-Tse Wu, Rutgers University	Chair: Paolo Mezzanotte, University of Perugia
Room: Texas A	Room: Texas C	Room: Seguin AB

15:40

Tu3A-1: Concurrent Vibration and Location Detection Using W-band On-chip Super-Regenerative Oscillator-Based Pulsed Radar

Authors: Donglin Gao, Rutgers University; Shuping Li, Rutgers University; Minning Zhu, Rutgers University; Chung-Tse Michael Wu, Rutgers University

Tu3C-1: A 23-30 GHz Low-phase-noise 5-Bit Voltage-Controlled Oscillator in 90-nm CMOS Process

Authors: Po-Yuan Chen, National Central University; Jun-Liang Chen, National Central University; Hong-Yeh Chang, National Central University

JP5-1: Single-Input Broadband Hybrid Doherty Power Amplifiers Design Relying on a Phase Sliding-Mode of the Load Modulation Scheme

Authors: Chenyu Liang; Patrick Roblin; Yunsik Hahn; Jose I. Martinez-Lopez; Hsiu-Chen Chang; Vanessa Chen

16:00

Tu3A-2: A Multi-layer Coil Magnetic Stimulation Device for autonomous function regulation

Authors: Po-Lei Lee, National Central University; Kuo-Kai Shyu, National Central University

Tu3C-2: Low Phase Noise 104 GHz Oscillator Using Self-Aligned On-Chip Voltage-Tunable Spherical Dielectric Resonator in 130-nm SiGe BiCMOS

Authors: Yu Zhu, Technische Universitaet Dresden; Georg Sterzl, University of Stuttgart; Jan Hesselbarth, University of Stuttgart; Tilo Meister, Technische Universitaet Dresden; Frank Ellinger, Technische Universitaet Dresden

JP5-2: A Broadband Asymmetrical Doherty Power Amplifier With Optimized Continuous Mode Harmonic Impedances

Authors: Alex Pitt; Gautam Jindal; Kevin Morris; Tommaso Cappello

16:20

Tu3A-3: Respiratory Dynamics of Thoracic and Abdominal Motion in Doppler Radar Measurements

Authors: Jannatun Noor Sameera, University of Hawaii at Manoa; Alexander Lee, University of Hawaii at Manoa; Victor Lubecke, University of Hawaii Manoa; Olga Boric-Lubecke, University of Hawaii at Manoa

Tu3C-3: A 34 GHz CMOS VCO with Transformer Tail-Node Filter and TSPC Frequency Divider in 22 nm FDSOI

Authors: Andre Engelmann, Friedrich-Alexander-Universität Erlangen-Nürnberg; Florian Probst, Friedrich-Alexander-Universität Erlangen-Nürnberg; Philip Hetterle, Friedrich-Alexander-Universität Erlangen-Nürnberg; Robert Weigel, Friedrich-Alexander-Universität Erlangen-Nürnberg

JP5-3: 3-Way Doherty Power Amplifiers: Design Guidelines and MMIC Implementation at 28 GHz

Authors: Anna Piacibello; Vittorio Camarchia; Paolo Colantonio; Rocco Giofrè

16:40

Tu3A-4: Enhancing Heart Failure Monitoring: Biomedical Radar-Based Detection of Cheyne-Stokes Respiration

Authors: Li Wen, Shanghai Jiao Tong University; Zhi Zhang, Shanghai General Hospital; Jinliang Wang, Shanghai General Hospital; Jiaqi Liu, Shanghai General Hospital; Shuqin Dong, Shanghai Jiao Tong University; Changzhan Gu, Shanghai Jiao Tong University; Junfa Mao, Shanghai Jiao Tong University

Tu3C-4: D-Band VCO with Uniformly Low Phase Noise versus Frequency and Temperature

Authors: Isabel Kraus, Ruhr University Bochum; Herbert Knapp, Infineon Technologies AG; Nils Pohl, Ruhr University Bochum

JP5-4: A Broadband Outphasing GaN Power Amplifier Based on Reconfigurable Output Combiner

Authors: Weiwei Wang; Shiping Li; Shichang Chen; Jialin Cai; Yuanchun Li; Xinyu Zhou; Giovanni Crupi; Gaofeng Wang; Quan Xue

17:00

Tu3C-5: Voltage-Controlled-Oscillator Using 8-shaped Transformer-coupled Transmission Line

Authors: Sheng-Lyang Jang, National Taiwan University of Science and Technology; Yi-Ping Hsieh, National Taiwan University of Science and Technology; Wen-Cheng Lai, Min Chi University of Technology, Taiwan

17:20

Tuesday, 23 January 2024 • Space Night

MTT-S Space Night

Organizer: Jan Budroweit, German Aerospace Center

Room: Texas B

17:30

MTT-S Space Night

As part of the Radio Wireless Week and its co-located conference "Space Hardware and Radio Conference" (SHaRC), we invite you to the 3rd MTT-S Space Night! The Space Night topic for 2024 is "Ground Stations and Advanced Technologies for Satellite Communications" and we will host a interactive panel discussion with experts from agency, academia and industry. They will provide us insights on recent developments for ground station technologies and satellite communication services, while attendees can interactively contribute to this panel session by asking questions, comment and participating to surveys using their smart phones (no applications required). All registered attendees are also welcome to join the Space Quiz which we will host for the time. Experience an interesting event with us in an entertaining atmosphere, enjoy the interactive panel discussions with experts with light snacks and drinks and don't miss the chance to win great prizes in our space night quiz!

Moderators:

- Marie Piasecki, NASA Glenn Reserach Center

- Jan Budroweit, DLR

Panelists:

- Václav Valenta, ESA

- Allen Katz, College of New Jersey, MACOM Fellow

- Benjamin Schoch, University Stuttgart

- Eyal Trachtman, Addvalue Technologies

19:30

Image credit: SHUTTERSTOCK.COM/Boris Rabtsevich

Wednesday, 24 January 2024 • Early Morning Sessions

RWS Session We1A	WiSNet Session We1B	SHARC Session We1C
Wireless Digital Signal Processing and Artificial Intelligence	**Wireless Sensing and Localization Concepts**	**Microwave Subsystems and Antennas for Space**
Chair: Ken Kolodziej, Massachusetts Institute of Technology Co-Chair: Markus Gardill, Brandenburg University of Technology	Chair: Paolo Mezzanotte, University of Perugia Co-Chair: Valentina Palazzi, University of Perugia	Chair: Jan Budroweit, DLR Group Co-Chair: Charlie Jackson, Earthlink
Room: Texas A	Room: Texas B	Room: Texas C

=== 8:00 ===

We1A-1: BeamCIM: A Compute-In-Memory based Broadband Beamforming Accelerator using Linear Embedding

Authors: Nael Mizanur Rahman, Georgia Institute of Technology; Sudarshan Sharma, Georgia Institute of Technology; Coleman B Delude, Georgia Institute of Technology; Wei Chun Wang, Georgia Institute of Technology; Justin Romberg, Georgia Institute of Technology; Saibal Mukhopadhyay, Georgia Institute of Technology

We1B-1: Distributed Radar Network with Polymer Microwave Fiber (PMF) Based Synchronization

Authors: A. Chaminda J Samarasekera, Johannes Kepler University Linz; Sergio Lopez Fernandez, Johannes Kepler University Linz; Reinhard Feger, Johannes Kepler University Linz; Richard Hüttner, Johannes Kepler University Linz; Frank Gruson, ZF Friedrichshafen AG; Siegfried Krainer, Infineon Technologies AG; Andreas Stelzer, Johannes Kepler University Linz

We1C-1: High-Performance Compact Diplexer Based on the Alternative Low-Cost AFSIW Technology

Authors: Maxime Le Gall, Exens Solutions; Anthony Ghiotto, Bordeaux INP; Issam Marah, Exens Solutions

---

We1A-2: Active Vector Modulator Design for Self-Interference Cancellation in STAR Systems

Authors: Marcus W Wolff, Massachusetts Institute of Technology, Lincoln La; Pierre-Francois W Wolfe, MIT Lincoln Laboratory; Kenneth E Kolodziej, Mit Lincoln Laboratory

We1B-2: A Three-Dimensional Localization System Based on Magnetic Fields and Induction

Authors: Lukas Messner, University of Innsbruck; Thomas Ussmueller, University of Innsbruck

We1C-2: Demonstration of GaN HEMT MMIC High-Power Amplifier for Lunar Proximity Communications

Authors: Rainee N Simons, NASA Glenn Research Center; Marie T Piasecki, NASA Glenn Research Center; Joseph A Downey, NASA Glenn Research Center; Bryan L Schoenholz, NASA Glenn Research Center

---

We1A-3: Modulation Recognition with Untrained Deep Neural Network for IoT and Mobile Applications

Authors: Jongseok Woo, Georgia Institute of Technology; Kuchul Jung, Georgia Institute of Technology; Saibal Mukhopadhyay, Georgia Institute of Technology

We1B-3: Secure Occupancy Sensing with Passive Radar for Spectrally Congested Spaces

Authors: Rachel Ma, Texas Tech University; Aaron B Carman, Texas Tech University; Changzhi Li, Texas Tech University

We1C-3: Mechanical Tuning of an Offset-fed Reflector Antenna

Authors: Taehak Lee, Yuhan University; Sang-Gyu Lee, Korea Aerospace Research Institute; Sang-Burm Ryu, korea Aerospace Research Institute

=== 9:00 ===

We1A-4: Predistortion of Charge Trapping Memory Effects in GaN based RF Power Amplifiers with Artificial Neural Networks

Authors: Patrick Jueschke, Nokia ; Georg Fischer, Friedrich-Alexander-Universität Erlangen-Nürnberg

We1B-4: A Digital Beamforming Approach for Indoor Passive Sensing

Authors: Aaron B Carman, Texas Tech University; Changzhi Li, Texas Tech University

We1C-4: Design and Characterization of a Multi-Channel ADS-B Antenna for Small Satellites

Authors: Jan Budroweit, DLR; Felix Eichstaedt, German Aerospace Center; Ferdinand Stehle, DLR e.V.

We1A-5: Transfer Learning Optimized PA Behavioral Modeling over 2D Operation States

Authors: Jose M Domingues, University of Aveiro; Hugerles S Silva, Universidade de Aveiro; Nuno Carvalho, Instituto De Telecomunicacoes; Arnaldo R Oliveira, Univ. de Aveiro - Inst. de Telecom.

We1B-5: Wind Turbines Structural Health Monitoring Using a FMCW Radar Mounted on a Drone

Authors: Victor G Rizzi Varela, Texas Tech University; Changzhi Li, Texas Tech University

We1C-5: Microlens Coupler from Integrated Photonic Circuit to Fiber Design for Space Application

Authors: Chengtao Xu, Embry-Riddle Aeronautical University, Daytona Beac; Jayaprakash B Shivakumar, Embry-Riddle Aeronautical University, Daytona Beac; Eduardo Rojas, Embry-Riddle Aeronautical University

=== 9:40 ===

Wednesday, 24 January 2024 • Late Morning Sessions

RWS Session We2A

Passive Components and Filters

Chair: Rashaunda Henderson, University of Texas at Dallas
Co-Chair: Bayaner Arigong, Florida A&M University

Room: Texas A

WiSNet Session We2B

Recent Developments of Smart Radar Sensors

Chair: Fabian Lurz, Otto von Guericke University
Co-Chair: Davi Rodrigues, University of Texas at El Paso

Room: Texas B

SHARC Session We2C

Space Communication Systems

Chair: Markus Gardill, Brandenburg University of Technology
Co-Chair: Vaclav Valenta, European Space Agency

Room: Texas C

10:10

We2A-1: A Planar Monopulse Comparator Network Design from Port-Transformation Rat-Race Coupler

Authors: Hanxiang Zhang, Florida State University; Powei Liu, Florida State University; Jonathan Casamayor, Florida State University; Saeed Zolfaghary Pour, Florida A&M University; Mitch Plaisir, Florida State University; Bayaner Arigong, Florida State Univ.

We2B-1: A Modular 61 GHz Vital Sign Sensing Radar System for Long-term Clinical Studies

Authors: Marvin Wenzel, Hamburg University of Technology; Dominik Langer, Hamburg University of Technology; Alexander Koelpin, Hamburg University of Technology; Fabian Lurz, Hamburg University of Technology

We2C-1: Towards Gbps Downlinks from Low-Cost Active Phased Arrays

Authors: Adam Gannon, National Aeronautics and Space Administration; James Downey, National Aeronautics and Space Administration; Bryan L Schoenholz, National Aeronautics and Space Administration

---

10:30

We2A-2: Compact Multilayer AFSIW Diplexer

Authors: Maxime Le Gall, Exens Solutions; Anthony Ghiotto, Bordeaux INP; Issam Marah, Exens Solutions

We2B-2: Displacement Motion Sensing with Asynchronous Bandpass Sampling Using a Single-Channel Dual-PLL SSB low-IF Doppler Radar

Authors: Fei Tong, Shanghai Jiao Tong University; Jingtao Liu, Shanghai Jiao Tong University; Changzhan Gu, Shanghai Jiao Tong University; Junfa Mao, Shanghai Jiao Tong University

We2C-2: Real-time Wideband Video Synchronization via an Analog QPSK Costas Loop in a Laboratory Demonstration of an E-Band Satellite Downlink

Authors: Janis Woermann, University of Stuttgart; Laura Manoliu, University of Stuttgart; Simon Haussmann, University of Stuttgart; Milos Krstic, IHP Microelectronics; Ingmar Kallfass, University of Stuttgart

---

10:50

We2A-3: Methodology to Accurately Replicate a Non-Planar Thin-Film Microstrip BEOL in 3D EM Simulation

Authors: Dominik Wrana, University of Stuttgart; Christopher M Groetsch, Keysight Technologies; Benjamin Schoch, University of Stuttgart; Lukas Gebert, University of Stuttgart; Thomas Ufschlag, Universität Stuttgart; Arnulf Leuther, Fraunhofer Institute for Applied Solid State Physics; Roger Lozar, Fraunhofer Institute for Applied Solid State Physics; Ingmar Kallfass, University of Stuttgart

We2B-3: Enhancing Multi-Subject Vital Sign Estimation by Utilizing the Generalized Sidelobe Canceller

Authors: Abdel-Kareem Moadi, University of Tennessee; Chandler J Bauder, University of Tennessee; Abdel-Hamid Djouadi, University of Tennessee; Paul Theilmann, MaXentric Technologies, LLC; Aly E Fathy, University of Tennessee

We2C-3: SDR based radio-frequency noise measurements

Authors: Giacomo Schiavolini, University of Perugia; Giulia Orecchini, Universiti of Perugia; Valentina Palazzi, University of Perugia; Luca Roselli, University of Perugia; Paolo Mezzanotte, University of Perugia; Guendalina Simoncini, University of Perugia; Anna Gregorio, University of Trieste; Federico Alimenti, University of Perugia

---

11:10

We2A-4: Planar-Magic-T-Based Dual-Band Bandpass Filters

Authors: Xi-Bei Zhao, Xidian University; Feng Wei, Xidian University; Li Yang, University of Alcala; Roberto Gomez-Garcia, University of Alcala

We2B-4: Stepped-Frequency PMCW-Radar Modulation Scheme for Automotive Applications

Authors: Moritz Kahlert, HELLA GmbH & Co. KGaA; Tai Fei, HELLA GmbH & Co. KGaA; Claas Tebruegge, HELLA GmbH & Co. KGaA; Markus Gardill, Brandenburg University of Technology

We2C-4: Pre-Flight Evaluation of a Multi-Channel ADS-B Receiver in a Stratospheric Balloon Mission

Authors: Felix Eichstaedt, German Aerospace Center; Ferdinand Stehle, German Aerospace Center; Jan Budroweit, German Aerospace Center

---

11:30

We2A-5: Multilayer Dual-Band Bandpass Filter Using Microstrip-to-Slotline Transitions and Transversal Signal-Interference Microstrip Lines

Authors: Li Yang, University of Alcala; Mohamed Malki, University of Alcalá; Roberto Gomez-Garcia, University of Alcala

We2B-5: Tracking Driver's Foot Movements Using mmWave FMCW Radar

Authors: Davi Rodrigues, University of Texas at El Paso; Changzhi Li, Texas Tech University

We2C-5: IDRS, a persistent, always-on connectivity for LEO spacecraft

Authors: Eyal J Trachtman, Addvalue Technologies

11:50

Wednesday, 24 January 2024 • Early Afternoon Sessions

RWS Session We3A	WiSNet Session We3B	Journal Paper Session JP6
Emerging Wireless Technologies	**Advanced Signal Processing and Machine Learning Concepts in Radar Sensing**	**Phase Shifters, Switches and Resonators**
Chair: Tejinder Signh, Dell Technologies Co-Chair: Hong-Yeh Chang, National Central University	Chair: Michael Brown, Los Alamos National Laboratory Co-Chair: Thomas Kurin, Friedrich-Alexander-Universität Erlangen-Nürnberg	Chair: Roberto Gómez García, IEEE MWTL
Room: Texas A	Room: Texas B	Room: Texas C

13:30

We3A-1: DDS-based Multiphase Local Oscillator Generator for Fast-Beam-Switching Phased-Array Antennas

Authors: Shuichi Inaguma, Ritsumeikan University; Koki Nagata, Ritsumeikan University; Hideyuki Nosaka, Ritsumeikan University

We3B-1: A Large-scale Movement Path Fitting Based Phase Compensation Algorithm for FMCW Radar Vital Sign Detection

Authors: Li Sun, Nanjing University of Science and Technology; Ge Bai, Nanjing University of Science and Technology; Changhao Luo, Nanjing University of Science and Technology; Shuaiming Huang, Nanjing University of Science and Technology

JP6-1: A K-Band Ultra-Wideband Binary Phase Shifter for Phase Modulating Applications in Radar

Authors: Michael C. Brown; Changzhi Li

We3A-2: Combined RF-Ultrasonic Wireless Powering System for Sensor Applications in Harsh Environment

Authors: Yufei Ma, University of Glasgow; Yunan Jiang, University of Glasgow; Chong Li, University of Glasgow

We3B-2: Deep Learning-based Person Detection on a Moving Robot

Authors: Jasmin Gabsteiger, Friedrich-Alexander-Universität Erlangen-Nürnberg; Timo Maiwald, Friedrich-Alexander-Universität Erlangen-Nürnberg; Thomas Kurin, Friedrich-Alexander-Universität Erlangen-Nürnberg; Christian Dorn, Technical University of Munich; Robert Weigel, Friedrich-Alexander-Universität Erlangen-Nürnberg; Fabian Lurz, Hamburg University of Technology

JP6-2: Ultrawideband Schiffman Phase Shifter Designed With Deep Neural Networks

Authors: Sensong An; Bowen Zheng; Hang Tang; Hang Li; Zhou Li; Yunxi Dong; Mohammad Haerinia; Hualiang Zhang

---

We3A-3: The Impact of Interference on Macrodiversity Gain in mmWaveCellular Networks

Authors: Enass F Hriba, Ohio Northern University; Marwan M Alkhweldi, Ohio Northern University

We3B-3: Gesture Recognition for FMCW Radar on the Edge

Authors: Maximilian Strobel, Infineon Technologies AG; Stephan Schoenfeldt, Infineon Technologies AG; Jonas Daugalas, Infineon Technologies AG

JP6-3: A Plasma-Switch Impedance Tuner with Microsecond Reconfiguration

Authors: Justin Roessler; Alden Fisher; Austin Egbert; Zach Vander Missen; Trevor Van Hoosier; Charles Baylis; Mohammad Abu Khater; Dimitrios Peroulis; Robert J. Marks

14:30

We3A-4: Modelling of 32-APSK Constellation Distortion and EVM in GaN Power Amplifiers From AM-AM and AM-PM curves

Authors: Gamal M Hegazi, Aethercomm Inc.

We3B-4: Device-Free Occupant Counting Using Ambient RFID and Deep Learning

Authors: Guoyi Xu, Cornell University; Edwin C Kan, Cornell University

JP6-4: Surface-Acoustic-Wave Devices Based on Lithium Niobate and Amorphous Silicon Thin Films on a Silicon Substrate

Authors: Yansong Yang; Liuqing Gao; Songbin Gong

We3A-5: A 100 GHz Varactor-less Fundamental VCO With 12% Tuning Range in 22nm FDSOI Technology

Authors: Nazmus Saquib, Rensselaer Polytechnic Institute; Ahmed Elmenshawi, Rensselaer Polytechnic Institute; Mona Hella, Rensselaer Polytechnic Institute

We3B-5: Resonate-and-Fire Spiking Neurons for Hand Gesture Label Refinement

Authors: Ahmed Shaaban, Infineon Technologies AG; Zeineb Chaabouni, Infineon Technologies AG; Maximilian Strobel, Infineon Technologies AG; Wolfgang Furtner, Infineon Technologies AG; Robert Weigel, Friedrich-Alexander-Universität Erlangen-Nürnberg; Fabian Lurz, Otto von Guericke University

15:10

Wednesday, 24 January 2024 • Short Courses

Short Courses - Room: Republic B

=== 13:30 ===

Characterization of MMIC DPAs - From Wafer Screening to System Level

Lecturer: Anna Piacibello, Politecnico di Torino

Anna Piacibello received the bachelor's and master's degrees in electronic engineering and the Ph.D. degree in electric, electronic and communication engineering from the Politecnico di Torino, Turin, Italy, in 2013, 2015, and 2019, respectively.

In 2017, she was a Visiting Researcher with the Centre for High Frequency Engineering, Cardiff University, Cardiff, U.K. She is currently an Assistant Professor with the Department of Electronics and Telecommunications, Politecnico di Torino. Her research interests include the design and characterization of microwave and millimeter-wave electronic circuits, mainly focusing on broadband and highly efficient power amplifiers.

She has been an Affiliate Member of the IEEE MTT-S Technical Committee TC-12 on Microwave High-Power Techniques since 2022. She was a recipient of the 2018 Young Engineer Prize awarded by the European Microwave Association.

Abstract:

This talk will give an overview of the characterization flow for MMIC Doherty PAs, from the initial wafer screening to the performance assessment at system level. Doherty PAs require specific procedures to test the correct operation of its Main and Auxiliary branches, which are biased in different classes and therefore respond differently to signal excitations. This reflects in the way in which DC, small signal and large signal measurements are devised and carried out during the characterization flow. Finally, a short overview of the different metrics that can be used to estimate the PA linearity in different scenarios will be given.

=== 14:30 ===

=== 15:40 ===

High Efficiency CMOS Power Amplifiers: Design Challenges and Outlook

Lecturer: Narek Rostomyan, Waveye, Inc.

Narek Rostomyan earned his M.Sc. in electrical engineering from the Technical University of Munich, Germany, in 2014, and completed his Ph.D. in electrical engineering, specializing in RF/mm-wave transceivers and high-efficiency CMOS SOI power amplifiers, at the University of California, San Diego (UCSD), in 2018.

During his early career, from 2013 to 2014, he contributed to RF front-end design of mm-wave signal generators at Rohde & Schwarz in Munich, Germany. Later, from 2018 to 2020, he played a pivotal role in developing the first generation of 76-81 GHz automotive radar transmitter and receiver chipsets at Metawave in Carlsbad, USA. From 2020 to 2022, he served as a senior principal engineer at IQ-Analog in San Diego, focusing on mm-wave, broadband front-ends, and high-speed mixed-signal IPs for RF sampling, high-speed ADCs/DACs in FinFET CMOS. Currently, Narek serves as the co-founder and Chief Innovation Officer (CIO) at Waveye, Inc. (Palo Alto, CA), leading efforts in next-generation mm-wave radar imaging and perception. His research interests encompass high power and efficiency CMOS integrated circuits and systems for wireless communication, radar, and sensing applications.

Abstract:

Power consumption of mm-wave communication and radar systems is a significant bottleneck for many applications due to size, cost, battery, and heatsinking constrains. Depending on the application, the RF front-end, and in particular the power amplifier, can be one of the main contributors to the overall power consumption. Additionally, both in communication and radar systems, the choice of the transmit waveform poses design constrains on the RF front-end architecture and achievable specifications.

It is well known that achieving high efficiency with on-chip silicon power amplifiers at mm-wave frequencies is quite challenging. Furthermore, in order to attain high linearity without resource-hungry digital pre-distortion, a significant amount of back-off from Psat/P1dB is usually necessary, which penalizes the average efficiency even more. On the other hand, the choice of the transmit waveform has even more impact on the overall system architecture and ultimately the power consumption for modern mm-wave radars.

In this short course, we will review various challenges and techniques that affect the efficiency and linearity of mm-wave PAs in silicon technologies. We will also review various commonly used efficiency improvement techniques. Practical design challenges will be emphasized. Finally, we will conclude with a step-by-step case study of a single stage Ka-band Doherty amplifier design.

=== 16:40 ===

Wednesday, 24 January 2024 • Late Afternoon Sessions

RWS Session We4A

Advancements in Wireless Sensing and Communication

Chair: Davi Rodrigues, University of Texas at El Paso

Room: Texas A

WiSNet Session We4B

Emerging Concepts for Wireless Sensors

Chair: Thomas Ussmueller, University of Innsbruck
Co-Chair: Vaclav Valenta, European Space Agency

Room: Texas B

15:40

We4A-1: Energy Constraints in Wireless Technologies - how to improve efficiency

Authors: Nuno Carvalho, Instituto De Telecomunicacoes

We4B-2: Robust Doppler Displacement Measurement Resolving the Uncertainty During Target Stationary Moment

Authors: Luigi Ferro, University of Messina; Graziella Scandurra, University of Messina; Changzhi Li, Texas Tech University; Emanuele Cardillo, University of Messina

---

We4A-2: Direct Sampling Receiver with an Adjustable Bandpass Filter for Use in Passive Radar with FM Radio

Authors: Marie Horlbeck, Friedrich-Alexander-Universität Erlangen-Nürnberg; Jonathan Fiedelak, Friedrich-Alexander-Universität Erlangen-Nürnberg; Benedict Scheiner, FAU; Robert Weigel, University Erlangen-Nuremberg; Fabian Lurz, Hamburg University of Technology

We4B-3: Investigation of a Simple and Versatile Concept for OFDM Radar Target Simulator Enhancement

Authors: Christoph Birkenhauer, Friedrich-Alexander-Universität Erlangen-Nürnberg; Georg Körner, Friedrich-Alexander-Universität Erlangen-Nürnberg; Patrick Stief, Friedrich-Alexander-Universität Erlangen-Nürnberg; Gerhard Hamberger, Rohde & Schwarz GmbH & Co. KG; Matthias Beer, Rohde & Schwarz GmbH & Co. KG; Christian Carlowitz, Friedrich-Alexander-Universität Erlangen-Nürnberg; Martin Vossiek, Friedrich-Alexander-Universität Erlangen-Nürnberg

---

We4A-3: Gesture Recognition to Control a Moving Robot With FMCW Radar

Authors: Timo Maiwald, Friedrich-Alexander-Universität Erlangen-Nürnberg; Jasmin Gabsteiger, Friedrich-Alexander-Universität Erlangen-Nürnberg; Robert Weigel, Friedrich-Alexander-Universität Erlangen-Nürnberg; Fabian Lurz, Otto von Guericke University

We4B-4: Phase Modulation Based TX Channel Calibration for MIMO Radar Systems

Authors: Simon Heining, Johannes Kepler University Linz; Reinhard Feger, Johannes Kepler University Linz; Thomas Faseth, Infineon Technologies Austria; Christoph Wagner, Silicon Austria Labs; Andreas Stelzer, Johannes Kepler University Linz

---

16:40

We4A-4: Parametric Classification of Recoverable Radar-Assessed Respiratory Rate Data

Authors: Mohammad Shadman Ishrak, University of Hawaii at Manoa; Jannatun Noor Sameera, University of Hawaii at Manoa; Olga Boric-Lubecke, University of Hawaii at Manoa; Victor M Lubecke, University of Hawaii Manoa

We4B-5: Passive Broadband Harmonic Sensor-Tag using Circular Disk Dipole Antenna

Authors: Nobuhiro Kuga, Yokohama National University; Iori Serizawa, Yokohama National University; Kun Xiao, Yokohama National University

---

We4A-5: Angular Dependency of Human Speech Recognition using Interferometry Radar

Authors: Christopher Williams, Texas Tech University; Changzhi Li, Texas Tech University

---

17:20

A Ku-Band Power Amplifier in 22nm FDSOI

Alexander Haag
Institute of Radio Frequency Engineering and Electronics
Karlsruhe Institute of Technology
Karlsruhe, Germany
alexander.haag@kit.edu

Ahmet Çağrı Ulusoy
Institute of Radio Frequency Engineering and Electronics
Karlsruhe Institute of Technology
Karlsruhe, Germany
ulusoy@kit.edu

Abstract—**This paper presents the design of an 18 GHz power amplifier (PA) in GlobalFoundries' 22 nm fully-depleted silicon on insulator (FD-SOI) technology. The PA features a single-ended cascode core design with a P_{sat} of 17 dBm and maximum PAE of 45%. The whole PA, including pads and ESD protection, occupies only 0.23 mm^2 and can be used as a baseline for more complex power-combined designs. Additionally, for the core design the performance of a common source, a cascode and a three-stack core are compared.**

Index Terms—**CMOS, 22nm, FDSOI, Silicon, Power Amplifier, PA, 18 GHz, Ku Band, High Efficiency, Efficient, High PAE**

I. INTRODUCTION

With the on-going scaling of silicon-based technologies excellent high frequency performance in conjunction with high integration densities for digital and mixed-signal components is possible. However, down-scaling beyond a certain threshold will deteriorate high frequency performance. A technology like 22 nm FDSOI is a suitable candidate for the purpose of adding multiple functionalities into the same chip, as the transistors show excellent performance for analog and digital circuits alike. Due to the aggressive scaling, nominal supply voltages below 1 V complicate the design of high frequency PAs significantly. A simple single core design appears to be insufficient, as most published designs have moved on to different and more complex design architectures [1]–[4]. There are two main methods to achieve improved performance.

Firstly, the stacking of transistors to increase output voltage swing, loadpull impedance and gain of the amplification stage. This method focuses on circumventing the limited supply voltage. However, useful stacking is only possible up to three to four transistors at millimeter wave frequencies, as stacking becomes less effective as more transistors are stacked. This is shown in [5] for a 45 nm CMOS SOI technology.

Secondly, it is common to adopt a differential design exploiting the possibility to utilize the neutralized differential pair topology and transformer based matching and combining. Transformers are a very common matching approach as their high matching efficiency, small footprint and DC feeding through center-taps promise superior performance. When the differential to single-ended conversion is omitted, a differential design will have an advantage over a single-ended design.

Only two publications for 22 nm FDSOI technologies could be identified by the authors that proposed a single-ended design [6], [9]. Both designs feature a two-stage three-stacked design. While [6] is matched using a combination of transmission lines, inductors and capacitors, [9] is completely LC based. Both designs target a wide operational bandwidth. For comparison this paper presents a PA designed by applying basic design principles to the 22 nm FDSOI technology.

II. DESIGN

The presented PA consists of a single-ended single core design, using a cascode topology. The amplifier core is then matched to $50\,\Omega$ using LC-based matching networks. The circuit is shown in Fig. 1.

The amplifier core consists of four identical unit cells connected in parallel. Each unit cell contains a common gate device stacked on top of a common source device in a cascode topology. This is mainly done to increase voltage swing for higher output power and output impedance for higher efficiency in the output matching network. Additionally, the isolation and gain are improved. At the input of each unit cell a parallel RC is placed for low frequency stabilization.

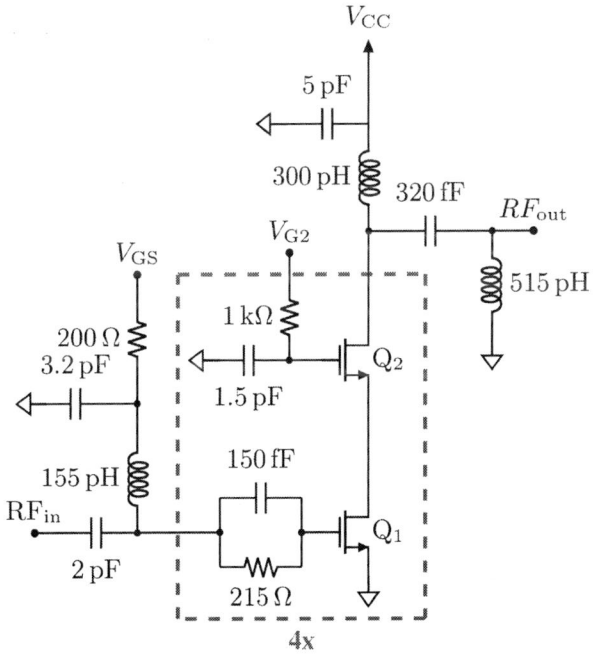

Fig. 1: Circuit schematic. Total transistor gate width is $Q_{1,total} = Q_{2,total} = 4$ x 67.2 µm. $V_{CC} = 1.6\,V$, $V_{G2} = 1.3\,V$ & $V_{GS} = 0.5\,V$.

979-8-3503-4331-1/24 $31.00 © 2024 IEEE

SiRF 2024

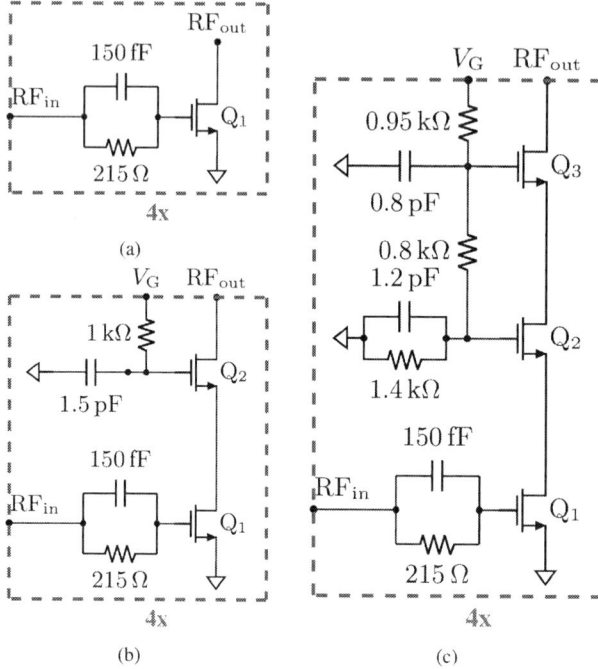

Fig. 2: Core schematics: a) common source ($V_{DD} = 0.8\,V$); b) cascode ($V_{DD} = 1.6\,V$); c) three stack ($V_{DD} = 2.4\,V$). $Q_{1,total} = Q_{2,total} = Q_{3,total} = 4 \times 67.2\,\mu m$.

The bias voltage V_{G2} for the common gate device is supplied through a large resistor, while V_{GS} is provided through the input matching network.

The cascode core was chosen due to the PAE advantage it offers compared to a common source or three stack core. The core schematics are shown in Fig. 2. All cores are evaluated for the same total transistor gate width and therefore approximately equivalent I_D per core. The performance is compared under ideal loading conditions and with a lossy matching network to $50\,\Omega$. The ideal load impedances are determined by loadpull simulations for optimal PAE. The common source, cascode core and three stack core are denoted by the indices 'A', 'B' and 'C', respectively, and include EM simulated parasitics. The results including the matching network loss are additonally denoted with the index 'PA'. The simulation results for the core comparison at 18 GHz are shown in Fig. 3.

The common source core achieves a saturated output power of 13 dBm, including matching network losses. This is 4 dB lower than the cascode core with nearly 17 dBm. Additionally, the common source core's PAE is halved by the matching network losses. This is due to the low loadpull impedance and therefore high matching network losses and secondly due to the low gain in compression (<5 dB). When the gain is low the required input power will negatively influence the achievable PAE. The three stack core, including network losses, achieves 19.9 dBm of saturated output power. However, even though the output network loss is about 0.7 dB for the three stack core and 1 dB for the cascode core, the PAE peaks for the cascode core and then declines for higher stacking. This confirms that

results previously shown in [5] can be applied for 22 nm FDSOI as well. A simulated peak PAE of 45 % is achieved for the cascode core including matching networks.

Due to the necessary parallelization the input and load impedance for the cascode design are at $Z_{in} = (10 - 5j)\,\Omega$ and $Z_{L,opt} = (10 + 10j)\,\Omega$, respectively. The load impedance was determined by loadpull simulation for optimal PAE. Since the core shows sufficient gain, the input matching network loss is not optimized, thereby slightly improving stability. The output matching network is a Pi-network and iteratively optimized using EM simulation to improve network efficiency and consequently PA performance. A chip photo is provided in figure 5.

III. MEASUREMENT RESULTS

For the measurement the IC is directly probed. The small signal measurement is performed using a 2-port LRRM calibration with a known calibration substrate. The results are shown in figure 4a. The small signal measurement shows very good agreement with simulation. A small signal gain of 20 dB

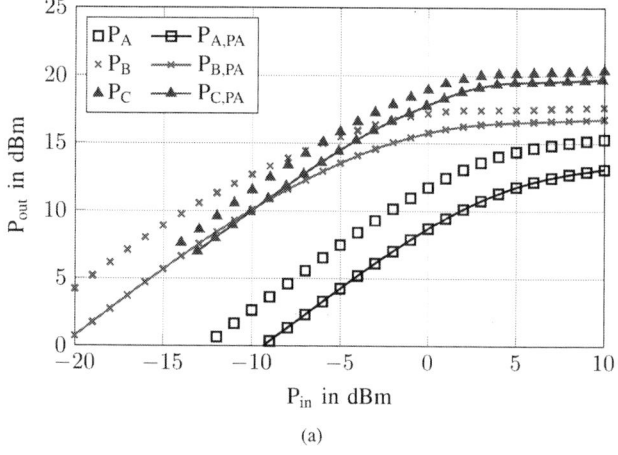

(a)

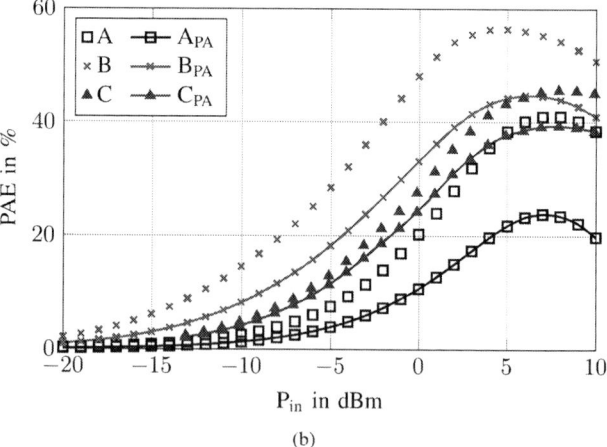

(b)

Fig. 3: Large signal performance comparison of a common source (A) core, a cascode (B) core and a 3-stack core (C) under ideal loading conditions and with lossy matching networks (indicated by 'PA') at 18 GHz: a) output power; b) PAE.

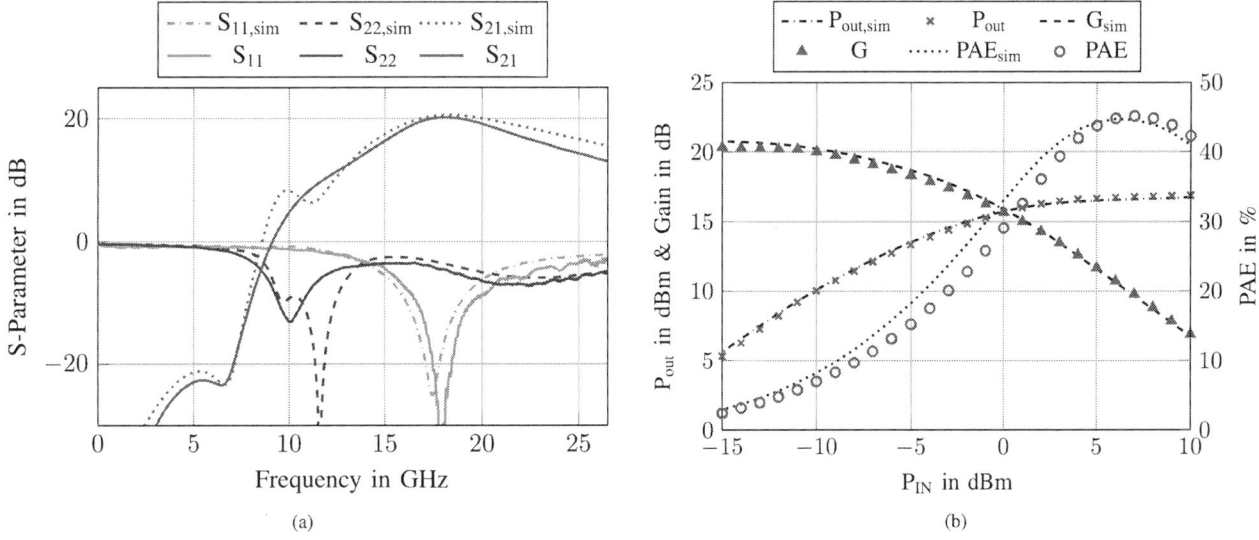

Fig. 4: Measurement results: a) small signal; b) large signal single tone at 18 GHz.

is measured and a 3 dB bandwidth is achieved from 15.4 GHz to 21.9 GHz, corresponding to a relative bandwidth of 35 %.

The large signal measurement is performed using a signal generator as the source and a power meter connected to the output. Cable losses are calibrated out. Probe loss is accounted for with the respective data sheet values. Peak performance is achieved at 17 - 18 GHz. The corresponding large signal measurement results are shown in figure 4b. A P_{sat} of 17 dBm at 45 % PAE is measured. At this point the amplifier is 8 dB compressed. At 1 dB compression the PA outputs 12 dBm at 11 % PAE.

Modulated measurements were performed for 64 QAM at a data rate of 1.2 Gbit s^{-1} and 256 QAM at a data rate of

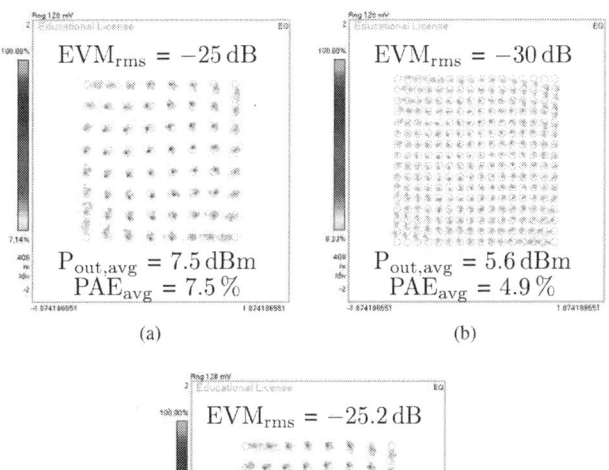

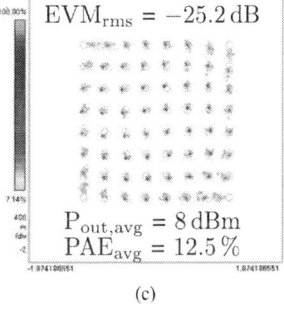

Fig. 6: Modulated measurement results: a) 200 MHz 64 QAM; b) 100 MHz 256 QAM; c) 200 MHz 64 QAM with reduced biasing.

0.8 Gbit s^{-1}. For 64 QAM an average output power, $P_{out,avg}$, of 7.5 dBm and a corresponding average PAE, PAE_{avg}, of 7.5 % are measured. This fits in well with state-of-the-art presented in table I. For the 256 QAM $P_{out,avg}$ is measured to be 5.6 dBm at an average PAE of 4.9 %. The constellation diagrams of the modulated measurements are presented in Fig. 6. During the modulated measurements different bias conditions were tested and biasing the PA at $V_{G2} = 1.15$ V & $V_{GS} = 0.35$ V improves the performance. The lower bias

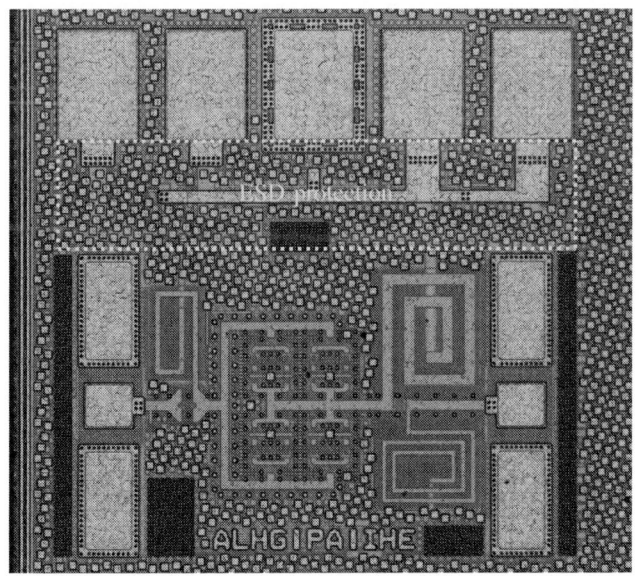

Fig. 5: Chip microphotograph. Chip area is 480 μm x 480 μm = 0.23 mm^2.

979-8-3503-4331-1/24 $31.00 © 2024 IEEE

TABLE I: Performance comparison with state of the art.

Reference	This work		[7]	[8]	[9]	[10]
Technology	22 nm FD-SOI		22 nm FD-SOI	22 nm FD-SOI	22 nm FD-SOI	22 nm FD-SOI
Architecture	Cascode, SE, LC-based		3 Stack Diff. Doherty, Transformer	2 Stack Diff., Transformer	3 Stack, SE, LC-based	2 Stack Diff., Transformer
f_C	18 GHz		28 GHz	28 GHz	24.5-34.5 GHz	24 GHz
P_{sat}	17 dBm		22.5 dBm	21 dBm	16.8 dBm	14.9 dBm
PAE_{max}	45 %		28.5 %	31.5 %	19.8 %	26.1 %
P_{1dB}	12 dBm		21.1 dBm	-	13 dBm	11.5 dBm
G_{lin}	20 dB		26.1 dB	16.5 dB	25 dB	15.7 dB
V_{DD}	1.6 V		2.4 V	1.6 V	2.4 V	1.6 V (estimated)
RF core power density	$642 \frac{mW}{mm^2}$		$889 \frac{mW}{mm^2}$	$420 \frac{mW}{mm^2}$ (estimated)	$200 \frac{mW}{mm^2}$	$206 \frac{mW}{mm^2}$ (estimated)
Core area	$0.08 \, mm^2$		$0.2 \, mm^2$	$0.61 \, mm^2$ (full chip)	$0.24 \, mm^2$	$0.31 \, mm^2$ (full chip)
Modulation	QAM 64		QAM 64	QAM 64	-	QAM 256
Data rate	$1.2 \, Gbit \, s^{-1}$		$2.4 \, Gbit \, s^{-1}$	$0.6 \, Gbit \, s^{-1}$	-	$3.2 \, Gbit \, s^{-1}$
EVM_{rms}	−25 dB		−25.1 dB	−28 dB	-	?
$P_{out,avg}$	7.5 dBm	8 dBm*	10.9 dBm	12.5 dBm	-	7 dBm
PAE_{avg}	7.5 %	12.5 %*	9.2 %	6.6 %	-	7.7 %

* measurement results with changed bias conditions ($V_{G2} = 1.15 \, V$ & $V_{GS} = 0.35 \, V$)

voltages reduce current consumption, allowing for an average PAE of 12.5 % for the same EVM. Compared to other designs in 22 nm FD-SOI similar values of output power and an increased peak PAE are achieved. Since only a single core is used the area consumption is reduced.

IV. CONCLUSION

A very compact (0.23 mm^2) and efficient power amplifier in 22 nm FD-SOI is presented. A simple LC-based design is chosen on purpose, highlighting that excellent performance can be achieved without the need to resort to differential and transformer-based designs. However, for a significant increase in output power a more sophisticated design in terms of stacking and combining is required.

ACKNOWLEDGMENT

The authors would like to thank GLOBALFOUNDRIES for providing silicon fabrication through the 22FDX university program.

REFERENCES

[1] M. Cui, C. Carta and F. Ellinger, "A 21-dBm 3.7 W/mm² 28.7% PAE 64-GHz Power Amplifier in 22-nm FD-SOI," in IEEE Solid-State Circuits Letters, vol. 3, pp. 386-389, 2020, doi: 10.1109/LSSC.2020.3023499.

[2] Z. Zong et al., "A 39GHz T/R front-end module achieving 25.6% PAEmax, 20dBm Psat, 5.7dB NF, and -13dBm IIP3 in 22nm FD-SOI for 5G communications," 2021 IEEE Radio Frequency Integrated Circuits Symposium (RFIC), 2021, pp. 23-26, doi: 10.1109/RFIC51843.2021.9490466.

[3] J. Mayeda, D. Y. C. Lie and J. Lopez, "Broadband Millimeter-Wave 5G CMOS Power Amplifiers With High Efficiency at Power Backoff and ESD-Protection in 22nm FD-SOI," 2021 IEEE International Midwest Symposium on Circuits and Systems (MWSCAS), 2021, pp. 899-902, doi: 10.1109/MWSCAS47672.2021.9531817.

[4] J. C. Mayeda, J. Lopez and D. Y. C. Lie, "Highly-Efficient Broad-band Medium Power Amplifier Design in 22nm CMOS FD-SOI for mm-Wave 5G," 2020 IEEE Texas Symposium on Wireless and Microwave Circuits and Systems (WMCS), 2020, pp. 1-4, doi: 10.1109/WMCS49442.2020.9172413.

[5] H. -T. Dabag, B. Hanafi, F. Golcuk, A. Agah, J. F. Buckwalter and P. M. Asbeck, "Analysis and Design of Stacked-FET Millimeter-Wave Power Amplifiers," in IEEE Transactions on Microwave Theory and Techniques, vol. 61, no. 4, pp. 1543-1556, April 2013, doi: 10.1109/TMTT.2013.2247698.

[6] X. Xu et al., "A 28 GHz and 38 GHz High-Gain Dual-Band Power Amplifier for 5G Wireless Systems in 22 nm FD-SOI CMOS," 2020 50th European Microwave Conference (EuMC), 2021, pp. 174-177, doi: 10.23919/EuMC48046.2021.9338181.

[7] Z. Zong et al., "A 28GHz Voltage-Combined Doherty Power Amplifier with a Compact Transformer-based Output Combiner in 22nm FD-SOI," 2020 IEEE Radio Frequency Integrated Circuits Symposium (RFIC), Los Angeles, CA, USA, 2020, pp. 299-302, doi: 10.1109/RFIC49505.2020.9218280.

[8] Z. Al-Husseini et al., "A 28 GHz 22FDX® PA with 31.5 % Peak PAE and Output Power of 21 dBm in CW, 18.5 dBm in QPSK, and 12.5 dBm in 64QAM," 2022 52nd European Microwave Conference (EuMC), Milan, Italy, 2022, pp. 349-352, doi: 10.23919/EuMC54642.2022.9924479.

[9] X. Xu et al., "A 21-39.5 GHz Power Amplifier for 5G Wireless Systems in 22 nm FD-SOI CMOS," 2019 IEEE Asia-Pacific Microwave Conference (APMC), 2019, pp. 1640-1642, doi: 10.1109/APMC46564.2019.9038202.

[10] J. Mayeda, C. Sweeney, D. Y. C. Lie and J. Lopez, "A 19.1 - 46.5 GHz Broadband Efficient Power Amplifier in 22nm CMOS FD-SOI for mm-Wave 5G," 2022 IEEE International Symposium on Circuits and Systems (ISCAS), Austin, TX, USA, 2022, pp. 1112-1116, doi: 10.1109/ISCAS48785.2022.9937729.

A W-Band Amplifier in FinFET Technology

Yuen-Sum Ng
Graduate Institute of Communication Engineering
National Taiwan University
Taipei, Taiwan
d06943004@ntu.edu.tw

Yunshan Wang
Graduate Institute of Communication Engineering
National Taiwan University
Taipei, Taiwan
d01942017@ntu.edu.tw

Huei Wang
Graduate Institute of Communication Engineering
National Taiwan University
Taipei, Taiwan
hueiwang@ntu.edu.tw

Abstract—This paper presents a W-band power amplifier (PA) operating between 68-82 GHz in 16 nm fin field effect transistor (FinFET) technology. The proposed PA is implemented using transformer coupling with device size and layout optimization. The 3-stage PA achieves a high power gain of 21.4 dB, beneficial from the transformer with shunt capacitance and series inductance matching technique. This PA provides 12.6 dBm saturated output power (P_{sat}) and 20% of power-added efficiency (PAE) under a low supply voltage of 0.8 V.

Index Terms—Power Amplifier, W-band, Fin Field Effect Transistor (FinFET), transformer, MMICs

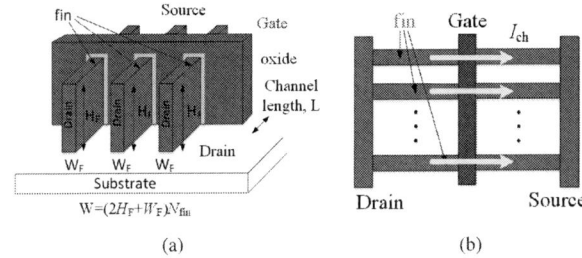

Fig. 1. (a) 3D structure of FinFET device [12] (b) Top view

I. INTRODUCTION

The FinFET technology is a three-dimensional structure device that has been used in mass production in different aspects of consumer electronics since 2013. This technology has the advantage of low leakages current when the short channel effect is severe in most advanced digital process technology [1]. In recent years, the millimeter wave application on CMOS has been emphasized because the requirement of the output power is not as demanding as those counterparts in sub-6 GHz application, but the research on the millimeter application using digital dominated FinFET technology has also not been fully investigated. In recent years, this topic has started to be a concern [2], [3] in order to integrate the phased-array system into system-on-chip (SoC), which reduces the power consumption of the chip-to-chip interface, and alter the phased-array system behavior in digital manner.

The E-band PA in [2], [3] have achieved sufficient output power used in the phased-array system at 71-75 GHz in 22 nm FinFET and 14 nm FinFET respectively under a low supply voltage of 1 V. It is proven that the FinFET technology is suitable in millimeter wave application in E-band or high frequency under phased-array System [4]. Other publications on E, W or D-band applications in planar MOSFET technology such as 65 nm [5]–[7], 40 nm [8], [9], and 28 nm [10], [11] have been investigated in recent years, that means the suitability of CMOS and FinFET in this spectrum for research.

This paper presents the design methodology of the FinFET in TSMC 16 nm FinFET process for W-band phased-array application. The P_{sat} of the PA is 12.6 dBm with 21.4 dB of gain and 20.2% of PAE at 78 GHz.

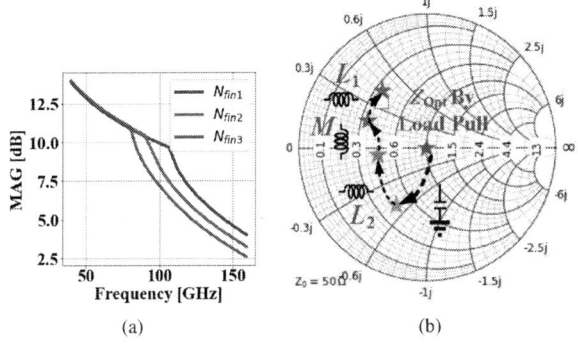

Fig. 2. (a) Device size selection by MAG simulation, where the values of N_{fin1}, N_{fin2}, N_{fin3} are 8, 10, and 12 respectively (b) Transformer matching with Shunt C

II. CIRCUIT DESIGN

The FinFET has the advantage of low leakage current which benefits the RF/mmW performance even though the gate resistance of FinFET is higher than the planar counterpart [13]. Fig. 1(a) and (b) shows the 3D structure and the top view of the FinFET devices. The transistor width W is equivalent to the following equation:

$$W = N_{fin}(2H_F + W_F) \quad (1)$$

where N_{fin} is the number of fin in the FinFET structure, H_F is the height of the fin and the W_F is the width of the fin. W_F and H_F are non-tunable parameters and the transistor width can only be adjusted by modifying the value of N_{fin}.

979-8-3503-4331-1/24 $31.00 © 2024 IEEE

SiRF 2024

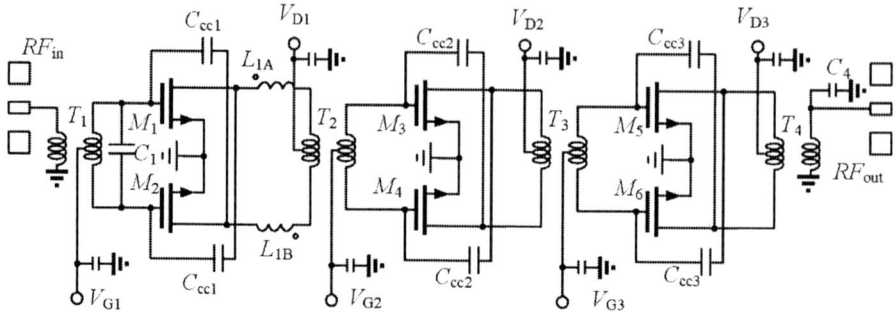

Fig. 3. Schematic of the proposed PA

TABLE I
CIRCUIT PARAMETERS OF PROPOSED PA

T_1	L_p :100 pH , L_s :83 pH , k:0.63	C_{cc1}	5 fF
T_2	L_p :56 pH , L_s :71 pH , k:0.63	C_{cc2}	12 fF
T_3	L_p :65 pH , L_s :57 pH , k:0.42	C_{cc3}	25 fF
T_4	L_p :90 pH , L_s :122 pH , k:0.62	L_1	30 pH
M_1, M_2	w:0.442 μm , nf:24 , m:1	C_1	20 fF
M_3, M_4	w:0.442 μm , nf:24 , m:2	C_4	40 fF
M_5, M_6	w:0.442 μm , nf:24 , m:4		

Note: L_p: Primary coil, L_s: Secondary coil, k: Coupling Coefficient, w: width, nf: finger number, m: no. of device in power cell

The device size of FinFET used in the proposed is chosen based on the optimization process under the combination of the number of fin ($N_{fini}, i = 1, 2, 3$) in the transistor, the number of fingers (nf), and maximum available gain (MAG) around W-band.

Fig. 2(a) shows the simulation of MAG value, where the values of N_{fin1}, N_{fin2}, N_{fin3} are 8, 10, and 12 respectively. The larger the number of N_{fin}, the wider the width of transistors, which degrades the MAG in high frequency, but the small value of the fin number does not provide enough MAG. It shows that N_{fin3} does not provide sufficient MAG around the W-band, and N_{fin1} covers the W-band, but MAG is not as good as N_{fin2}. Finally, a value of N_{fin2} which is equivalent to the transistor width of 0.442 μm and nf of 24 is chosen as the device sizes. The simulated f_t/f_{max} of the chosen device is 332/235 GHz respectively.

The three-stage PA architecture with simultaneous conjugate matching at the 1st stage, power matching at 2nd and 3rd stages [14] are adopted. First, the layout of all the transistor stages is built and simulated by Cadence EMX. Second, the load-pull simulation of 2nd and 3rd stages are performed by Cadence Spectre-RF to find out the optimum impedances for power delivery. For the 1st stage, simultaneous conjugate matching impedances, Γ_{Ms} and Γ_{ML}, are calculated with simulated S-parameter results.

The complete schematic of the PA is shown in Fig. 3, with the circuit parameter shown in Table I. The differential common source architecture is adopted in each power cell, with the capacitance cross-coupling technique to stabilize the potential instability in the differential cells.

There are compromises of the size of transformers, as large transformers exhibit good coupling effects but low self-resonant frequencies, which should be located away from the interested band. The PA is designed in the following procedure. First, the impedances of differential cells together with the cross-coupling capacitance are found by using the Cadence EMX simulator. Transformers are designed to match the cells after the input and output impedance of the whole differential power cells are known. Finally, series inductance or shunt capacitance are used to optimize those matching networks. Fig. 2(b) shows the matching strategy in the output transformer with shunt capacitance fine-tuning.

III. MEASUREMENT RESULTS

The proposed PA is fabricated in 16 nm FinFET technology. The chip micrograph is shown in Fig. 4 with a core area of 0.09 mm^2. A 0.8 V of supply voltage is used in all the stages. The gate bias of the 1st, 2nd and 3rd stages are 0.5, 0.5, and 0.6 V respectively.

The measurement setup is shown in Fig. 5(a) and the corresponding setup photo is shown in Fig. 5(b). Fig. 6 shows the large signal simulation and measurement result of the proposed PA. The PA achieves a P_{sat} of 12.6 dBm, with a gain of 21.4 dB and PAE of 20.2% at 78 GHz. Fig. 7 shows the S-parameter result which matches with the simulation. The 3-dB bandwidth of the S_{21} is 68-82 GHz.

The proposed PA achieves comparable P_{sat}, gain, and PAE with other FinFET publications [2], [3] as shown in Table. II. The proposed work with 80% of supply voltage achieves a similar level of output power. Although other planar

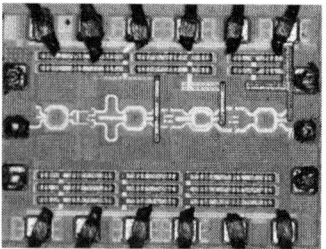

Fig. 4. Micrograph of the proposed PA

979-8-3503-4331-1/24 $31.00 © 2024 IEEE

TABLE II

COMPARISON TABLE OF THE PROPOSED WORK AND THE PREVIOUS PUBLICATION

	This Work	[2]	[3]	[10]	[9]	[5]	[15]	[6]
Process	**16 nm FinFET**	22 nm FinFET	14 nm FinFET	28 nm CMOS	40 nm CMOS	65 nm CMOS	28 nm FDSOI	65 nm CMOS
Architecture	**1 way 3 stages X_f w Sh-C# and Ser-L##**	1 way 2 stages X_f matching	1 way 3 stages X_f matching	1 way 2 stages X_f matching	2 ways 2 stages X_f matching	8 ways 1 stage TL matching	1 way 3 stage X_f FC matching	1 way 3 stage TL matching
Supply Voltage [V]	**0.8**	1	1	0.9	0.9	1	1	1
Meas. Freq [GHz]	**78**	75	71	79	78.5	79	77	77
3-dB BW [GHz]	**68-82**	62-86	65-75*	60-81.5*	70.3-85.5	71-82*	76-78*	74.3-86.2
Gain [dB]	**21.4**	16.6	16.7	17	18.1	24.2	26.5	24.4
OP_{1dB} [dBm]	**8**	5.7	2	8.25	17.8	16.4	10	12.1
P_{sat} [dBm]	**12.6**	12.8	7.4	12.3	20.9	19.3	13.5	15.4
Peak PAE [%]	**20.2**	26.3	8.9	13.8	22.3	19.2	14.5	10.4
Core Area [mm^2]	**0.09**	0.054	0.1	0.043**	0.19	0.855	0.14	0.433**

* Observation from the figures in the literature ** Observation from the figure excluding the PADs in the literature # Shunt capacitance ## Series inductance

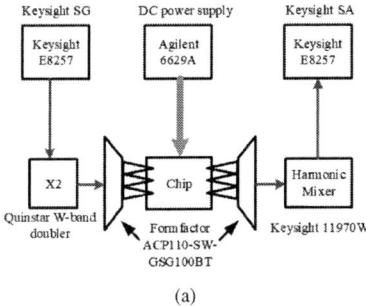

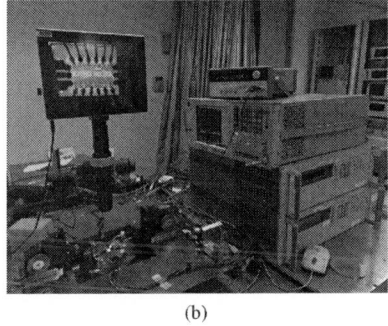

Fig. 5. (a) Measurement setup (b) photo

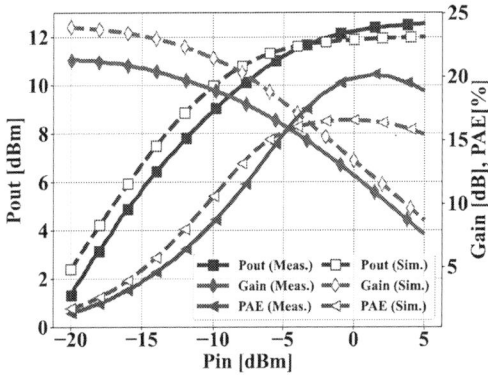

Fig. 6. Measured and Simulated Large Signal Measurement at 78 GHz

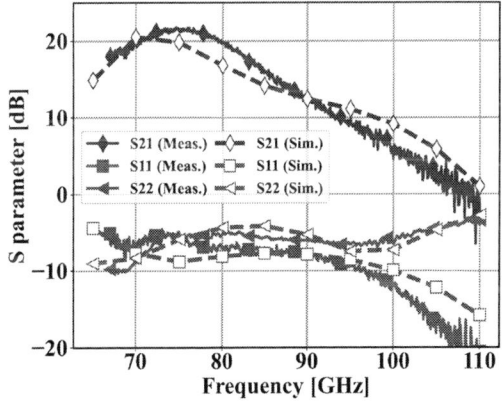

Fig. 7. Measured and Simulated S parameter

counterpart works of 65 nm [5]–[7], 40 nm [8], [9], and 28 nm [10], [11] achieve better output power and PAE, their supply voltages are 0.9 to 1.2 V or more than one power combining path. And the core area of the proposed work is relatively smaller compared with the planar counterpart works, which is more suitable to be integrated into SoC. The gain of the 3-stage proposed PA is lower than those of counterparts in [5], [15], and [6] because of less no. of combined paths, smaller transistors size, SOI technology, and lower supply voltage. The PAE of our work is lesser than [2] and [9] because the proposed work has fewer power channels and more stages.

The measured PAE is around 4 percentage points higher than the simulated PAE (∼16%) because the DC power consumption is lower than expected due to process variation.

The DC power consumption of measurement is around 80% of the simulation result. It is also in the acceptable range of process variation.

IV. CONCLUSION

This paper presents the 3-stage PA at 78 GHz implemented in 16 nm FinFET technology. The design methodology of using a transformer as a matching network together with shunt capacitance and series capacitance is used in both the gain and power matching. The proposed PA deliveries a P_{sat} of 12.6 dBm, with a gain of 21.4 dB and PAE of 20.2% at 78 GHz with 3-dB bandwidth in S_{21} of 68-82 GHz. The core area of the PA is 0.09 mm^2 which is small and comparable to other planar counterparts.

ACKNOWLEDGMENT

The authors wish to thank Taiwan Semiconductor Manufacturing Company (TSMC), Hsinchu, Taiwan for providing the university shuttle program in 16-nm FinFET technology, Taiwan Semiconductor Research Institute (TSRI) for providing measurement service, Cadence Design Systems for providing temporary university licenses and Amazon Web Service and Chunghwa Telecom for providing cloud computing services.

REFERENCES

[1] Cadence. [Online]. Available: https://resources.system-analysis.cadence.com/blog/msa2021-using-finfets-vs-mosfets-for-ic-design

[2] S. Callender, S. Pellerano, and C. Hull, "An E-Band Power Amplifier With 26.3 % PAE and 24-GHz Bandwidth in 22-nm FinFET CMOS," *IEEE Journal of Solid-State Circuits*, vol. 54, no. 5, pp. 1266–1273, 2019.

[3] S. Callender, S. Pellerano, and C. Hull, "A 73GHz PA for 5G phased arrays in 14nm FinFET CMOS," in *2017 IEEE Radio Frequency Integrated Circuits Symposium (RFIC)*, 2017, pp. 402–405.

[4] S. Pellerano, S. Callender *et al.*, "9.7 a scalable 71-to-76ghz 64-element phased-array transceiver module with 2? direct-conversion ic in 22nm finfet cmos technology," in *2019 IEEE International Solid- State Circuits Conference - (ISSCC)*, 2019, pp. 174–176.

[5] K.-Y. Wang, T.-Y. Chang, and C.-K. Wang, "A 1V 19.3dBm 79GHz power amplifier in 65nm CMOS," in *2012 IEEE International Solid-State Circuits Conference*, 2012, pp. 260–262.

[6] Y.-H. Hsiao, Y.-C. Chang *et al.*, "A 77-GHz 2T6R Transceiver With Injection-Lock Frequency Sextupler Using 65-nm CMOS for Automotive Radar System Application," *IEEE Transactions on Microwave Theory and Techniques*, vol. 64, no. 10, pp. 3031–3048, 2016.

[7] W. Wu, R. Chen *et al.*, "A Compact W-Band Power Amplifier With a Peak PAE of 21.1% in 65-nm CMOS Technology," *IEEE Microwave and Wireless Technology Letters*, vol. 33, no. 6, pp. 703–706, 2023.

[8] D. Zhao and P. Reynaert, "A 40-nm CMOS E-Band 4-Way Power Amplifier With Neutralized Bootstrapped Cascode Amplifier and Optimum Passive Circuits," *IEEE Transactions on Microwave Theory and Techniques*, vol. 63, no. 12, pp. 4083–4089, 2015.

[9] D. Zhao and P. Reynaert, "An E-Band Power Amplifier With Broadband Parallel-Series Power Combiner in 40-nm CMOS," *IEEE Transactions on Microwave Theory and Techniques*, vol. 63, no. 2, pp. 683–690, 2015.

[10] A. Medra, V. Giannini *et al.*, "A 79GHz variable gain low-noise amplifier and power amplifier in 28nm CMOS operating up to 125 °C," in *ESSCIRC 2014 - 40th European Solid State Circuits Conference (ESSCIRC)*, 2014, pp. 183–186.

[11] M. Vigilante and P. Reynaert, "20.10 A 68.1-to-96.4GHz variable-gain low-noise amplifier in 28nm CMOS," in *2016 IEEE International Solid-State Circuits Conference (ISSCC)*, 2016, pp. 360–362.

[12] B. Razavi, *Design of analog CMOS integrated circuits / Behzad Razavi.*, ser. McGraw-Hill series in electrical and computer engineering. Boston, MA: McGraw-Hill, 2006.

[13] B. Philippe and P. Reynaert, *Mm-wave circuit design in 16nm FinFET for 6G applications / Bart Philippe and Patrick Reynaert.*, ser. Analog Circuits and Signal Processing Ser. Cham, Switzerland: Springer, 2022.

[14] G. Gonzalez, *Microwave Transistor Amplifiers: Analysis and Design.* Prentice Hall, 1997.

[15] C. Nocera, G. Papotto *et al.*, "A 13.5-dBm 1-V Power Amplifier for W-Band Automotive Radar Applications in 28-nm FD-SOI CMOS Technology," *IEEE Transactions on Microwave Theory and Techniques*, vol. 69, no. 3, pp. 1654–1660, 2021.

A D-Band 28 nm CMOS-Bulk Power Amplifier with 12.8 dBm Output Power and 31.3 GHz 3 dB Bandwidth

Pascal Stadler
Ruhr University Bochum
Inst. of Integrated Systems
Bochum, Germany
pascal.stadler@rub.de

Hakan Papurcu
Ruhr University Bochum
Inst. of Integrated Systems
Bochum, Germany
hakan.papurcu@rub.de

Justin Romstadt
Ruhr University Bochum
Inst. of Integrated Systems
Bochum, Germany
justin.romstadt@rub.de

Nils Pohl
Ruhr University Bochum
Inst. of Integrated Systems
Bochum, Germany
Fraunhofer FHR
Wachtberg, Germany
nils.pohl@rub.de

Abstract—We present a 2-way, 4-stage power amplifier (PA) in TSMC's 28 nm CMOS-bulk technology. The D-Band PA consists of three capacitively-neutralized, common-source (CS) gain stages in conjunction with a cascode output stage. All stages are realized differentially with the interstage match, DC-block and bias voltages provided via the use of transformers. The PA achieves a saturated maximum output power of 12.8 dBm, small signal gain of above 36 dB, and 3 dB bandwidth of 31.3 GHz covering the range of 106-137.3 GHz. Its power consumption of 286 mW is derived from a dual-supply of 0.9/1.8 V with a current of 119 mA/99.5 mA, respectively. A maximum PAE of 6 % was determined from these values. The total PA takes up 0.32 mm^2 including the in- and output pads.

Index Terms—CMOS, CMOS Bulk, D-band, Power Amplifier, Power Combiner, PA, Radar, Transformer, 28nm

I. INTRODUCTION

In the wake of a rapidly increasing demand for high bandwidth circuitry, which was recently magnified by the interest in frequency bands above 100 GHz by the FCC (USA), Ofcom (UK) and CEPT/ETSI (Europe) [1], the cumulative attention of industry and researchers alike is directed towards mmWave frequencies. This attention enables fields such as imaging, distance/velocity measurements or near-field communications to thrive as a result of an increase in resolution with a rise in frequency [2]–[6]. Yet higher frequencies also necessitate effects like an enlarged free space attenuation to be overcome, which is why the need to provide ample power in the form of power amplifiers (PAs) also grows simultaneously.

While CMOS, especially its bulk variety, is not particularly known for high-power devices [8], it does have economic implications in mass production which makes it stand out from its competitors. Prospects with digital circuitry particularly are of high importance to create low-cost, compact and efficient solutions on a single silicon die. This is also aided by technology scaling since not only f_T of the transistors increase but also a significant decrease in the silicon footprint is achieved. As the focus of this work lies in designing a PA suitable for D-band radar systems with the possibility

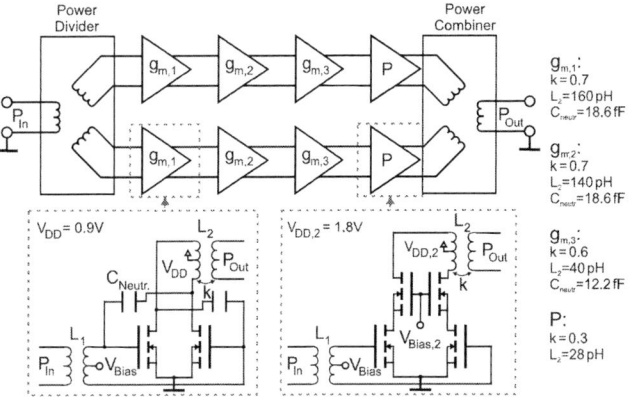

Fig. 1. Block diagram of the power amplifier and design of its various stages.

for digital interconnects, the choice of technology fell on TSMCs 28nm technology. The target for this PA was to provide a high bandwidth to cover the ISM- (122-123 GHz) and potential future automotive band (134-141 GHz) [7] along with a saturated output power high enough to produce a sufficient link budget for a CMOS radar system in regards to the free space loss.

II. CIRCUIT ARCHITECTURE AND DESIGN CONSIDERATIONS

Fig. 1 depicts a simplified schematic of the 2-way, 4-stage PA. Power-combining, i.e. connecting $n \in \mathbb{N}^{>1}$ PAs in parallel, is used as a means to ideally increase P_{Sat} of a saturated PA by $3 \cdot \log_2(n)$ dB with the downside of multiplying the power consumption by n. According to [9], a current combining scheme amounts to a lower susceptibility to parasitics, making it more appropriate for higher-frequency circuits. A figure displaying the 3D-model of the current combiner used in this work is given in Fig. 2. To ease measurement requirements a single-ended output was opted for. If the outer transmission lines (GND) are used instead of the inner one

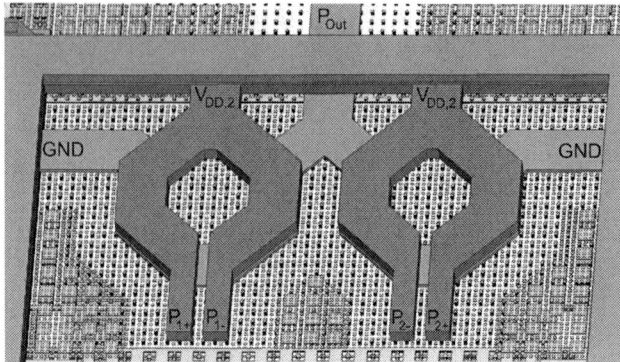

Fig. 2. 3D-model of the utilized power-combiner including filling structures.

(P_{Out}), the output can be adapted to be differential. The widest possible line widths were chosen based on the optimal output impedance lacking in capacitance. This lack could not be ameliorated by utilizing dedicated MOM capacitances, due to them generally being present in the lower metal layers. Their low metal thickness (high resistance) in conjunction with the proximity to the semi-conducting bulk deteriorates the transmission of the signal, which is why they were avoided as DC-blocks or matching elements in this work.

The PA is power-combined at its cascode output stages. By virtue of its common-gate stage, C_{DG} is mitigated, alleviating the demand for neutralization capacitances. A higher supply voltage compared to a common-source (CS-) stage is also leveraged to create a larger power swing. This stacking of transistors together with higher supply voltages is a common way to rectify the low breakdown voltages of FETs induced by transistor scaling. Caused by a reduction in channel length, C_{Gate} decreases, raising f_T. Simultaneously, a lower R_{DS} and smaller depletion area inhibits the voltage handling capabilities, making the transistor more sensitive to changes in potential. Exceeding technology-specific thresholds, the use of additional wells is required to not introduce a current into the substrate and retain sub-breakdown voltages. The benefit of reducing C_{Gate} will be compensated or even reversed as a result of bulk-well and FET-well capacitances. Hence, this work does not employ a high degree of transistor-stacking or circuit structures like current mirrors.

Using transmission line matching, as is typically the case for e.g. SiGe or SOI, is infeasible for the highly capacitive input impedance of CMOS-bulk transistors. The expected transmission line lengths in the order of $\frac{\lambda}{4}$ for a sufficient matching would add notably to the occupied area and introduce significant losses. Alongside the presence of low Q-capacitances, simultaneous provision of the load and supply of bias-potentials via the center-tap AC-grounds, this culminates in the prevalent use of transformers in analog bulk technologies. By varying the transformers' width, diameter and turn ratio, an optimum between the input match of the current stage and the load impedance of the previous stage was achieved. The antiproportional behavior of the Q- (losses) to the k-factor

(transmission) was also considered therein. To avoid substrate swelling, transformers usually require a specific metal density, which is supplied by means of filling structures. These can lead to large Q-factor reductions and have been taken into account when simulating transformers as can be seen in Fig. 2.

In order to make power combining for this PA viable, gain stages have to drive the output stages into saturation. Despite cascodes providing a higher gain per unit-stage, the drawback of a doubled power consumption along with the increase in supply voltage from 0.9 V to 1.8 V becomes evident. 3 common-source gain stages are therefore used alongside cross-coupled capacitances in the range of the transistors' C_{GD} to induce negative feedback. This in turn produces a peaked output power in the CS-stage, trading gain for stability. With careful design considerations, these peaks can be arranged to produce a wideband, amplified output signal.

III. MEASUREMENT RESULTS

A. Measurement Setup

D-band and W-band measurements were conducted with a Keysight PNA-X and Infinity GSG probes from FormFactor for the specific frequency range. A Virginia Diodes Inc (VDI) WR 6.5-/Keysight 1mm-extender along with a VDI Erickson PM5b/Keysight U8489A power meter were used to measure the output power under power-calibrated conditions, respectively. All of the probes, waveguides, tapers and pads have been de-embedded in the subsequent measurement results.

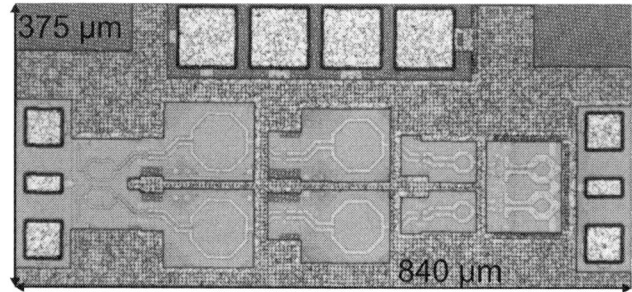

Fig. 3. Chip micrograph showing the 840 μm x 375 μm (0.32 mm^2) PA

B. Measurement Results

Fig. 3 shows a chip micrograph of the PA. It takes up a core area of 690.6 μm x 224.1 μm (0.15 mm^2) that increases to 840 μm x 375 μm (0.32 mm^2) with pads and draws 119 mA from a 0.9 V and 99.5 mA from a 1.8 V supply resulting in a total power consumption of 286 mW. Compared to simulations, the power consumption is roughly 30% lower and varied slightly over the frequency range. This suggests a shift in bias potential along with the presence of self-biasing. Concerning the former, the bias network is likely the origin by reason of it being made up of MOSFETs arranged as diodes together with physically small resistors. The resulting process variations and mismatch effects have been estimated

979-8-3503-4331-1/24 $31.00 © 2024 IEEE

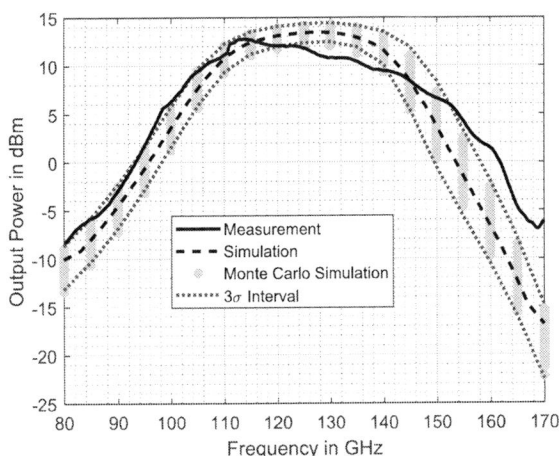

Fig. 4. Measured and simulated P_{Out} over the frequency at $P_{In} = 0\,dBm$. Monte Carlo simulations are given as grey point clusters with its 3σ intervals as red, dotted lines.

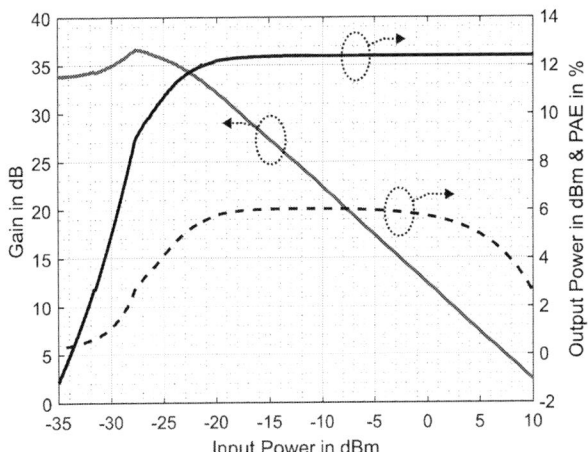

Fig. 5. 115 GHz measurement of gain, PAE and P_{Out} over the input power.

by Monte Carlo simulations (gray point clusters) and plotted with the simulated and measured output power in Fig. 4. Although the measured data set fits reasonably well with the simulation, larger discrepancies, unexplained by the Monte Carlo simulations, can be seen for frequencies >120 GHz. These might result from modeling related issues since the provided models were only validated up to 110 GHz. The widening discrepancy between simulation and measurement for frequencies past 135 GHz can be partially traced back to the devices' susceptibility to variations according to the large output power variation of the Monte Carlo simulation. Ultimately, a maximum output power of 12.8 dBm alongside a 3 dB bandwidth of 31.3 GHz spanning from 106-137.3 GHz with an almost seamless transition between the different extenders/frequency bands is present.

Indications of self-biasing can be perceived when inspecting the course of the PAs gain in Fig. 5. Instead of having a flat gain response at input powers below -27.5 dBm, a peaking behavior can be seen. The input power for the peaked gain may correlate to an optimal bias potential of the circuit. It could also be an artifact of the measurement setup because the utilized extension module fails to output accurate power levels for the low input powers required to drive the PA in the linear region even under power calibrated conditions. Nevertheless, the PA is driven in saturation rather efficiently with a PAE of 6% even for input powers as low as -17 dBm and provides a maximum gain above 36 dB.

When contrasted with state-of-the-art PAs of the 28 nm technology in Table I, the PA of this work exhibits both the highest output power as well as the highest gain. Moreover, a bandwidth on par with its competitors is exhibited. Yet its shortcoming lies within its efficiency when contrasted with the other PAs. If other technology nodes and CMOS technologies are considered, a lower P_{sat} is demonstrated while still retaining a competitive gain and bandwidth.

TABLE I
COMPARISON ON RECENTLY PUBLISHED CMOS D-BAND PAS

Ref.	Node (nm)	P_{sat} (dBm)	BW (GHz)	Gain (dB)	P_{DC} (mW)	PAE (%)	Core Area (mm²)
This Work	28	12.8	31.3	>36	286	6	0.15
[10]	28	8	21.8	22.7	78.5	6.6	0.027
[11]	28	11.8	41	21.9	140	10.9	0.24
[12]	28	6.9*	26.8	19	34.7	10.6*	0.027
[13]	28	6.3*	20#	18*	-	10*	-
[14]	22+	14.7	33	35.7	152	11.3	0,024
[15]	40	13.4	38.5	16	104.8	9.4	0.14

* Simulated, # Small signal BW, + SOI

IV. CONCLUSION

This article presented a fully differential, wideband 2-way, 4-stage power amplifier in TSMCs 28 nm bulk technology with a maximum saturated output power of 12.8 dBm, gain of above 36 dB, PAE of 6% and 3 dB-bandwidth of 31.3 GHz. 3 common-source, neutralized gain stages are used to drive a cascode output stage into saturation that is then power-combined to further increase the output power. Discrepancies between simulation and measurements were elucidated upon and deduced to be the effect of a shift in bias potential due to process variations and self-biasing. The resulting non-linear power progression is of minor concern due to the high gain and early saturation of the PA. The PA will thus be improved upon or used in future works to provide steady output power to antennas in radar systems.

REFERENCES

[1] S. Kueppers, T. Jaeschke, N. Pohl and J. Barowski, "Versatile 126–182 GHz UWB D-Band FMCW Radar for Industrial and Scientific Applications," LSENS, vol. 6, no. 1, pp. 1-4, Jan. 2022.

[2] T. T. Braun, J. Schöpfel, P. Kwiatkowski, C. Schweer, K. Aufinger and N. Pohl, "Expanding the Capabilities of Automotive Radar for Bicycle Detection With Harmonic RFID Tags at 79/158 GHz," TMTT, vol. 71, no. 1, pp. 320-329, Jan. 2023.

[3] S. Koch, M. Guthoerl, I. Kallfass, A. Leuther and S. Saito, "A 120–145 GHz Heterodyne Receiver Chipset Utilizing the 140 GHz Atmospheric Window for Passive Millimeter-Wave Imaging Applications," JSSC, vol. 45, no. 10, pp. 1961-1967, Oct. 2010.

[4] M. Kucharski, A. Ergintav, W. A. Ahmad, M. Krstic, H. J. Ng and D. Kissinger, "A Scalable 79-GHz Radar Platform Based on Single-Channel Transceivers," TMTT, vol. 67, no. 9.

[5] H. -J. Song and T. Nagatsuma, "Present and Future of Terahertz Communications," TTHZ, vol. 1, no. 1, pp. 256-263, Sept. 2011.

[6] R. Appleby and R. N. Anderton, "Millimeter-Wave and Submillimeter-Wave Imaging for Security and Surveillance," JPROC, vol. 95, no. 8, pp. 1683-1690, Aug. 2007.

[7] A. Filippi, V. Martinez and M. Vlot, "Spectrum for Automotive Radar in the 140 GHz Band in Europe,", EuRAD, 2022, pp. 1-4.

[8] Stadler, Pascal, et al. "An overview of state-of-the-art D-band radar system components." Chips 1.3 (2022): 121-149.

[9] Q. J. Gu, Z. Xu and M. -C. F. Chang, "Two-Way Current-Combining W -Band Power Amplifier in 65-nm CMOS," TMTT, vol. 60, no. 5, pp. 1365-1374, May 2012.

[10] X. Tang et al., "Design of D-Band Transformer-Based Gain-Boosting Class-AB Power Amplifiers in Silicon Technologies," TCSI, vol. 67, no. 5, pp. 1447-1458.

[11] J. Zhang, T. Wu, L. Nie, S. Ma, Y. Chen and J. Ren, "A 120–150 GHz Power Amplifier in 28-nm CMOS Achieving 21.9-dB Gain and 11.8-dBm Psat for Sub-THz Imaging System,", ACCESS, vol. 9, pp. 74752-74762, 2021.

[12] S. Park et al., "A D-Band Low-Power and High-Efficiency Frequency Multiply-by-9 FMCW Radar Transmitter in 28-nm CMOS," JSSC, vol. 57, no. 7, pp. 2114-2129, July 2022.

[13] S. Park et al., "A 135-155 GHz 9.7%/16.6% DC-RF/DC-EIRP Efficiency Frequency Multiply-by-9 FMCW Transmitter in 28 nm CMOS," 2021 IEEE Radio Frequency Integrated Circuits Symposium (RFIC), Atlanta, GA, USA, 2021, pp. 15-18,

[14] X. Tang, J. Nguyen, G. Mangraviti, Z. Zong and P. Wambacq, "Design and Analysis of a 140-GHz T/R Front-End Module in 22-nm FD-SOI CMOS," JSSC, vol. 57, no. 5, pp. 1300-1313, May 2022.

[15] H. S. Son, T. H. Jang, S. H. Kim, K. P. Jung, J. H. Kim and C. S. Park, "Pole-Controlled Wideband 120 GHz CMOS Power Amplifier for Wireless Chip-to-Chip Communication in 40-nm CMOS Process," TCSII, vol. 66, no. 8, pp. 1351-1355, Aug. 2019.

The role of Varactor, the Nonlinear Semiconductor, for Next Generation of Intelligent and Reconfigurable Radio Nodes

Najme Ebrahimi
Northeastern University
USA
University of Florida
USA
n.ebrahimi@northeastern.edu

Abstract—This paper highlights the significance of intelligent and reconfigurable radio nodes, incorporating the nonlinear and tunable semiconductor, varactors, in shaping the future of wireless communication. It presents their use in frequency generations at different bands, including IoT to mm-wave (MMW), in intelligent reflecting surfaces (IRS) for reflecting incident waves with beamforming, and enhanced performance of a machine learning (ML)-based system for number of tag/sensor detection and channel estimations. The focus is on the recent developed technique on a nonlinear ring resonator (NRR) capable of generating harmonic and subharmonic dual bands at 2.4 GHz and 4.8 GHz, extending its application to MMW frequencies at 60 GHz and 120 GHz. Additionally, the utilization of IRS using the proposed ring-based resonator is demonstrated, showcasing their capability in achieving 300-degree phase shifting. The paper also explores the integration and enhancement of the proposed dual band, tunable and reconfigurable circuit with ML-based tag collision recovery technique, enabling multi-tag and sensor detection. This leads to a remarkable 10 dB improvement in signal-to-noise ratio (SNR) and overall throughput of the system. The presented research contributes insights and advancements for the future intelligent communication and sensing technologies employing tunable and reconfigurable circuit and devices.

Keywords— IoT/RFID, mm-wave (MMW), Varactor, Nonlinear phenomena, Intelligent reflecting surfaces (IRS)

I. Introduction

The integration of intelligent and reconfigurable Internet of Things (IoT) sensors and radio nodes, along with Radio Frequency Identification (RFID) tags, plays a vital role in shaping the next generation of wireless communication, smart devices, Machine-to-Machine (M2M) communication, and environmental sensing [1-14]. With an increasing number of connected sensors and nodes within networks, the need for optimized algorithms and network structures arises to dynamically adjust system parameters. To address this, a reconfigurable front-end is required for efficiently tuning and adapting to evolving system requirements, including critical adjustments for interference mitigation, number of tag/sensor and channel estimation, beamforming, and the integration of intelligent reconfigurable surfaces (IRS), Fig. 1.

Tunable components, such as varactors and inductance, enhance the intelligence and reconfigurability of electronic systems. The adjustability of these components allows for fine-tuning of circuit parameters, such as resonant frequencies, time constants, phase delay, and impedance matching. Varactors

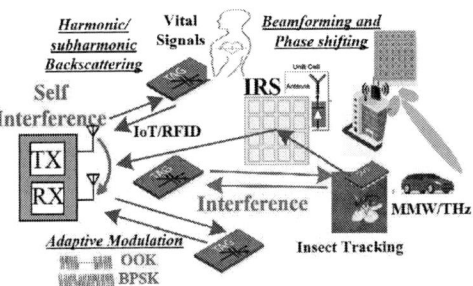

Fig 1: Next generation of intelligent and reconfigurable wireless network for interference mitigation, beamforming and the IRS, number of tag/sensor detection and channel estimation.

particularly play a pivotal role in modern wireless communications, offering nonlinear frequency generation capabilities [1-9], adaptability to harmonic and subharmonic (multi-band) frequency generation for interference mitigation and frequency diversity [1], and advanced phase shifting and beamforming techniques [6-7], [10-11]. Notably, for the latter application in IRS, varactors control the phase and amplitude of reflected electromagnetic waves, actively shaping the radiated wavefront, and enabling efficient beamforming, [10-11].

This paper focuses on the fundamental principles and functionality of varactors as nonlinear semiconductors within reconfigurable radio nodes. It explores their applications for frequency generation and phase shifting, ranging from IoT (2.4 GHz) to MMW and high-frequency bands (60 GHz/120GHz). The concept of utilizing the newly developed nonlinear ring resonator (NRR) as a standing wave resonator for efficient harmonic and subharmonic frequency generation is investigated [1-2]. The integration of the developed NRR unit cell with the controllable components, varactors/diodes, into an array of IRSs will be studied for phase shifting and reflective beamforming. Moreover, the paper details the incorporation of the multi-band circuitry within an ML-aided algorithm to enhance the efficiency in number of tag/sensor and channel estimation, aiming to pave the way for next-generation systems.

II. Fundamentals of Varactor, the Nonlinear Semiconductor, for Frequency and Phase Generation

The nonlinear electronic components such as Metal-Oxide-Semiconductor Field-Effect (MOSFET) varactors or PIN diodes have been utilized for frequency and phase generation,

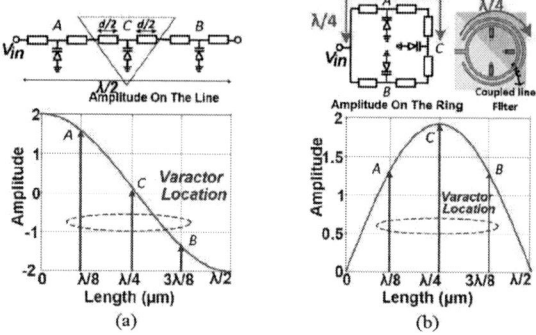

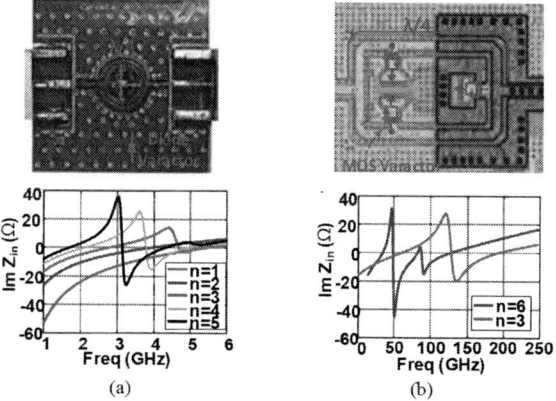

Fig. 2: Periodic transmission line for frequency generation (a) travelling wave in $\lambda/2$ line requiring multiple stages, b) proposed standing wave resonator in ring achieving maximum voltage swings with minimum number of stages.

Fig. 3: The implemented dualband, bidirectional NRR circuit with their resonance behaviour and input impedance for different number of unit-cells, (a) PCB for RFID and IoT at 2.4GHz/4.8 GHz using silicon varactor [1-2], (b) Chip for 60GHz/120 GHz band using 90nm SiGe (developed using GlobalFoundries)

leveraging their distinctive nonlinear characteristics [1,7,8,9,10]. The discrete component such as PIN diode is suitable for low-frequency applications, like IoT and RFID for harmonic and subharmonic backscatter communications [1-2]. The FET-based varactors, or Schottky diodes are preferred for high-frequency applications where silicon-based integration is preferred. In both cases, the variable capacitance depends on the inherent junction capacitance of the technology, the reverse biasing voltage, and the input voltage swing. At the onset of the operation, the capacitance of the varactor, $C_v(t)$, is related to the applied voltage/bias, v_m, that can be estimated using a Taylor series expansion. The capacitance range of a MOS varactor for 90nm SiGe technology provides the steepest nonlinearity slope in the sub-threshold region. On the other hand, the subharmonic generation relies on having a suitable circuitry with an adequate pump signal, which is necessary to generate the negative resistance from varactor's equivalent series resistant. This is essential for establishing the oscillation condition required for subharmonic generation. To achieve harmonic or subharmonic generation, it is essential to have an efficient circuitry capable of delivering the required voltage swing and pump signal across each nonlinear component. However, these matching and termination circuitry can be space-consuming for low-frequency applications and introduce losses in higher frequency applications.

To enhance the conversion efficiency of the harmonic/subharmonic generator, a parametric approach utilizing periodic/artificial transmission lines [8-9] have been proposed. This circuit involves cascading multiple stages of a non-linear transmission line (NLTL) with a Bragg frequency (f_{bragg}) set by the equivalent inductance of the T-line ($d/2$ in Fig. 2(a)) and the capacitance of the varactor (Fig. 2(a)). This configuration creates a low pass filter characteristic, which can be optimized along with the number of periodic cells (n) to achieve the improved output power and conversion loss. The circuit operates based on the traveling wave concept, resulting in waveform generation through constructive superposition from multiple varactors (designated as 'n') aligned with the propagation direction, Fig. 2(a). The prior work demonstrates optimum number of 9 stage for 2nd harmonic generation of RFID tag at 3.5 GHz, [9], and 10 stages for 24 GHz frequency divider applications [8].

A. Proposed Harmonic and Subharmonic Frequency Generation from IoT sensors towards MMW Applications

In this talk, we will present our latest innovative design that effectively minimizes the number of periodic and cascaded cells to $n=3$, while generating bidirectionally harmonics and subharmonics signals [1-2]. This architecture will be employed to applications spanning from IoT/RFID, 2.4 GHz to MMW, 60 GHz frequency bands. The developed frequency generation circuitry is based on a novel nonlinear ring resonator (NRR) operating on standing wave resonation to optimize the voltage swing at the varactors' nodes. Fig. 2(b) depicts the standing-wave voltage distribution along the $\lambda/2$ transmission line in ring configurations for different nodes. In a ring transmission line with a half-wavelength, it results in an open end at its center, node C, at the quarter-wavelength from input. At the center of the ring, the voltage experiences maximum swing because of the constructive and destructive interference of the incident and reflected waves along the transmission line, forming a standing wave pattern. In addition, at the locations with $\lambda/8$ distance from the input demonstrate significant swings, making them suitable for varactor placement. An additional significant advantage of periodic capacitive loaded configurations is their capacity to achieve a compact design due to the slow-wave property enabled by a larger propagation constant.

The proposed circuitry can be optimized for bidirectional operation, serving as both harmonic and subharmonic generators for use in a backscatter system. A dual-band system can enhance backscattering communication and localization performance by mitigating the self-interference issue at the reader, through separating the transmitting and receiving bands [1,9]. Additionally, the dual band system facilitates reconfigurable configurations by introducing frequency diversity to the system, see Sec. III. The sustainable dual band resonance conditions occur by creating zero, short impedance, or pole, infinite impedance, at excitation ports that can be transformed to each other using the inherent quarter-wavelength of the proposed NRR. Fig. 3(a) demonstrates the printed circuit board (PCB) implementation of the NRR at 2.4/4.8 GHz band using Skyworks Silicon PIN diode presented in [1-2] and Fig. 3(b) presents the new developed chip

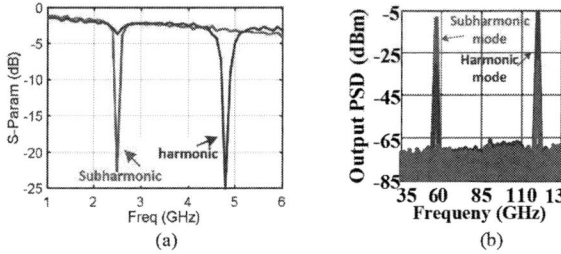

Fig. 4: Dual-band harmonic and subharmonic generation tag/sensor: (a) measured S-parameter response of subharmonics and harmonics of IoT tag, (b) output power spectral density (PSD) for MMW tag (Sim-EM-extracted).

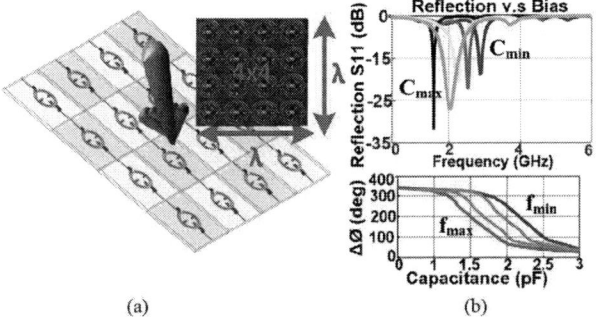

Fig. 5: (a) Conceptual illustration of scalable array of proposed NRR resonators for IRS implementation (b) top: simulated tunability of S_{11} and the reflection waveform, bottom: simulated achieved phase shift range.

implementation (using 90nm SiGe technology) of the harmonic generation at 60/120 GHz band, to be presented in this talk. The input impedance for the two NRR for various numbers of periodic unit cells, denoted as n are also plotted in Fig. 3(a) and (b). The results indicate that with a minimum of $n=3$, it is possible to achieve two resonance conditions for the two frequency bands, 2.4/4.8 GHz and 60/120 GHz bands. This configuration leads to a zero impedance (short impedance) at the subharmonic frequency and an infinite impedance (pole) at the harmonic bands. The designated boundary conditions and the quarter wavelength of the proposed NRR also creates a bandpass filter (BPF) based on the coupled line theory to apply/extract harmonic/subharmonic signals at the input port while transferring the zero and pole impedance nodes at each port. Fig. 4(a) demonstrates the measurement results of the S-parameter, S_{11}, at both ports, for IoT tag while Fig. 4(b) illustrates the Em-extracted output power spectral density (PSD) for the MMW tag using 90nm SiGe. The results verify the dual band operation of subharmonic and harmonic generation with near 15 to 25 dB signal rejection at each mode for MMW and IoT bands, respectively.

B. Application of Phase shifting and Beamforming for RIS Surface

Intelligent reflecting surfaces (IRS) are a type of smart surface that composed of an array of reconfigurable reflecting unit cells [10-11]. By integrating controllable components like varactors or PIN diodes into these unit cells, IRSs can be programmed for reflective beamforming [10]. Common designs for IRSs include the use of metallic patches integrated with varactor diodes and the split-ring-resonators (SRR) [11]. The SRR can

be externally driven by a magnetic field aligned with the axis of the ring to resonate within a narrow frequency band, that can be modelled as LC resonant tanks. The developed NRR frequency generation in Sec. II. A can be also employed as a tunable SRR, Fig. 5(a). The inclusion of varactors/diodes in the circular ring can enable the tuning of its resonant frequency or phase response. For the incident signal, by varying the bias voltage across the varactors, the effective electrical length of the circular ring can be modified, allowing for the adjustment of the phase shift it introduces to the reflected wave. This enables accurate modulation of the reflected wave, facilitating beamforming by switching different capacitance values.

Fig. 5(b) demonstrates the S_{11} and reflection characteristics of the proposed NRR at 2.4 GHz at various biasing states and with different tunable capacitance values, offering the capability of adjusting the transmitted waveform at different resonance frequencies. The tuning range of the NRR is 20% that can provide near 300° phase shift range through tuning the varactor from minimum to maximum capacitance values, $\Delta C \approx$ 1 pF. Once an IRS is accurately modelled to characterize its reflective phase response, the properties of the surface such as beamforming can be easily managed using the controllable unit.

III. TUNABLE VARACTOR AND FREQUENCY DIVERSITY IN ML-AIDED MUTI-TAG COLLISION RECOVERY TECHNIQUE

A significant obstacle faced by future multi-node connectivity and sensing is dealing with collisions that arise when numerous RFID tags or sensors try to communicate with the reader simultaneously. Various collision recovery techniques have been proposed, including randomizing the transmission procedure by frame slotted ALOHA (FSA) protocol [12], exploring joint-decoding of multiple tags, utilizing antenna arrays with blind source separation techniques [13].

Estimating the number of tags involved in a collision is a crucial step in collision recovery algorithms. In this work, initially, the number of colliding tags is estimated accurately using ML tools, Fig. 6(a). In [14], It has been observed that the number of tag/sensors can be extracted from tag collision using received signal strength (RSS) method, where in a collision involving S tags 2^S power states will be present, creating M-QAM like constellation points ($M= 2^S$). The received signal creates constellation points containing both in-phase and quadrature (I/Q) components. An example of the constellation points in the I/Q plane are plotted in Fig. 7(a), for 4 tags in a single band (only harmonic or subharmonic mode). By observing the amplitudes of both I/Q components, 16 distinct clusters corresponding to different numbers of colliding tags are identified. The I/Q cluster points for 4 tags using the proposed dual band NRR tag modes as shown in Fig. 6(b), resulting the enhancement of distinction between the constellation points, illustrated in Fig. 7(b). A simple yet effective deep learning model is then employed to find the experienced channel coefficients, enabling the separation of each tag's signal from the received signal using maximum likelihood decoding in two bands [3]. The proposed approach uses four different feed-forward neural network models, each trained with 4 fixed additional symbols, to estimate the channel coefficients for the

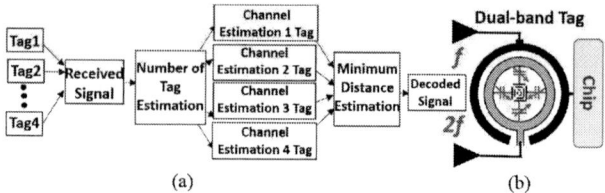

(a)

(b)

Fig. 6: (a) The workflow diagram of the proposed ML-aided collision recovery algorithm for number of tags up to 4. (b)The proposed dual band NRR tag integrated with the detection algorithms.

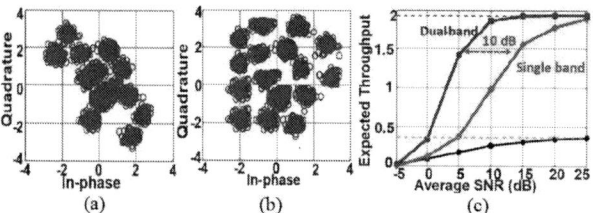

(a)   (b)   (c)

Fig. 7: The received constellation points containing both in-phase and quadrature (I/Q) components for 4 simultaneous tags, (a) for single band, (b) for proposed dual-band NRR, (c) enhancement of the throughput of the proposed dual band NRR tags in the proposed tag collision recovery algorithms.

given number of tags. These additional symbols are orthogonal to each other and aid in extracting the channel coefficients from the received signal corrupted by additive white Gaussian noise (AWGN). In this measurement, both training and testing are done on synthetic data generated by Universal Software Radio Peripheral. The channel model assumes on-off keying modulation at the tags, a single-antenna receiver at the reader, additive noise, and leakage between the Tx and Rx antennas at the receiver. With the estimated number of tags and the corresponding channel coefficients, a minimum distance decoder is applied to separate the transmitted signals of the passive tags, Fig. 6. This process allows the system to recover individual tag signals from the received signal, even in cases of collisions. The proposed method in combination with the proposed dualband tag circuit (Fig. 6(b)) achieves a throughput boost from 0.368 to 1.837 and can decode up to 4 tags [3]. Therefore, employing the proposed dual-band NRR tag into this proposed ML-aided algorithm in [3] enhance the required signal to noise ratio (SNR) by 10 dB to achieve the targeted throughput, Fig. 7(c). A dual-band system significantly enhances collision recovery techniques in passive UHF-RFID due to increased estimation accuracy, improved signal separation, and leveraging different channel characteristics. Frequency diversity allows separate processing of signals from different bands, reducing interference and improving tag signal accuracy. Utilizing multiple frequency bands also enables leveraging variations in channel characteristics to refine the estimation and recovery process further.

IV. CONCLUSION:

In conclusion, the integration of intelligent and reconfigurable radio nodes with varactors holds tremendous promise for advancing wireless communication technologies.

Varactors, with their ability to dynamically adjust circuit parameters and facilitate nonlinear frequency generation, provide essential adaptability, frequency diversity and interference mitigation in the ever-expanding network of connected devices. Their significance spans from IoT and RFID to MMW and high-frequency bands. Furthermore, the incorporation of varactors in IRS empowers precise beamforming, adding another layer of versatility. Additionally, the use of dual-band and reconfigurable varactors in ML-aided collision recovery techniques paves the way for improved tag identification, signal separation, and channel estimation. Overall, the advancements in tunable semiconductor technology are needed to shape the next generation of reconfigurable and intelligent Internet of Things system and wireless communication paradigm.

REFERENCES

[1] P. Pahlavan, S. Z. Aslam and N. Ebrahimi, "A Novel Dual-band and Bidirectional Nonlinear RFID Transponder Circuitry," 2022 IEEE/MTT-S International Microwave Symposium - IMS 2022, 2022, pp. 44-47.

[2] P. Pahlavan, N. Ebrahimi, "Dual-band Harmonic and Subharmonic Frequency Generation Circuitry for Joint Communication and Localization Applications Under Severe Multipath Environment" *arXiv:2110.15363*

[3] T. Akyıldız, R. Ku, N. Harder, N. Ebrahimi and H. Mahdavifar, "ML-Aided Collision Recovery for UHF-RFID Systems," 2022 IEEE International Conference on RFID (RFID), 2022, pp. 41-46.

[4] N. Ebrahimi, H. -S. Kim and D. Blaauw, "Physical Layer Secret Key Generation Using Joint Interference and Phase Shift Keying Modulation," in IEEE Transactions on Microwave Theory and Techniques, vol. 69, no. 5, pp. 2673-2685, May 2021.

[5] N. Ebrahimi, K. Sarabandi, J. F. Buckwalter, "A 71-76 / 81-86 GHz, E-band, Phased Array Transceiver Module With Image Selection Weaver Architecture for Low EVM Variation" 2020 IEEE Radio Frequency Integrated Circuits Symposium (RFIC), 2020, pp. 95-98.

[6] N. Ebrahimi and J. F. Buckwalter, "A High-Fractional-Bandwidth, Millimeter-Wave Bidirectional Image-Selection Architecture with Narrowband LO Tuning Requirements," in IEEE Journal of Solid-State Circuits, (JSSC), vol. 53, no. 8, pp. 2164-2176, Aug. 2018.

[7] N. Ebrahimi, P. Wu, M. Bagheri and J. F. Buckwalter, "A 71–86-GHz Phased Array Transceiver Using Wideband Injection-Locked Oscillator Phase Shifters," in IEEE Transactions on Microwave Theory and Techniques (TMTT), vol. 65, no. 2, pp. 346-361, Feb. 2017.

[8] Lee, Wooram, and Ehsan Afshari. "Distributed parametric resonator: A passive CMOS frequency divider." IEEE journal of solid-state circuits 45.9 (2010): 1834-1844.

[9] Yu, Fan, Keith G. Lyon, and Edwin Chihchuan Kan. "A novel passive RFID transponder using harmonic generation of nonlinear transmission lines." IEEE transactions on microwave theory and techniques 58.12 (2010): 4121-4127

[10] Venkatesh, S., Lu, X., Saeidi, H. et al. A high-speed programmable and scalable terahertz holographic metasurface based on tiled CMOS chips. Nat Electron 3, 785–793 (2020).

[11] Q. Wu and R. Zhang, "Towards smart and reconfigurable environment: Intelligent reflecting surface aided wireless network," IEEE Commun., vol. 58, no. 1, pp. 106–112, Nov. 2019

[12] F. Schoute, "Dynamic frame length ALOHA," IEEE Trans. Commun., vol. 31, no. 4, pp. 565–568, Apr. 1983.

[13] C. Huang, C. Zhang, J. Yang, B. Sun, B. Zhao, and X. Luo, "Reconfigurable metasurface for multifunctional control of electromagnetic waves," Advanced Optical Materials, vol. 5, no. 22, p. 1700485, Sept. 2017.

[14] R. S. Khasgiwale, R. U. Adyanthaya, and D. W. Engels, "Extracting information from tag collisions," in Proc. IEEE Int. Conf. RFID, Orlando, FL, Apr. 2009, pp. 131–138.

979-8-3503-4331-1/24 $31.00 © 2024 IEEE

A Compact, High Tuning Accuracy and Enhanced Linearity 37-43 GHz Digitally-Controlled Vector Sum Phase Shifter

Mehran Hazer Sahlabadi
Electrical and Computer Engineering dept.
University of Waterloo
Waterloo, Canada
mhazersa@uwaterloo.ca

Hang Yu
Electrical and Computer Engineering dept.
University of Waterloo
Waterloo, Canada
h78yu@uwaterloo.ca

Jingjing Xia
Electrical and Computer Engineering dept.
University of Waterloo
Waterloo, Canada
jingjing.xia@uwaterloo.ca

Slim Boumaiza
Electrical and Computer Engineering dept.
University of Waterloo
Waterloo, Canada
slim.boumaiza@uwaterloo.ca

Abstract—This paper presents a compact digitally controlled vector sum phase shifter (DC-VSPS) capable of precise gain and phase control across a wide bandwidth. The DC-VSPS utilizes a pair of differential variable gain amplifiers (VGAs) designed to optimize gain and phase tuning accuracy. Moreover, it incorporates differential transformers-based output combiner and input quadrature hybrid, enhancing integration, bandwidth, and matching performance. A 6-bit DC-VSPS prototype was successfully designed and fabricated using the 45 nm silicon-on-insulator (SOI) CMOS technology, occupying a minimal core area of 0.084 mm^2. Experimental results demonstrate an excellent tuning range of $360°$ for phase control with a resolution of 6 bits and 14 dB for gain control with 1 dB resolution. Furthermore, comprehensive measurements indicate excellent performance over the frequency range of 35 - 43 GHz, with total root-mean-square (RMS) phase error, total RMS gain error, and group delay variation all within $1.5°$, 0.24 dB, and ±12 picoseconds, respectively. The prototype also exhibits input-referred 1 dB gain compression power levels exceeding 4 dBm and input-output return losses better than 9 dB across the entire bandwidth.

Index Terms—Active phase shifter, millimetre wave (mm-wave), Digitally controlled vector sum phase shifter

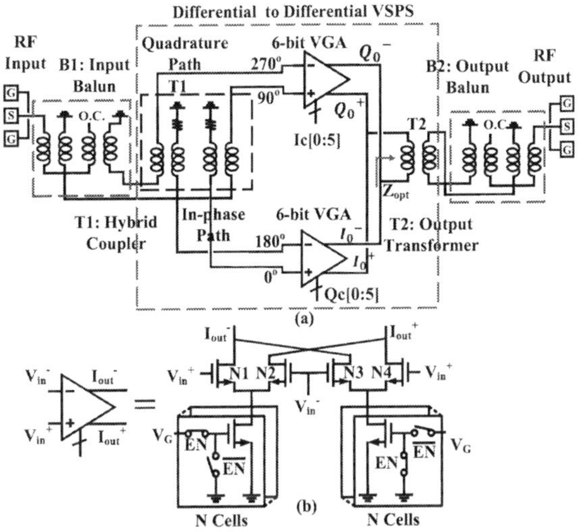

Fig. 1. (a) Block diagram of the proposed VSPS. (b) VGA schematic.

I. INTRODUCTION

The advent of the millimeter wave (mm-wave) spectrum has brought forth exciting opportunities for the deployment of cutting-edge wireless and satellite communication systems, capable of supporting ultra-high-speed links in the high-Gigabits per second (Gbps) range. However, the challenges posed by high path losses at mm-wave frequencies necessitate innovative solutions to improve coverage range and communication performance. Large-scale phased array antenna (LSPAA) systems have emerged as a pivotal technology, leveraging their ability to direct radiation in focused directions, thereby enhancing communication range and efficiency. The variable phase and gain block (VPGB) plays a crucial role in LSPAA systems, as its phase resolution and accuracy directly influence critical aspects such as beam directivity, steering resolution, and pointing accuracy. Given the complexity of LSPAA systems, multiple radio frequency (RF) signal paths are required, with each path incorporating a dedicated VPGB. Thus, it is imperative for the VPGB to possess key characteristics, including compact size, high phase and gain tuning range, low insertion loss (IL), wide bandwidth, and maximal tuning accuracy, all while minimizing DC power consumption and non-linear distortions.

Conventionally, the VPGB is realized through either a cascade of active/passive phase shifter (PS) and variable gain amplifier (VGA) or attenuator (Att) [1], [2], an active VGA-based vector sum phase shifter (VSPS) [3]–[9] or a passive Att-based VSPS [10], [11], [12]. However, each of these configurations has its own drawbacks: the cascade of

979-8-3503-4331-1/24 $31.00 © 2024 IEEE

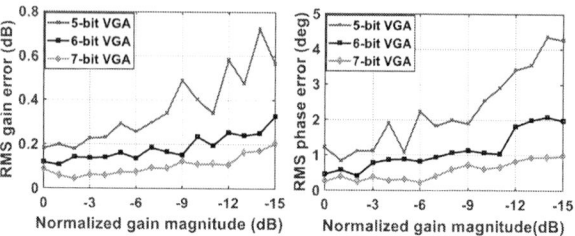

Fig. 2. Calculated DC-VSPS (a) gain error and (b) phase error versus the normalized gain magnitude and the VGA bit resolution.

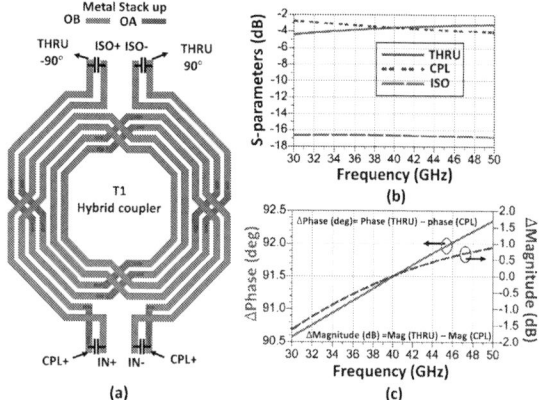

Fig. 3. Transformer based hybrid 90° coupler (a) layout view, (b) S-parameters (dB), and (c) phase and gain difference between the through and coupled ports.

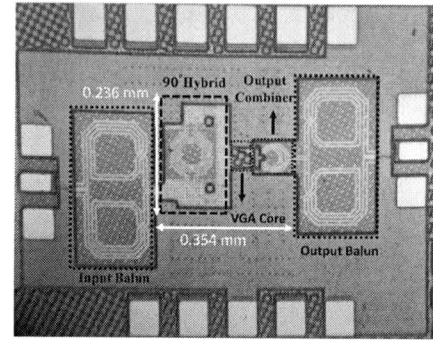

Fig. 4. Micro-photograph of the DC-VSPS prototype.

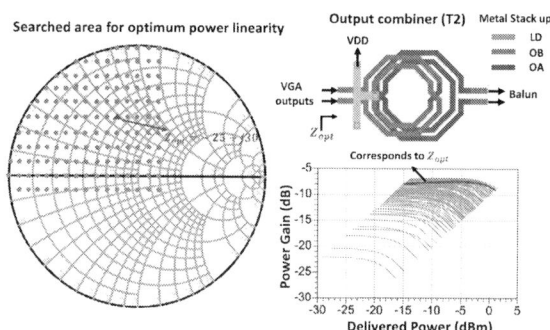

Fig. 5. VSPS power gain vs input power for different load termination.

components is area-intensive, the active VSPS suffers from poor linearity and unidirectionality, and the passive VSPS exhibits high insertion loss (IL).

To address the pressing requirements of small chip area, high tuning accuracy over the entire gain and phase range and frequency bandwidth, and linearity in mm-wave LSPAA applications, this work proposes a compact digitally-controlled VSPS (DC-VSPS). The proposed DC-VSPS offers precise phase and gain tuning accuracy and enhanced linearity.

II. CIRCUIT DESIGN

Fig. 1(a) illustrates the simplified block diagram of the proposed DC-VSPS, comprising a 90° folded transformer-based hybrid coupler (T1) inspired by [13], two 6-bit digitally-controlled VGAs (DC-VGA), and a transformer-based output current combiner (T2). The layout and full EM simulation results of the designed 90° hybrid coupler is depicted in Fig. 3. As illustrated in Fig. 3, the hybrid exhibits an insertion loss of about 0.5 dB at 40 GHz and an isolation and return loss at all ports of better than 16.5 dB, and 13 dB, respectively, over the frequency range of 30 to 50 GHz. Moreover, the phase and gain imbalance between the through (THRU) and coupled (CPL) ports are less than 1.8° and 0.4 dB, respectively, from 37 to 43 GHz. For testing purposes, wide band baluns are included at the input and output (B1 and B2). The operation principle of the VSPS is as follows: the hybrid coupler divides

the RF input signal into in-phase (I_{RF}) and quadrature-phase (Q_{RF}) signals with equal magnitude and a 90° relative phase difference. These I_{RF} and Q_{RF} signals are then weighted by two separate VGAs, controlled by signals (I_C) and (Q_C) respectively. The combination of the VGA output signals results in the final output signal, which represents the magnitude-scaled and phase-adjusted version of the input RF signal.

A. VGA Design

Fig. 1(b) provides a detailed schematic of the proposed 6-bit VGA. The chosen VGA core topology incorporates a single trans-conductance stage with multiple digital control cells (63 equally sized unit cells) for current steering. Inspired by [6], the VGA topology is modified in the digital control unit cells to minimize area and layout complexity. The trans-conductance stages (N1-N4) are implemented using n-type transistors, each with a gate width and length of 30 μm and 45 nm, respectively. To achieve an acceptable balance between RF performance and design complexity, a study is conducted to determine the number of VGA unit cells for the desired gain and phase accuracy of the proposed DC-VSPS. Assuming an ideal VSPS constructed with ideal VGA, splitter, and combiner, the RMS gain (dB) and phase (deg) errors are calculated for a gain tuning range of up to 15 dB with 1 dB resolution while varying the number of VGA bits between 5 and 7.

Fig. 2 demonstrates that the gain and phase errors of the VSPS degrade with fewer VGA bits and lower gain states.

To achieve a 6-bit VSPS with 6-bit phase resolution, a gain tuning range (GTR) of $\sim$ 15 dB with 1 dB gain resolution, and gain/phase errors less than half of the gain/phase resolution, it is essential to have at least 6-bit VGA control. Consequently, the optimal number of VGA control bits is carefully determined to be 6 bits, striking the right balance between RF performance and design complexity.

B. VSPS Linearity

To assess the linearity performance of the VSPS, a load pull analysis was conducted, as illustrated in Fig. 5, sweeping the load impedance presented to the output terminals of VGA-I and VGA-Q. The load-pull was performed at the maximum gain state and 0° phase state, as this configuration is expected to yield the worst linearity performance, with the first and second VGAs set to maximum and minimum gains, respectively. The optimal load impedance (Z_{opt}) was chosen to be equal to 25+j30, to allow for an IP1dB of 5 dBm at 40 GHz. To realize this optimal impedance, a vertically stacked transformer-based output current combiner (T2) was implemented. The transformer was designed using the two topmost thick copper layers available in the 45nm SOI-CMOS process to minimize loss and parasitic effects. Electromagnetic (EM) simulations of the output VSPS combiner transformer revealed primary and secondary self-inductances of 300 pH and 370 pH, respectively, and quality factors of 19.6 and 19.7, respectively. These design considerations were made to enhance the linearity and overall performance of the VSPS.

III. MEASUREMENT RESULTS

Fig. 4 displays the fabricated 45nm SOI-CMOS DC-VSPS micro-photograph. The core chip area measures $0.084mm^2$, excluding the baluns which were added for testing and measurement purposes. This highlights the compactness and efficiency of the implemented design.

Fig. 6(a) illustrates the measured constellation diagram (in blue) alongside the gain and phase settings (in red) chosen within the desired tuning ranges of 14 dB (with 1 dB gain step) and 360° (with 5.6° phase step), respectively, at 40 GHz. Fig. 6(b) depicts the measured S-parameters of the largest gain circle acquired while configuring the DC-VSPS settings at 39 GHz fixed (only calibrated at 39 GHz) to maintain a gain of -8 dB, with the phase varying between 0° and 360°, it also illustrates the simulated average S-parameters (in dashed line). Notably, as depicted in Fig. 6(b), the proposed DC-VSPS demonstrator exhibited an excellent IL variation, remaining below ±0.23 at 40 GHz and ±2.3 across 35 - 43 GHz. Additionally, it sustained input and output return losses below -17 dB and -9 dB, respectively, over the same frequency range. Fig. 6(c) illustrated the relative phase response on the left which is achieved by calibration per 1 GHz and the group delay on the right which is acquired by keeping the DC-VSPS settings fixed at 39 GHz. The excellent performance of the DC-VSPS is further validated in Fig. 6(c), where a group delay variation of ±12 ps or lower is consistently maintained for all phase settings at a gain of -8 dB and within the desired

TABLE I
PERFORMANCE COMPARISON WITH PREVIOUS WORKS.

Reference	This work	[3]	[6]	[7]	[8]	[9]
PS type	VSPS	VSPS	VSPS	VSPS	VSPS	VSPS
Tech.	45 nm SOI	40 nm CMOS	45 nm SOI	65 nm CMOS	65 nm CMOS	65 nm CMOS
Freq. (GHz)	35 - 43	23.8 - 30.4	27 - 33	21 - 25	30 - 32.5	18 - 26
Avg IL (dB)	-8	-4.9	-5.8	-3.5	-2.8	-6.2
Gain range/ res. (dB)	14/1	-	-	17.8/ 0.56	17.3/4	-
RMS gain error (dB)	<0.24[1,2]	0.13[2,3]	0.15[3]	0.17[1]	0.4	1
RMS phase error (°)	<0.73[2,3] <1.5[1,2]	0.5[2]	0.8[3]	1.4[1]	3.5[1]	2.08
Phase range (°)	360	360	360	360	360	360
Phase res. (°)	5.6	0.61	5	2.8	22.5	3.75
IP1dB (dBm)	4.1	1@28GHz	2.2	-	-1[4]	-1.8
DC Power (mW)	27	23	25	6.6	18	16
Core area (mm^2)	0.084	0.19	0.27	0.134	0.21	0.157

[1] Total RMS error. [2] Frequency dependant calibration.
[3] for max gain circle. [4] Simulation

frequency range. This characteristic is evident from the flatness observed in the relative phase shift versus frequency plot within the same figure.

Based on the measured constellation diagram, Fig. 6(d) shows the RMS phase and gain error of the DC-VSPS at different gain states at 40 GHz. The RMS phase/gain error is 0.66°/0.13 dB for the maximum gain circle (-8 dB) and increases to 2.1°/0.3 dB for the smallest gain circle (-22 dB) at 40 GHz. Fig. 6(e) illustrates the total RMS phase and gain error for all the gain circles (-8 dB to -22 dB with a gain step of 1 dB) and all the phase states (from 0° to 360° with a 5.6° phase step) on each circle versus frequency with calibration done per 1GHz. The plot confirms that the DC-VSPS can maintain a total RMS phase/gain of less than 1.5°/0.24 dB over 35-43 GHz. Furthermore, Fig. 6(f) presents the measured IP1dB of the DC-VSPS for all the phase states at the maximum gain circle (-8 dB) over the frequency range 37-43 GHz. The results indicate that the DC-VSPS can achieve an IP1dB measurement greater than 4.1 dBm at the maximum gain circle (-8 dB) within the frequency range of 37-43 GHz.

Table I provides a performance comparison between the presented mm-wave DC-VSPS and state-of-the-art VSPS. The proposed DC-VSPS demonstrates competitive gain and phase tuning accuracies, a large gain tuning range, excellent linearity, a broad bandwidth, and a compact chip size when compared to the other VSPSs listed in Table I.

IV. CONCLUSION

In this paper, a digitally controlled vector sum phase shifter (DC-VSPS) specifically designed to cater to the demands of mm-wave LSPAA applications. The DC-VSPS harnesses the benefits of a pair of differential variable gain amplifiers (VGAs) and differential transformers-based passives, resulting in excellent integration, bandwidth, and matching performance. The fabricated DC-VSPS prototype, implemented

979-8-3503-4331-1/24 $31.00 © 2024 IEEE

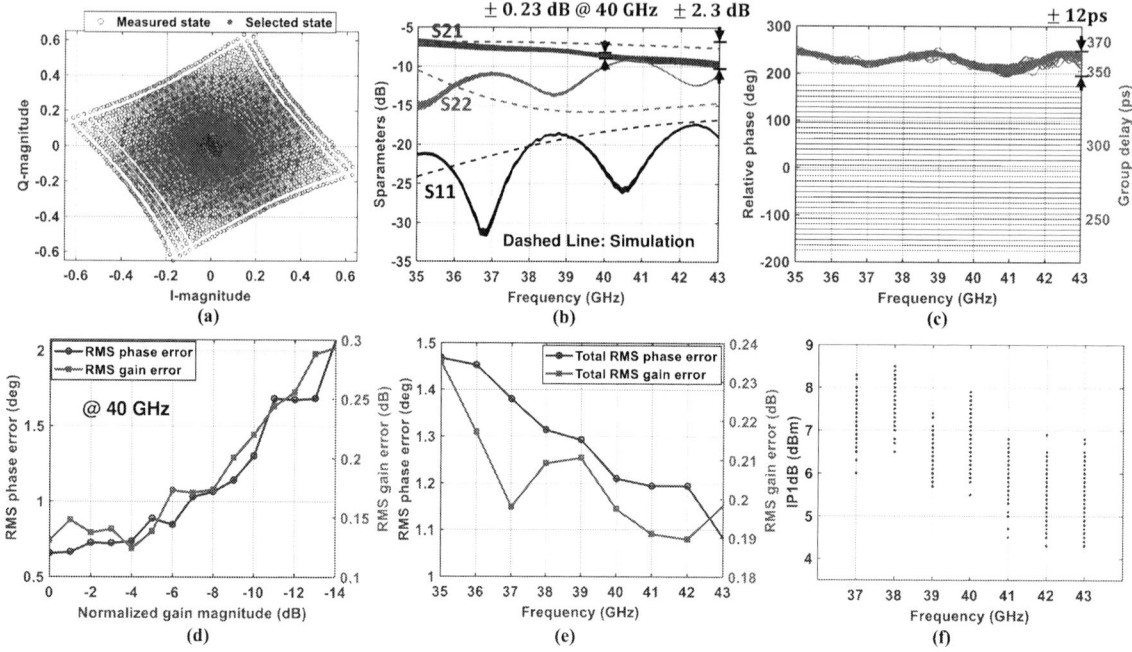

Fig. 6. (a) Measured constellation diagram of the fabricated DC-VSPS's and the selected states at 40 GHz. (b) Measured S-parameters for all the selected phase states on the max gain circle (-8 dB) and simulated (in dashed line) average S-parameters for max gain over the bandwidth. (c) Relative phase and group delay performance response over the bandwidth. (d) RMS phase and gain error versus normalized gain magnitude at 40 GHz. (e) Total RMS phase and gain error versus frequency. (f) Measured IP1dB for the maximum gain circle (-8 dB) and all the phase states from 37 - 43 GHz.

using 45 nm SOI-CMOS technology, occupies a mere core area of 0.084 mm^2. Experimental findings showcase excellent results, with total root-mean-square (RMS) phase error, total RMS gain error, and group delay variation all comfortably within 1.5°, 0.24 dB, and ±12 pico-seconds, respectively, across the frequency range of 35 - 43 GHz. Furthermore, the DC-VSPS exhibits a remarkable tuning range of 360° with a 6-bit resolution for phase control, along with a gain control range of 14 dB with 1 dB resolution. In addition to its excellent phase and gain control capabilities, the prototype demonstrates outstanding performance in terms of input-referred 1 dB compression points, exceeding 4 dBm, and input-output matching, which achieves better than 9 dB over the entire bandwidth.

ACKNOWLEDGEMENT

The authors would like to thank CMC Microsystems for providing access to CAD tools and semiconductor fabrication services. This work was financially co-supported by Skyworks Solutions, Natural Sciences and Engineering Research Council of Canada, and Ontario Research Fund.

REFERENCES

[1] C.-N. Chen et al., "38-GHz phased array transmitter and receiver based on scalable phased array modules with endfire antenna arrays for 5G MMW data links," *IEEE Trans. on Microw. Theory Techn.*, vol. 69, no. 1, pp. 980–999, 2021.

[2] W. Zhu et al., "A 24–28-GHz four-element phased-array transceiver front end with 21.1%/16.6% transmitter peak/OP1dB PAE and subdegree phase resolution supporting 2.4 Gb/s in 256-QAM for 5-G communications," *IEEE Trans. on Microw. Theory Techn.*, vol. 69, no. 6, pp. 2854–2869, 2021.

[3] L. Zhang, Y. Shen, L. de Vreede, and M. Babaie, "A 23.8–30.4-GHz vector-modulated phase shifter with two-stage current-reused variable-gain amplifiers achieving 0.23° minimum rms phase error," *IEEE Solid-State Circuits Letters*, vol. 5, pp. 150–153, 2022.

[4] T. Wu et al., "A 60-GHz variable gain phase shifter with 14.8-dB gain tuning range and 6-bit phase resolution across -25 °C-110 °C," *IEEE Trans. on Microw. Theory Techn.*, vol. 69, no. 4, pp. 2371–2385, 2021.

[5] J. Park, S. Lee, D. Lee, and S. Hong, "9.8 a 28GHz 20.3%-transmitter-efficiency 1.5°-phase-error beamforming front-end IC with embedded switches and dual-vector variable-gain phase shifters," in *IEEE Int. Solid-State Circuits Conf. (ISSCC)*, 2019, pp. 176–178.

[6] J. Xia and S. Boumaiza, "Digitally assisted 28 GHz active phase shifter with 0.1 dB/0.5° RMS magnitude/phase errors and enhanced linearity," *IEEE Trans. Circuits Syst. II, Exp. Briefs*, vol. 66, no. 6, pp. 914–918, 2019.

[7] S. Wang, J. Park, and S. Hong, "A K-band variable-gain phase shifter based on Gilbert-cell vector synthesizer with RC–RL poly-phase filter," *IEEE Microw. Wireless Compon. Lett.*, vol. 31, no. 4, pp. 393–396, 2021.

[8] J. Park, G. Jeong, and S. Hong, "A Ka-band variable-gain phase shifter with multiple vector generators," *IEEE Trans. Circuits Syst. II, Exp. Briefs*, vol. 68, no. 6, pp. 1798–1802, 2021.

[9] F. Qiu, H. Zhu, W. Che, and Q. Xue, "A K-band full 360° phase shifter using novel non-orthogonal vector summing method," *IEEE J. Solid-State Circuits*, vol. 58, no. 5, pp. 1299–1309, 2023.

[10] J. Jang, B. Kim, C.-Y. Kim, and S. Hong, "79-GHz digital attenuator-based variable-gain active vector-sum phase shifter with high linearity," *IEEE Microw. Wireless Compon. Lett.*, vol. 28, no. 8, pp. 693–695, 2018.

[11] P. Gu, D. Zhao, and X. You, "Analysis and design of a cmos bidirectional passive vector-modulated phase shifter," *IEEE Trans. Circuits Syst. I Reg. Papers*, vol. 68, no. 4, pp. 1398–1408, 2021.

[12] Y. Tian et al., "A 26-32GHz 6-bit bidirectional passive phase shifter with 14dBm IP1dB and 2.6° RMS phase error for phased array system in 40nm CMOS," in *2023 IEEE/MTT-S International Microwave Symposium - IMS 2023*, 2023, pp. 195–198.

[13] J. S. Park and H. Wang, "A transformer-based poly-phase network for ultra-broadband quadrature signal generation," *IEEE Trans. on Microw. Theory Techn.*, vol. 63, no. 12, pp. 4444–4457, 2015.

A Novel High Q-factor Structure of Digitally Tunable Capacitor for High RF Power Handling Applications

Wonwoo Seo
Department of Electrical and Electronic Engineering
Hanyang University
Ansan, Republic of Korea
sww@hanyang.ac.kr

Sunghyuk Kim
Department of Electrical and Electronic Engineering
Hanyang University
Ansan, Republic of Korea
kimsh96@hanyang.ac.kr

Byunghun Ko
Department of Electrical and Electronic Engineering
Hanyang University
Ansan, Republic of Korea
qudgsn34@hanyang.ac.kr

Jehwan Lee
Department of Electrical and Electronic Engineering
Hanyang University
Ansan, Republic of Korea
wpghks2005@hanyang.ac.kr

Yongbae Choi
Department of Electrical and Electronic Engineering
Hanyang University
Ansan, Republic of Korea
yongbaec@hanyang.ac.kr

Taejoo Sim
Department of Electrical and Electronic Engineering
Hanyang University
Ansan, Republic of Korea
taejoosim@hanyang.ac.kr

Junghyun Kim
Department of Electrical and Electronic Engineering
Hanyang University
Ansan, Republic of Korea
junhkim@hanyang.ac.kr

Abstract— **This paper presents a novel high quality-factor (Q-factor) structure of a digitally tunable capacitor (DTC) for high RF power handling applications, such as reconfigurable power amplifiers and reconfigurable antenna circuits. Conventional DTC structures handle high power through series-stacked FETs, which directly degrades the Q-factor. To overcome this structure, a parallel connected structure that effectively uses the stacked-FETs in terms of Q-factor and power handling capability are proposed. All the fabricated circuits were measured at 2 GHz. At an ON-state capacitance of 1.5 pF, the Q-factor was improved by up to 30.9 % over a conventional structure. In addition, the OFF-states power handling capability of both conventional and proposed structures achieved an input RF power of 40 dBm. However, the improvement of the proposed structure has capacitance limitations depending on the process. All the conventional and proposed structures were implemented with a 0.13-μm partially depleted silicon-on-insulator (PD-SOI) CMOS process and verified.**

Keywords—Digitally tunable capacitor (DTC), high Q-factor, high power, reconfigurable component, SOI CMOS, tunable capacitor.

I. INTRODUCTION

Since the introduction of the fifth generation (5G) mobile communication, mobile handsets have been required to cover many operating frequency bands, from sub-6 GHz bands to millimeter wave bands. However, as mobile devices become smaller, the size of the transceiver module also has to be reduced. Consequently, several methods have been introduced to reduce the form factor. One of these methods is reconfigurable circuit. And most of this method include tunable components. However, due to the quality factor (Q-factor) limitations of tunable components, efficiency of the

This work was supported in part by the 2022 Hanyang University Project in Avago Technologies, U.S. Inc. and in part by the Technology Innovation Program and 20019457, Development of SoC RF Front End Module for 5G C-V2X Communication funded By the Ministry of Trade, Industry & Energy (MOTIE, Korea).

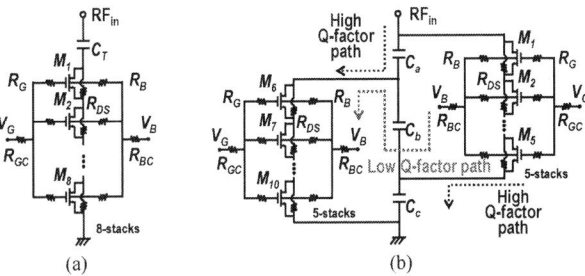

Fig. 1. (a) Conventional and (b) proposed 1-bit conventional tunable capacitor structure.

reconfigurable circuits is lower than that of conventional circuits. Therefore, many literatures study high Q-factor digitally tunable capacitor (DTC) structures. However, most of the literatures only focuses on controlling the total gate width of transistors [1], [2], [3], and rearranging the positions of the capacitors and transistors [4]. As shown in Fig. 1(a), most studies use capacitors and stacked-FETs connected in series. To overcome the limitations of conventional structures, this paper proposes a parallel connection structure as shown in Fig. 1(b). To verify the proposed structure, conventional and proposed structure were fabricated and measured.

This paper is organized as follows. In Section II, we introduce the circuit design and operating principles of the conventional and proposed tunable capacitor structures. The simulation and implementation measurement results are presented in Section III. Finally, Section IV concludes this paper.

II. CIRCUIT DESIGN AND OPERATING PRINCIPLE

To implement a novel tunable capacitor structure, two stacked-FETs and three capacitors were used. Three capacitors were connected in series, and each stacked-FETs was connected in parallel to two capacitors, as shown in Fig. 1 (b). The conventional structure uses 8 FETs (8 stacks × 1) to handle the

979-8-3503-4331-1/24 $31.00 © 2024 IEEE

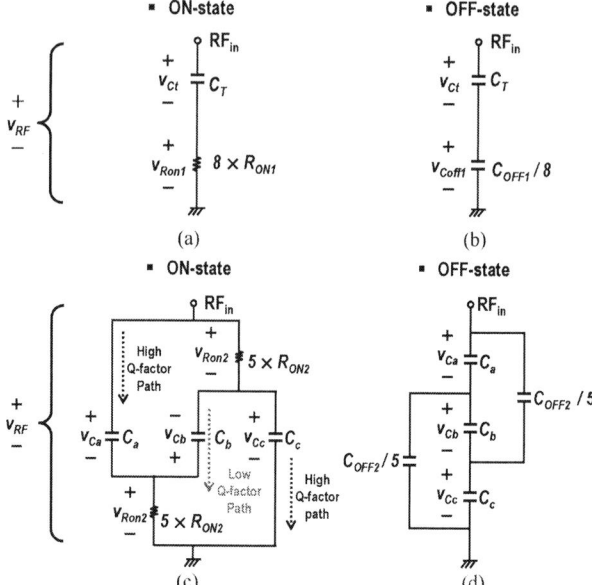

Fig. 2. Simple equivalent circuits of the 1-bit conventional tunable capacitor structure ((a) ON-state and (b) OFF-state) and the 1-bit proposed tunable capacitor structure ((c) ON-state and (d) OFF-state).

40 dBm, and the proposed structure uses 10 FETs (5 stacks × 2). To achieve the same transistor chip area for both the conventional and the proposed structures, the transistors total gate widths of the proposed structures are narrower than those of the conventional one. Thus, the ON-resistance (R_{ON2}) of the proposed structure was higher than the ON-resistance (R_{ON1}) of the conventional structure ($R_{ON1} < R_{ON2}$). To equalize the ON-state capacitance of the conventional and proposed structures, the target ON-state capacitor value (C_T) of the conventional structure is the sum of the capacitor values (C_a, C_b, and C_c) of the proposed structure ($C_T \approx C_a + C_b + C_c$). The values of C_a and C_c are the same value. Since the value of C_b is the trade-off point between the Q-factor and the power handling capability.

A. ON-state Operating Principle

The capacitance of the conventional structure is C_T during the ON-state, as shown in Fig. 2(a). The capacitance of the proposed structure is $C_a + C_b + C_c$ because the capacitors (C_a, C_b, and C_c) are connected in parallel during the ON-state, as shown in Fig. 2(c).

The conventional structure during the ON-state uses 8 stacked R_{ON1} in series, as shown in Fig. 2(a). However, in the proposed structure, C_a and C_c paths include only 5 stacks R_{ON2} in series, and C_b uses 10 stacks R_{on2} in series, as shown in Fig. 2(c). Thus, decreasing the value of C_b increases the Q-factor of the proposed tunable capacitor; however, there are two limitations. One limitation is that the sum of the capacitor values (C_a, C_b, and C_c) is C_T and the other limitation is that the ratio between capacitors (C_a, C_b, and C_c) affects the power handling capability.

In the ON-state of the conventional and proposed structures, most of the voltage swing is endured by the capacitors (C_T or C_a, C_b, and C_c) rather than the stacked-FETs with a small ON-

resistance. Thus, during the ON-state, stacked-FETs have to handle a very small RF peak voltage. And the power handling capability depends on the physical structure and material of the capacitors (C_T, C_a, C_b, and C_c).

B. OFF-state Operating Principle

The capacitance of the conventional structure is $C_T \parallel (C_{OFF1}/8)$, as shown in Fig. 2(b). This capacitance is very low because of the small OFF-capacitance (C_{OFF1}) of the transistor. In the proposed structure, because a C_{OFF2} is much smaller than C_a, C_b, and C_c. The capacitance of the proposed structure is roughly $C_a \parallel C_b \parallel C_c$, as shown in Fig. 2(d). However, in practical case, the conventional structure uses the minimum capacitance $C_{off1}/8$, by connecting C_T in series, and the proposed structure uses the minimum capacitance $C_{off2}/5$, by connecting capacitances (C_a, C_b, and C_c) in parallel. Thus, total OFF-capacitance of the proposed structure is higher than conventional one.

In the OFF-state of the conventional structure, the OFF-capacitance (C_{OFF1}) is much smaller value than that of the capacitor (C_T). Thus, most of the RF peak voltage (v_{RF}) is applied to the stacked-FETs rather than to the capacitor (C_T). As a result, the applied voltage of the transistor (v_{Coff1}) is $v_{Coff1} \approx v_{RF}$. In the OFF-state of the proposed tunable capacitor, because the value of the OFF-capacitance (C_{OFF2}) is much smaller value than that of the capacitors (C_a, C_b, and C_c), we simply neglect the OFF-capacitance (C_{OFF2}) path. Because of the voltage dividing rule, the voltages applied to C_a, C_b, and C_c are,

$$v_{Ca} = \frac{c_b c_c}{c_a c_b + c_b c_c + c_c c_a} v_{RF} \quad (1)$$

$$v_{Cb} = \frac{c_c c_a}{c_a c_b + c_b c_c + c_c c_a} v_{RF} \quad (2)$$

$$v_{Cc} = \frac{c_a c_b}{c_a c_b + c_b c_c + c_c c_a} v_{RF} \quad (3)$$

Therefore, the upper stacked-FETs has to handle is,

$$v_{Ca} + v_{Cb} = \frac{c_b c_c + c_c c_a}{c_a c_b + c_b c_c + c_c c_a} v_{RF} \quad (4)$$

and lower stacked-FETs has to handle is,

$$v_{Cb} + v_{Cc} = \frac{c_c c_a + c_a c_b}{c_a c_b + c_b c_c + c_c c_a} v_{RF} \quad (5)$$

Thus, total stacked-FETs in proposed tunable capacitor has to handle is,

$$v_{Ca} + 2v_{Cb} + v_{Cc} = v_{RF} + v_{Cb} \quad (6)$$

The v_{RF} in (6) depend on the RF power rather than the capacitance ratio and structures. And the v_{Cb} decreased as increasing the C_b from (2). In other words, increasing the C_b alleviates the RF peak voltage that total stacked-FETs in proposed tunable capacitor has to handle. And harmonic levels are also improved as power handling capability performance improves. However, we have already mentioned that increasing

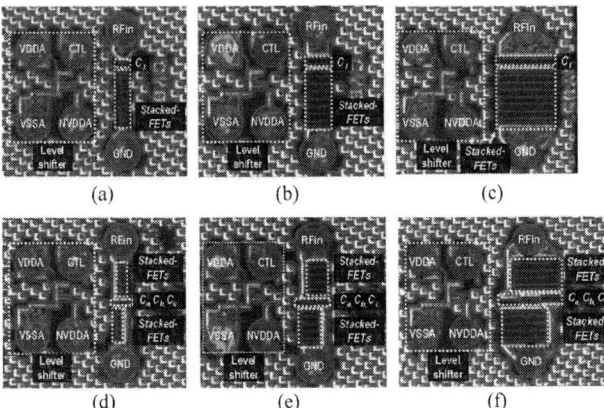

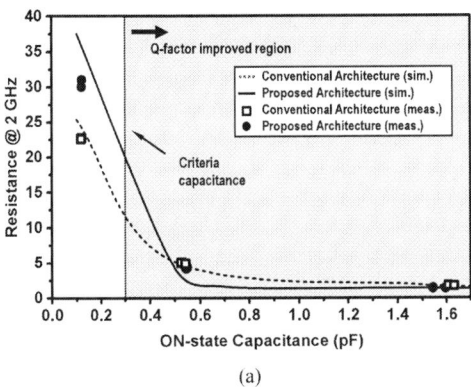

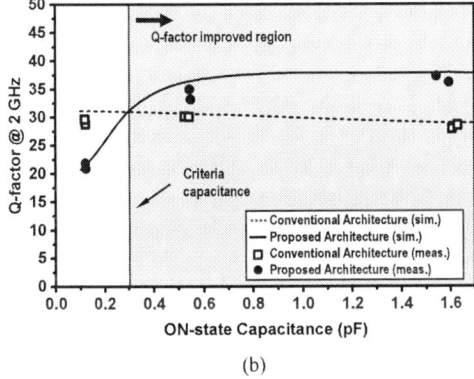

Fig. 3. Photograph of (a) 0.1 pF, (b) 0.5 pF, (c) 1.5 pF conventional tunable capacitor structures and (d) 0.1 pF, (e) 0.5 pF, (f) 1.5 pF proposed tunable capacitor structures.

Fig. 4. Simulation and measurement results of the (a) resistance and (b) Q-factor versus ON-state capacitance of the tunable capacitor structure.

the C_b decreases Q-factor of the ON-state proposed structure. And total capacitances limited by the conventional target capacitance ($C_T \approx C_a + C_b + C_c$). In practical case, Due to size limitations, the lower Q-factor design, the more $C_{off2}/5$ is not sufficiently smaller than the capacitances (C_a, C_b, and C_c) to neglect the OFF-state transistor path in proposed structure. Therefore, the capacitance ratio was optimized according to the power handling capability and Q-factor simulation results.

III. MEASUREMENT RESULTS

To compare the Q-factor of the conventional and proposed structures, three ON-state capacitance values (0.1 pF, 0.5 pF, and 1.5 pF) of the 1-bit tunable capacitor structure were fabricated using the 0.13-μm PD-SOI CMOS process and the photograph is shown in Fig. 3. All circuits include the level shifter, logic ESD protection blocks, and the conventional or proposed tunable capacitor structure. All measurements were performed at 2 GHz using a probe tip on a multi-layer printed circuit board (PCB). Two samples of the same design were measured and are plotted in Fig. 4.

A. Capacitance and Q-Factor

Fig. 4 shows the resistance and Q-factors of the conventional and proposed tunable capacitor structure that have only 1-bit and not multi-bits. The Q-factor of conventional circuits is targeted at 30, and proposed circuits are designed in the same area as the total transistor chip area. All the Q-factors and capacitances were de-embedded by an open-short pattern of the PCB. At an ON-state capacitance of 1.5 pF, the Q-factors of the proposed and conventional tunable structures are 37.3 and 28.5 respectively, at 2 GHz. The Q-factor of the proposed structure was 30.9 % higher than the Q-factor of the conventional structure. At a capacitance of 0.5 pF, the Q-factors of the proposed and conventional tunable structures are 35.0 and 30.1 at 2 GHz, respectively. The Q-factor of the proposed structure was 16.3 % higher than the Q-factor of the conventional structure. At a capacitance of 0.1 pF, the Q-factors of the proposed and conventional tunable structures are 21.9 and 29.6, respectively. The Q-factor of the proposed structure is 26.0 % lower than conventional structure. As shown in Fig. 4, above 0.3 pF capacitance, the proposed tunable capacitor structure has a higher Q-factor than that of the conventional tunable capacitor structure. However, under 0.3 pF capacitance, the conventional tunable capacitor structure has a higher Q-factor than that of the proposed tunable capacitor structures. Therefore, there is a criteria capacitance that the Q-factor improves depending on the process.

B. Power Handling Capability

The power handling capability of all the conventional and the proposed tunable capacitor structures is 40 dBm of input power at 2 GHz RF power in the OFF-state. All the conventional and proposed structures handle input RF power of 41 dBm (limitation power of the measurement setup device) at 2 GHz RF power in the ON-state. As shown in Fig. 5, the power handling capability of the proposed tunable structure (1.5 pF) in ON- and OFF-state were simulated and measured.

IV. CONCLUSION

To verify the proposed tunable capacitor structure, conventional and proposed structures were fabricated and measured. The proposed tunable structure achieved a higher Q-factor than that of the conventional tunable structure above 0.3 pF. At capacitance of 1.54 pF, the proposed structure achieved 29.7 % increase in Q-factor compared with that of the conventional one. However, below 0.3 pF, the proposed structure shows a lower Q-factor than that of the conventional one. Both the conventional and proposed tunable capacitor structures used the same active transistor chip area and achieved

979-8-3503-4331-1/24 $31.00 © 2024 IEEE 23

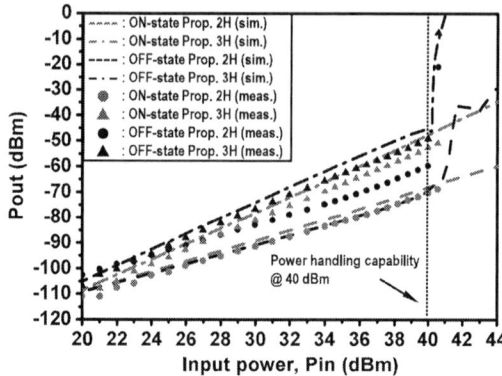

Fig. 5. Simulation and measurement results of the 2nd (2H) and the 3rd (3H) harmonic levels in the ON-state and OFF-state of the tunable capacitor structure (1.5 pF) according to input power at 2 GHz

TABLE I. COMPARISON WITH THE STATE OF THE ART

Performance	[2] 2014	[3] 2013	[5] 2013	This Work (Proposed structure)		
Frequency (GHz)	1	1	2	2	2	2
Number of Bits	6	5	4	1	1	1
Cmin. / Cmax (pF)	1 / 10	1.2 / 7.2	1 / 4.7	0.04 / 0.12	0.12 / 0.54	0.28 / 1.54
Tuning Ratio	10	6	4.7	3	4.5	5.5
Step Size (pF)	0.16	0.19	0.25	0.08	0.42	1.26
Q-factor @ C_{min} / C_{max}	15 / 88	61 / 41	50 / 15	16 / 22	42 / 35	62 / 37
Power Handling Capability (dBm)	36	36	> 35	> 40	> 40	> 40
H2 / H3 (dBm)	− 42 / − 45[b]	− 59 / − 47[c]	− 57 / − 52[a]	− 86 / −72[a]	− 80 / − 71[a]	− 78 / − 71[a]
Technology	SOI	SOI	SOI	SOI	SOI	SOI

[a] 32 dBm, [b] 34 dBm, [c] 36 dBm of the input power

an 40 dBm of the input RF power at 2 GHz in the OFF-state. As shown in Table 1, the proposed structure shows good performance in terms of Q-factor and power handling capability. To verify the proposed structure idea and Q-factor improved criteria capacitance, only 1-bit proposed tunable capacitor structures were simulated and measured in this paper. However, by connecting the several 1-bit proposed tunable capacitor structures in parallel, it is also possible to achieve multi-bits and a higher maximum capacitance of 10 pF, as in reference [2], and [3].

REFERENCE

[1] F. Carrara, C. D. Presti, F. Pappalardo and G. Palmisano, "A 2.4-GHz 24-dBm SOI CMOS Power Amplifier With Fully Integrated Reconfigurable Output Matching Network," *IEEE Transactions on Microwave Theory and Techniques*, vol. 57, no. 9, pp. 2122-2130, Sept. 2009.

[2] A. Thomas, W. Bakalski, W. Simbürger and R. Weigel, "A high quality factor bulk-CMOS switch-based digitally programmable RF capacitor," *2014 Asia-Pacific Microwave Conference*, Sendai, Japan, pp. 58-60, Nov. 2014.

[3] B. -K. Kim, T. Lee, D. Im, D. -K. Im, B. Kim and K. Lee, "Design methodology of tunable impedance matching circuit with SOI CMOS tunable capacitor array for RF FEM," *2013 Asia-Pacific Microwave Conference Proceedings*, Seoul, Korea (South), pp. 7-9, Nov. 2013.

[4] D. Nicolas et al., "Study of SOI CMOS tunable capacitor architectures and application to antenna aperture tuning," *2016 IEEE MTT-S International Microwave Symposium*, San Francisco, CA, USA, pp. 1-4, May 2016.

[5] D. Im and K. Lee, "Highly Linear Silicon-on-Insulator CMOS Digitally Programmable Capacitor Array for Tunable Antenna Matching Circuits," *IEEE Microwave and Wireless Components Letters*, vol. 23, no. 12, pp. 665-667, Dec. 2013.

Compact D-Band Passive Phase Shifters with Fine and Coarse Control Steps in BiCMOS-55nm

Lorenzo Piotto
Dept. of Electrical, Computer and Biomedical Engineering
University of Pavia, Italy

Guglielmo De Filippi
Dept. of Electrical, Computer and Biomedical Engineering
University of Pavia, Italy

Mahmoud M. Pirbazari
Dept. of Electrical, Computer and Biomedical Engineering
University of Pavia, Italy

Andrea Mazzanti
Dept. of Electrical, Computer and Biomedical Engineering
University of Pavia, Italy

Abstract—Programmable phase shifters are key building blocks in emerging Sub-THz phased array transceivers. This work proposes different structures to implement compact passive phase shifters in D-Band (110 - 170 GHz). Tunable band-pass filters and variable-delay transmission lines are evaluated as phase shifters with fine control resolution over a limited range. To extend the range, 0°/180° and 0°/90° phase shifters are also investigated and implemented with a hybrid coupler and a switched λ/4 transmission line. Experimental results are presented on a test chip in a 9 metal BiCMOS technology with 55 nm MOS transistors used as switches. By combining the different structures, the full 0°/360° programmable range is covered from 120 GHz to 170 GHz with a size of only 0.13 mm². The estimated insertion loss is aligned with previous works, but the remarkable footprint reduction facilitates the adoption in dense sub-THz phase array systems.

Index Terms—D-band, BiCMOS, phase shifter, phased-array

I. INTRODUCTION

The demand for ultra-high-capacity wireless links in the network infrastructure of 5G and beyond is driving the development of integrated circuits above 100 GHz, where ultra-wide bands are available [1]. Within this scenario, D-Band (110 GHz – 170 GHz) is quickly gaining interest, and G-Band (140 GHz – 220 GHz) may follow. On the other hand, digital CMOS technologies keep getting closer to their intrinsic limitations (f_{max}) in these bands. Whilst more expensive semiconductors are available, they bring penalties in terms of cost, scalability and lack of integration of digital functions. SiGe BiCMOS fills the gap with f_t / f_{max} in the 350-500 GHz range and a roadmap to push the HBT frequency limits even further [2].

A key block in sub-THz applications is the phase shifter (PS), which enables active phased arrays for beam-forming and beam-steering, relaxing Power Amplifiers requirements and improving the SNR. Active PSs are based on the vector interpolation principle, typically characterized by high resolution but limited linearity, even at moderately high power consumption [3]. On the contrary, passive PSs offer excellent linearity at the expense of an increased area occupation and unavoidable insertion loss. To date, a very limited number of works explored programmable passive PSs in sub-THz bands [4]–[7], particularly in silicon. Switched transmission lines (TLINEs) were demonstrated in SiGe [5] and compound-semiconductor technologies [6], but they require a large area that does not fit a sub-THz phased-array system, with antenna

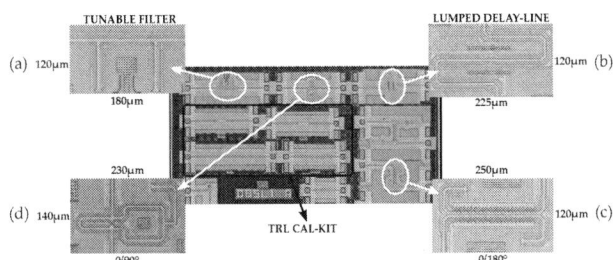

Fig. 1. Chip photo of the realized passive phase shifters.

spacing of 1 mm or less. A PS which combines reflective-type hybrid couplers and an active stage for phase inversion was proposed at 120 GHz in SiGe BiCMOS, proving low insertion loss with fine phase control resolution but over a narrow bandwidth [4].

This work investigates passive D-band PSs leveraging the availability of good MOS switches and a 9-level metal stack, with two thick top layers, in the SiGe BiCMOS-55nm of STMicroelectronics. Fig. 1 shows the photograph of the realized test chip, which includes also a TRL cal-kit used to de-embed pads and access lines from measurements. The PSs are classified into fine (1) and coarse (2) phase-control steps. The former feature a digitally-controlled phase shift in small steps but, to limit the insertion loss, the maximum range is limited. The latter provide a 1-bit phase control (nominally 0°/90° and 0°/180°). By cascading fine- and coarse-control PSs, the full 0°/360° can be covered with fine control resolution and limited insertion loss. Design and measurements of the PSs are presented in Sec. II and Sec. III, followed by a discussion and conclusion in Sec. IV.

II. PHASE SHIFTERS WITH FINE CONTROL STEPS

Two different approaches are investigated for PSs with a fine control step. The first is based on 4th order band-pass filters with tunable center frequency. Such kind of response may be achieved through magnetically coupled resonators, shown in Fig. 2a. An analysis of these networks is presented in [8]. If the components (R, L, C, k) are sized such that the two pairs of complex conjugate poles provide a flat in-band response, the phase of the transfer function is fairly linear across frequency, with a variation of π radians within the -3 dB bandwidth, $BW_{-3dB} = f_H - f_L$. Therefore, $d\phi/df \approx \pi/BW_{-3dB}$. If the filter center frequency, f_0, is shifted by Δf_0 by varying the net-

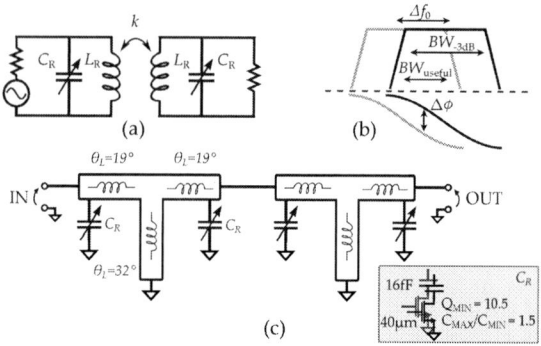

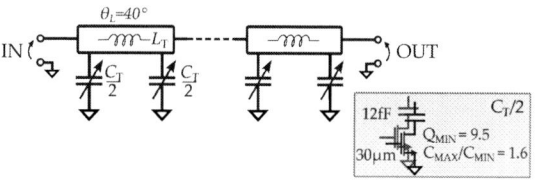

Fig. 2. Coupled-resonator band-pass filter (a) and transfer function (amplitude/phase) variation by tuning the center frequency (b). Schematic of the implemented network (c).

Fig. 3. Fine-resolution phase shifter realized as a transmission line periodically loaded by digitally-tuned capacitors.

work capacitors (gray plot in Fig. 2b), the signal experiences a phase shift variation:

$$\Delta\phi \approx \frac{\pi}{BW_{-3dB}} \cdot \Delta f_0 \qquad (1)$$

However, to avoid attenuation when the network transfer function is translated, the useful bandwidth (BW_{useful}) is lower than BW_{-3dB}. Looking at Fig. 2b, $BW_{useful} = BW_{-3dB} - \Delta f_0$ and making use of (1):

$$BW_{useful} = BW_{-3dB}\left(1 - \frac{\Delta\phi}{\pi}\right) \qquad (2)$$

From (2), given a target BW_{useful}, the wider is the desired maximum phase variation, the larger must be BW_{-3dB}. On the other hand, $BW_{-3dB} \propto 1/(RC)$, thus an upper bound on BW_{-3dB} (and hence $\Delta\phi$) is finally set by the minimum variable capacitors that can be reliably implemented to achieve the required tunability of f_0. To extend the attainable phase shift, multiple replicas of the networks can be cascaded. Two identical band-pass cells are used in the implemented PS, giving twice the phase shift range with a marginal bandwidth reduction. The schematic is shown in Fig. 2c. The coupled inductors of Fig. 2a are replaced by their equivalent T-network, implemented with short TLINEs that provide lower loss compared to magnetically-coupled circular coils. The TLINEs are shielded microstrips with the signal routed in the topmost metal layer and display a quality factor $Q = 35$ at 165 GHz. The network center frequency is digitally tuned by a bank of four switched MOM capacitors. The switches are nMOS with W/L = 40 um / 55 nm, implemented in triple well to limit the bulk parasitic capacitance. Each MOM-switch combination features C_{MAX} = 16 fF with C_{MAX}/C_{MIN} = 1.5 and minimum

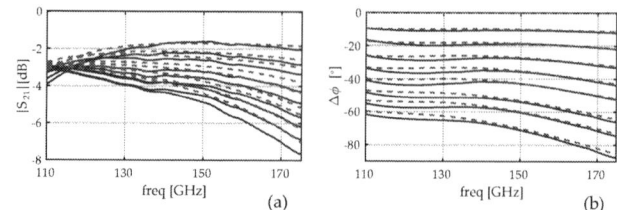

Fig. 4. Measurements (solid-red) and simulations (dashed-gray) of forward gain (a) and phase response (b) of the tunable band-pass filter.

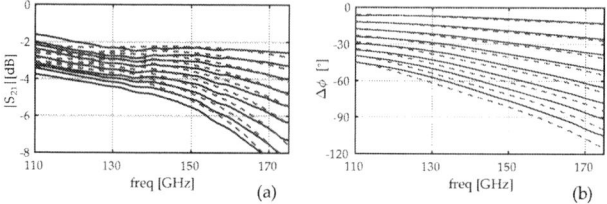

Fig. 5. Measurements (solid-red) and simulations (dashed-gray) of forward gain (a) and phase response (b) of the tunable-delay line.

quality factor Q_{MIN} = 10.5. The band-pass cells are sized for a center frequency of 150 GHz and 90 % fractional bandwidth. The capacitors bank shifts the center frequency by ±10 % thus, using (2), each cell introduces roughly 35° of programmable phase shift while the cascade of the two cells provides 70°.

The alternative fine-step PS investigated consists of a transmission line with tunable propagation delay. The schematic is shown in Fig. 3. Short sections of TLINEs, which approximate inductors of value L_T, are interleaved with digitally switched capacitors C_T. The overall network, composed of N sections, behaves as a low pass filter. The cut off frequency, defined as the frequency where the input impedance becomes purely reactive (i.e. no active power can be injected into the line) is $\omega_T = 2/\sqrt{L_T C_T}$. For a signal at angular frequency ω_0 sufficiently lower than ω_T, the network well approximates a TLINE with Z_0 and group delay τ_D given by:

$$Z_0 = \sqrt{\frac{L_T}{C_T}} \qquad \tau_D = N \cdot \sqrt{L_T C_T} \qquad (3)$$

The group delay, and hence the network phase shift ϕ, are tuned by capacitors C_T. If each capacitor is switched between $C_{T,MIN}$ and $C_{T,MAX}$, the variation of τ_D and of the network phase shift (in radians) are:

$$\Delta\tau_D = N\sqrt{L_T C_{T,MIN}}\left(\sqrt{\frac{C_{T,MAX}}{C_{T,MIN}}} - 1\right) \quad \Delta\phi = \omega_0 \Delta\tau_D$$
$$(4)$$

The variable capacitors C_T are realized with a bank of four digitally switched MOM capacitors of 12.5 fF, giving $C_{T,MAX}$ = 50 fF. The nMOS are sized with W/L = 30 um / 55 nm, leading to $C_{T,MAX}/C_{T,MIN}$ = 1.6 and Q_{MIN} = 9.5. The length of each piece of TLINE between consecutive capacitors is set to approximate L_T = 50 pH such that ω_T is greater than 2π 200 GHz. From (4), the delay variation of one cell (N=1) is $\Delta\tau$ = 0.34 ps, corresponding to a phase shift of 17.5° at 150 GHz. To reach a total phase shift variation comparable with the tunable band-pass filter, the structure is implemented with four (N=4) cascaded $C_T/2$-L_T-$C_T/2$ cells .

979-8-3503-4331-1/24 $31.00 © 2024 IEEE

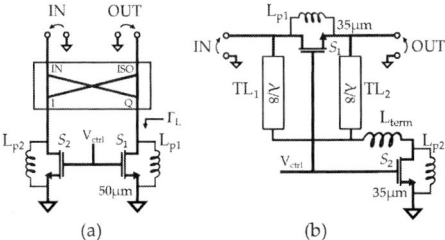

(a) (b)

Fig. 6. Coarse-resolution phase shifters to realize 180° (a) and 90° (b) phase steps.

The chip photos of the fine-control PSs are shown in Fig. 1a and Fig. 1b with core size of $180 \times 120\,\mu m^2$ and $225 \times 120\,\mu m^2$, respectively. The measured forward gain and phase of the tunable band-pass filters are plotted in Fig. 4 as solid lines and compared with simulation (dashed). The programmable phase shifting range is $> 60°$ at 110 GHz and 85° at 170 GHz with phase steps between 10° and 13°. The insertion loss is $< 4\,dB$ at the minimum phase shift and rises to 7 dB at 170 GHz for the maximum phase shift. The insertion loss normalized to the achievable phase shift is 65 mdB/° at 110 GHz and 80 mdB/° at 170 GHz. Fig. 5 reports the results for the variable-delay TLINE. The phase plot in Fig. 5b shows an almost linear phase shift over frequency, as a result of the constant group delay rather than constant phase shift. The insertion loss is around 2 dB at the minimum phase shift and rises to 4 dB at 110 GHz and 8.3 dB at 170 GHz at the maximum phase shift of 100°. The normalized insertion loss is 77 mdB/° at 110 GHz, and 83 mdB/° at 170 GHz, comparable to what measured on the the tunable filter. The return loss, not shown, is $< 10\,dB$ at all the phase settings for both the tunable filter and delay line.

III. PHASE SHIFTERS WITH COARSE CONTROL STEPS

The physical size and insertion loss of the structures presented in the previous Section increase proportionally to the desired range of programmable phase shift. Thus, to cover a wider range while maintaining a fine control resolution, it is more convenient to cascade those structures with networks that provide a large phase shift in a single coarse step, limiting the insertion loss and silicon area. A 0°/180° and a 0°/90° PSs are proposed in this section.

In [6], 0°/180° are achieved by using two hybrid couplers and two SPDT switches in a compound semiconductor technology. The structure proves low insertion loss (also thanks to the excellent performance of HEMTs) and a flat phase across frequency, but requires large chip size. On the other hand, while being desirable, a flat phase response is not mandatory if the block is to be cascaded with fine-control PSs, because its phase error versus frequency can be corrected by the latter. The implemented 0°/180° PS, shown in Fig. 6a, uses only one hybrid coupler and two grounded switches, leading to a very compact size. The circuit works as a reflective-type PS. The input signal, S_{IN}, is equally split at the I/Q ports of the coupler. If an identical impedance, Z_L, with non-zero reflection coefficient (Γ_L) loads the I/Q ports, the reflected signals sum

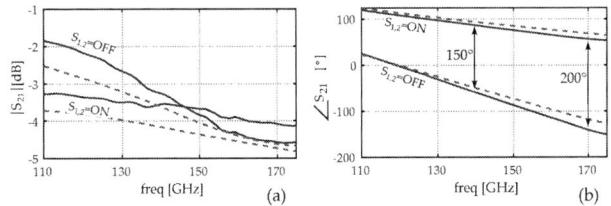

Fig. 7. 0/180° forward gain (a) and relative phase shift (b) compared against simulation (dashed-gray).

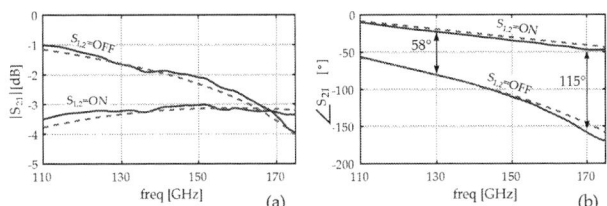

Fig. 8. 0/90° forward gain (a) and relative phase shift (b) compared against simulation (dashed-gray)

up at the ISO port, giving the output signal S_{OUT}. Neglecting losses, the following relation holds:

$$\left| \frac{S_{OUT}}{S_{IN}} \right| = |\Gamma_L|^2 \qquad \angle \frac{S_{OUT}}{S_{IN}} = -90° + \angle \Gamma_L \qquad (5)$$

The I/Q ports are terminated by the switches $S_{1,2}$, implemented with nMOS transistors, yielding ideally $\Gamma_L = \pm 1$ in the OFF and ON state, respectively. Using (5), S_{IN} is transferred to the ISO port with the same magnitude but $\pm 90°$ phase, corresponding to a 180° relative phase variation. The finite channel resistance and the parasitic capacitance of the nMOS switches introduce signal loss and deviation from the ideal 180° phase. The transistors are sized with W/L $= 50\,\mu m$ / 55 nm to achieve $r_{on} = 7\,\Omega$, while the OFF-state parasitic capacitance is resonated out by the parallel inductors $L_{p1,2} = 40\,pH$, implemented with bent TLINEs stubs visible on the chip layout in Fig. 1c. The hybrid coupler is realized with $230\,\mu m$-long coupled lines. The core size of the 0°/180° PS is $250\,\mu m \times 120\,\mu m$.

A 0°/90° phase shifter can be implemented by selecting the I or Q output of a 90° hybrid coupler, but with an intrinsic insertion loss of 3dB [6]. To avoid this loss, the proposed 0°/90° PS leverages a $\lambda/4$-long TLINE with a bypass switch. The schematic is drawn in Fig. 6b. Inductors $L_{p1,2} = 60\,pH$, in parallel to the nMOS switches $S_{1,2}$, resonate with transistors parasitic capacitances to rise the impedance and improve isolation when $S_{1,2}$ are OFF. In this condition, L_{term} is floating and the signal flows through TL_1 and TL_2, experiencing the 90° phase shift of the $\lambda/4$ total length. When $S_{1,2}$ are ON, S_1 shorts the input to the output, thus the phase shift is ideally 0°. In this situation, TL_1 and TL_2 are shunted and the top port is connected to the input/output. Not to compromise the insertion loss and reflection coefficient, the bottom port of $TL_{1,2}$ is terminated by $L_{term} = 25\,pH$ (grounded on one side by S_2). In fact, TL_1 and TL_2 behave like a single $\lambda/8$ TLINE with half the characteristic impedance, $Z_0/2$. L_{term} is thus sized for an impedance of $+jZ_0/2$, such that a high impedance

979-8-3503-4331-1/24 $31.00 © 2024 IEEE 27

TABLE I
MEASUREMENTS SUMMARY AND COMPARISON

	[4]	[5]	[6]			This work			
						Tunable Filter	Tunable TLINE	0°/90°	0°/180°
Technology	120 nm SiGe	130 nm SiGe	GaAs			55 nm SiGe			
Frequency [GHz]	116 - 128	110 - 170	110 - 170			110 - 170	110 - 170	130 - 170	140 - 170
Phase control range [°]	360	260 @ 110 GHz 405 @ 170 GHz	180±2	87±2	47±4	62 @ 110 GHz 84 @ 170 GHz	47 @ 110 GHz 100 @ 170 GHz	58 @ 130 GHz 115 @ 170 GHz	145 @ 140 GHz 200 @ 170 GHz
Max Insertion Loss [dB]	8*	24 @ 110 GHz 23 @ 170 GHz	5 @ 110 GHz 6.5 @ 170 GHz	6	6.5	4 @ 110 GHz 7 @ 170 GHz	4 @ 110 GHz 8.3 @ 170 GHz	3.2	3.5 @ 140 GHz 4.5 @ 170 GHz
Phase Resolution [°]	11.25	17 @ 110 GHz 27 @ 170 GHz	-	-	-	7 @ 110 GHz 13 @ 170 GHz	7 @ 110 GHz 13 @ 170 GHz	-	-
Area [mm²]	0.19	1.17	0.94	0.56	1.25	0.022	0.027	0.032	0.030
Normalized Insertion Loss † [mdB/°]	-	92 @ 110 GHz 57 @ 170 GHz	28 @ 110 GHz 36 @ 170 GHz	69	15	65 @ 110 GHz 83 @ 170 GHz	77 @ 110 GHz 83 @ 170 GHz	60 @ 130 GHz 28 @ 170 GHz	24

* Including buffer/s   † Insertion Loss/$\Delta\phi$

(ideally open circuit) is seen at the top port of TL_1, TL_2. The size of the nMOS switches is set to W/L=35 μm / 55nm as a compromise between insertion loss and phase error, caused by channel resistance and device capacitances, respectively. Inductors $L_{p1,2}$ and L_{term}, visible on the chip photo in Fig. 1, are realized with TLINEs. The core size of the 0°/90° phase shifter is 230 μm x 140 μm.

The simulated and measured magnitude and phase response of the 0°/180° and 0°/90° PSs are plotted in Fig. 7 and Fig. 8, respectively. The 0°/180° PS provides the 180° phase inversion at 160 GHz. The phase error from 180° is within 30° from 140 GHz to above 170 GHz. The insertion loss in this band is below 4.5 dB, which corresponds to < 24 mdB/°, remarkably lower than what achieved by the fine-control PSs. The 0°/90° PS introduces the 90° relative phase difference at 160 GHz with a deviation lower than 30° from 130 GHz to above 170 GHz. The insertion loss is within 2 - 4 dB which corresponds < 60 mdB/°. The return loss for the two PSs, not shown, is below -10 dB.

IV. DISCUSSION AND CONCLUSIONS

The measured results presented in the previous Sections are summarized in Table I and compared with the few other integrated PSs operating in D-band. The insertion loss of the proposed PSs is aligned with previous works, with area occupation which is significantly reduced, simplifying the adoption in dense Sub-THz phased array systems. Moreover, the implemented structures offer high flexibility by providing different phase shift control range and resolution. By cascading fine- and coarse-step PSs, a large programmable phase shift range can be covered maintaining the fine resolution in the phase control. As an example, Fig. 9 shows all the possible magnitude and phase responses achieved by cascading the measured S-parameters of 3 tunable band-pass filters with the 0°/90° and the 0°/180° PSs. Looking at Fig. 9b, the chain covers more than 360° with a fine resolution from 120 GHz to 175 GHz. The corresponding insertion loss for a relative phase shift programmed from 0° to 360°, marked with red dots in Fig. 9, is between 12 dB and 16 dB at 120 GHz and between 14 dB and 20 dB at 175 GHz. The maximum loss, normalized to the 360°, is below 55 mdB/°, still comparable or better than the other PSs in Table I. The footprint of

the five cascaded block is extremely small. The estimated silicon area, of only 0.13 mm², remains remarkably lower than the area occupation of previously reported PSs in D-band.

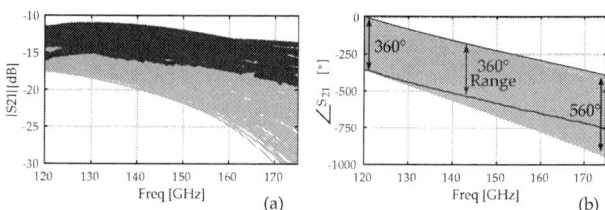

Fig. 9. S_{21} amplitude (a) and phase (b) response of the cascaded fine and coarse phase shifters

ACKNOWLEDGMENTS

This work received funding from the Commission of the European Union within the H2020 DRAGON project (Grant Agreement No. 955699) and KDT SHIFT project (Grant Agreement No. 1010962).

REFERENCES

[1] T. Maiwald et al., "A Review of Integrated Systems and Components for 6G Wireless Communication in the D-Band," *Proceedings of the IEEE*, vol. 111, no. 3, pp. 220–256, 2023.

[2] T. Zimmer et al., "SiGe HBTs and BiCMOS Technology for Present and Future Millimeter-Wave Systems," *IEEE Journal of Microwaves*, vol. 1, no. 1, pp. 288–298, 2021.

[3] D. d. Rio, I. Gurutzeaga, R. Berenguer, I. Huhtinen, and J. F. Sevillano, "A Compact and High-Linearity 140–160 GHz Active Phase Shifter in 55 nm BiCMOS," *IEEE Microwave and Wireless Components Letters*, vol. 31, no. 2, pp. 157–160, 2021.

[4] R. B. Yishay and D. Elad, "D-Band 360° Phase Shifter with Uniform Insertion Loss," in *2018 IEEE/MTT-S International Microwave Symposium - IMS*, 2018, pp. 868–870.

[5] A. Karakuzulu, M. H. Eissa, D. Kissinger, and A. Malignaggi, "Broadband 110 - 170 GHz True Time Delay Circuit in a 130-nm SiGe BiCMOS Technology," in *2020 IEEE/MTT-S International Microwave Symposium (IMS)*, 2020, pp. 775–778.

[6] D. Müller, S. Diebold, S. Reiss, H. Massler, A. Tessmann, A. Leuther, T. Zwick, and I. Kallfass, "D-Band digital phase shifters for phased-array applications," in *2015 German Microwave Conference*, 2015, pp. 205–208.

[7] M. De Wit and P. Reynaert, "An F-band active phase shifter in 28nm CMOS," in *2017 IEEE MTT-S International Microwave Symposium (IMS)*, 2017, pp. 965–968.

[8] A. Mazzanti and A. Bevilacqua, "Second-Order Equivalent Circuits for the Design of Doubly-Tuned Transformer Matching Networks," *IEEE Transactions on Circuits and Systems I: Regular Papers*, vol. 65, no. 12, pp. 4157–4168, 2018.

A 4.5 dBm SiGe Doubler-Amplifier Chain Covering the Entire D-Band

Matthias Möck
Institute of Radio Frequency
Engineering and Electronics
Karlsruhe Institute of Technology
Karlsruhe, Germany
matthias.moeck@kit.edu

İbrahim Kağan Aksoyak
Institute of Radio Frequency
Engineering and Electronics
Karlsruhe Institute of Technology
Karlsruhe, Germany
ibrahim.aksoyak@kit.edu

Ahmet Çağrı Ulusoy
Institute of Radio Frequency
Engineering and Electronics
Karlsruhe Institute of Technology
Karlsruhe, Germany
cagri.ulusoy@kit.edu

Abstract—We present an ultra-wideband doubler-amplifier chain with an output power 3-dB bandwidth of 110–186 GHz (46.2 %), covering the entire D-band frequencies (110–170 GHz). The chain consists of a push-push frequency doubler with a following buffer amplifier and is fabricated in a 130-nm SiGe BiCMOS technology, occupying a core area of only 0.11 mm². The circuit demonstrates a peak output power of 4.5 dBm at 140 GHz, while doubler and amplifier consume 86 mW in total.

Index Terms—SiGe, wideband, frequency doubler, amplifier, millimeter-wave, signal generation, D-band

I. INTRODUCTION

Recently, the D-band $(110 - 170\,\text{GHz})$ has attracted great interest for applications such as high-precision imaging or wideband communications. These applications set stringent requirements on the circuits, particularly on the signal generation chain. The design of low-noise voltage-controlled oscillators (VCOs) with wide tuning range, as required e.g. in fine-resolution radars, becomes very challenging at higher frequencies [1]. Thus, a low-frequency VCO with cascaded frequency multiplication chain is generally preferred [2].

Wideband frequency multiplication can be realized by using a distributed configuration [3] or resistive matching [4], both at the expense of an increased power loss due to the employed resistors. In [5], a marchand balun with broadband compensation technique is proposed to achieve wideband frequency

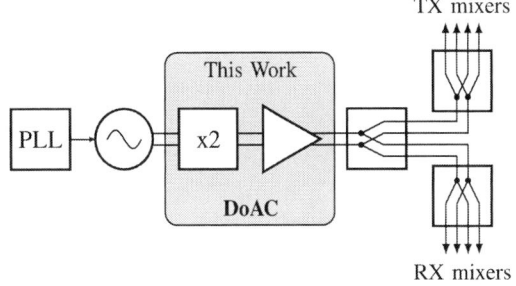

Fig. 2. Block diagram of LO path in I/Q-TRX system.

doubling. However, a relatively lower conversion gain of $-8\,\text{dB}$ is observed due to the absence of a buffer amplifier. The performance summary of frequency doublers at D-band frequencies in Fig. 1 highlights the difficulty of generating a high output power over a broad frequency bandwidth.

In this paper, we present a doubler-amplifier chain (DoAC) that demonstrates a wideband operation thanks to a careful co-design of the doubler and amplifier sub-circuits. By treating the sub-circuits as two stages with cascaded gain peaking and employing a broadband output matching network for the amplifier, we achieve a relative 3-dB bandwidth of 46.2 % covering the entire D-band frequencies.

II. CIRCUIT DESIGN

A. System Considerations

The proposed DoAC is to be used in a broadband D-band I/Q radar transceiver (TRX). The block diagram of the TRX's LO path is given in Fig. 2. The V/W-band signal is generated through a phase-locked VCO. After the DoAC, the D-band signal is distributed to the TX and RX I/Q mixers through differential power splitters, which demonstrate a simulated excess loss of $1\,\text{dB}$ each. Note that the quadrature phases for the I/Q mixers are provided by a quadrature coupler in the RF path, which is not shown in Fig. 2. From simulations, the intended LO drive of the RX mixers is found to be around $-5\,\text{dBm}$. Accounting for the splitters' loss, this leads to a target differential output power of $\approx 3\,\text{dBm}$ for the DoAC.

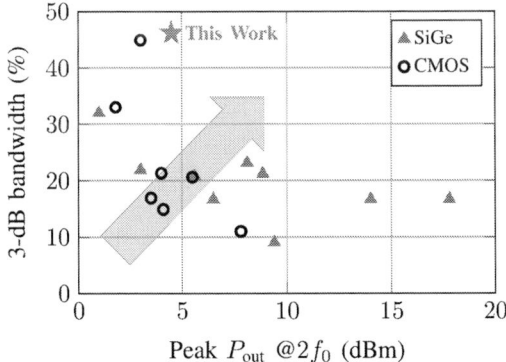

Fig. 1. Performance summary of state-of-the-art silicon-based D-band frequency doublers [5]–[17].

979-8-3503-4331-1/24 $31.00 © 2024 IEEE

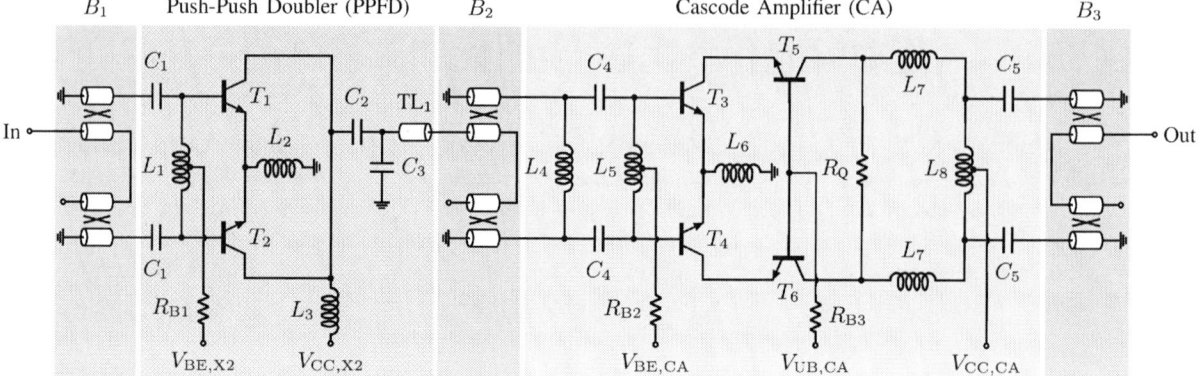

Fig. 3. Schematic of the DoAC. Baluns B_1 and B_3 are only for measurement purposes, thus their loss is de-embedded.

B. Design of the Doubler-Amplifier Chain (DoAC)

The schematic of the DoAC is shown in Fig. 3. A push-push frequency doubler (PPFD) is chosen over a Gilbert cell due to its improved efficiency and compact nature. Based on 2nd-harmonic load-pull optimizations, the transistors T_1 and T_2 are biased in Class-B and have $N_f = 8$ fingers to maximize the 2nd-harmonic output power. Small emitter degeneration inductance L_2 is added, resulting in a more favorable even-mode impedance. Harmonic reflectors can be used to boost the conversion gain [7], but have been avoided due to their narrow bandwidth. The PPFD's single-ended output signal is converted to a differential signal through marchand balun B_2.

A fully-differential cascode amplifier (CA) is added to meet the TRX's LO power requirements. A conservative device size of $N_f = 5$ is selected for T_{3-6} to keep the power consumption low while still satisfying saturated output power requirements. A special emphasis is on the design of the output matching network (OMN) formed by L_7, L_8 and C_5. To present the optimum load-pull impedance over a broad bandwidth, the OMN's quality factor is reduced by adding a differentially connected shunt resistor, $R_Q = 600\,\Omega$. Through this approach, the CA's 3-dB bandwidth is extended to $45\,\text{GHz}$, which is larger than most of the bandwidth values reported in state-of-the-art D-band amplifiers [18].

The simulation results in Fig. 4 illustrate the co-design process aiming to maximize the bandwidth. The simulation is performed over the whole DoAC, with $P_{\text{out,PPFD}}$ representing the input power of the CA. The PPFD's maximum output power ($P_{\text{out,PPFD}}$) exhibits a peak at $120\,\text{GHz}$. Conversely, the CA gain reaches its maximum at $160\,\text{GHz}$, coinciding with a local minimum of $P_{\text{out,PPFD}}$. This cascaded peaking results in a flatter frequency profile, contributing to an expanded bandwidth for the combined maximum output power ($P_{\text{out,DoAC}}$) of the DoAC. By leveraging this co-design approach, a remarkable enhancement in bandwidth is achieved.

The simulated maximum conversion gain and fundamental rejection ratio (FRR), both with and without the CA, are shown in Fig. 5. The simulations indicate that adding the CA buffer

results in up to $10\,\text{dB}$ improvement in conversion gain. Also, the FRR of the buffered DoAC remains better than $34\,\text{dBc}$ over the entire D-band.

Since phase noise (PN) is one of the main performance-limiting factors in radar systems, the DoAC's PN contribution is analyzed as well. Simulation results (not shown here) confirm a PN increase of $6\,\text{dB}$, which is in agreement with the theoretical expectation for a doubler.

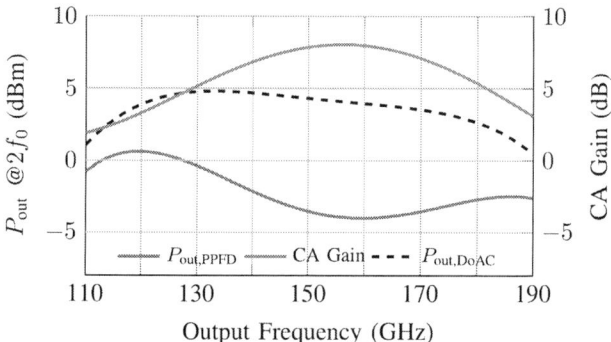

Fig. 4. Simulations showing cascaded gain peaking technique.

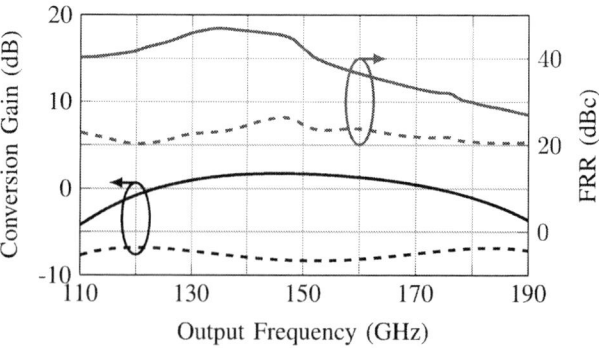

Fig. 5. Simulated peak conversion gain and fundamental rejection with (solid lines) and without buffer (dashed lines).

979-8-3503-4331-1/24 $31.00 © 2024 IEEE

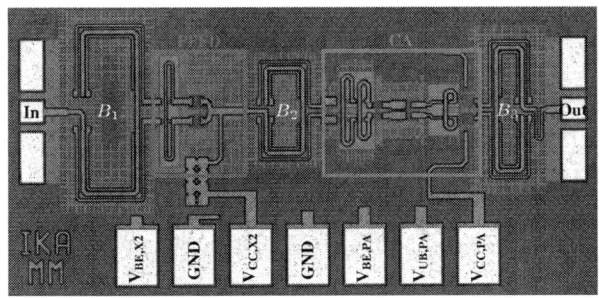

Fig. 6. Photograph of the fabricated doubler-amplifier chain. The chip area is $990 \times 470\,\mu m^2$ including the DC and RF pads.

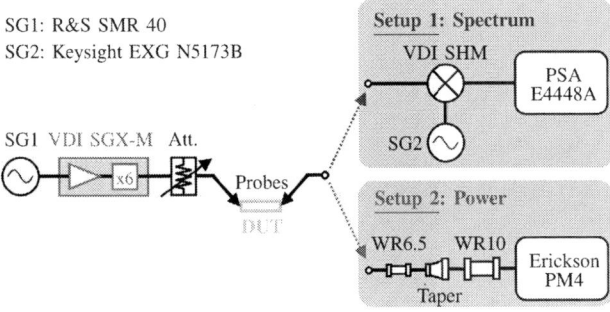

SG1: R&S SMR 40
SG2: Keysight EXG N5173B

Fig. 7. Measurement setup.

III. EXPERIMENTAL RESULTS

The chip was fabricated in IHP's 130-nm SiGe BiCMOS technology SG13G2 with f_T/f_{MAX} of $300/500\,GHz$ and featuring a seven metal layer aluminium back-end-of-line. The die micrograph is given in Fig. 6. The total chip area including pads is $990 \times 470\,\mu m^2$. The core area (PPFD, B_2, CA) measures only $560 \times 190\,\mu m^2$. The marchand baluns B_1 and B_3 are also taped out individually each in a back-to-back (B2B) configuration. Neither B_1 nor B_3 are part of the TRX in Fig. 2, thus the measured S-parameters of their B2B breakouts are used to de-embed their losses. B_2 is not de-embedded.

For the DoAC measurements, the setup in Fig. 7 is used. The input signal is applied through VDI SGX-M signal generator (SG) extension modules. At the output, VDI's D-band down-conversion mixer WR6.5X6SHM is connected to a spectrum analyzer (PSA) to measure the output signal frequency and verify single-tone spectrum content. For 2nd-harmonic power measurements, down-conversion mixer and PSA are replaced by an Erickson PM4 power meter.

The maximum conversion gain and maximum output power across output frequency are plotted in Fig. 8 and Fig. 9, respectively. For both plots, the input power is swept and the maximum value is reported for each frequency point. A peak conversion gain of $3.1\,dB$ is measured at $0\,dBm$ of input power. Even though the measurements were limited to $110 - 170\,GHz$, good agreement between simulations and measurements implies a 3-dB bandwidth ranging from $123 - 173\,GHz$ (33.8 %).

As shown in Fig. 9, the DoAC demonstrates a measured peak output power of $4.5\,dBm$ with a 3-dB bandwidth of $110 - 186\,GHz$ (46.2 %). As it can be seen in Fig. 10, the peak output power is achieved at an input power level of $4\,dBm$, at which doubler and amplifier draw $29\,mA/17\,mA$ from their $1.2\,V/3\,V$ supplies, respectively, leading to a total DC power consumption of $86\,mW$. The measured DC-RF efficiency ($\eta_{dc} = P_{out}/P_{dc}$) peaks at 3.4 % and remains above 2 % over the full D-band frequency range.

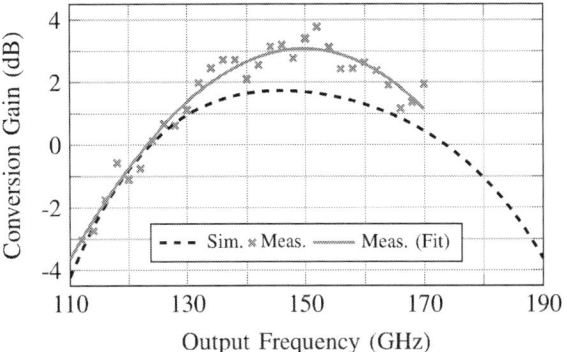

Fig. 8. Maximum conversion gain vs. frequency.

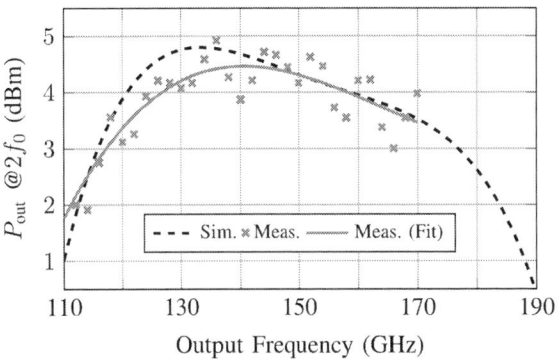

Fig. 9. Maximum output power vs. frequency.

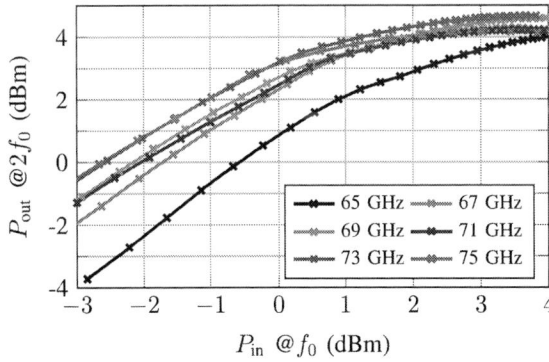

Fig. 10. Measured output power vs. input power for various input frequencies f_0.

979-8-3503-4331-1/24 $31.00 © 2024 IEEE

TABLE I
STATE-OF-THE-ART SILICON-BASED D-BAND FREQUENCY DOUBLERS.

Ref.	Technology	Type	f_c (GHz)	BW_{3dB}* (%)	Peak P_{out} (dBm)	P_{in}† (dBm)	CG_{max} (dB)	P_{dc}† (mW)	Area§ (mm^2)
[5]	65 nm CMOS	x2	122.5	44.9	3	11	-8	**22.8**	0.24
[6]	55 nm CMOS	x2	118.5	11.0	7.8	10	-0.5	88	**0.21**
[7]	22 nm FDSOI	x2	135	14.9	4.1	10	-3.8	24.7	0.33
[12]	55 nm SiGe	x2 + Buffer	130	23.1	8.1	**-5**	13.1†	97.5	n.a.
[13]	120 nm SiGe	x2 + 3-stage PA	120	16.7	**17.8**	-2	19.8†	600	1.92
[14]	130 nm SiGe	x2	115.5	32.0	1#	1	3.1	69	0.60
[15]	130 nm SiGe	x2 + Buffer	**154**	20.8	5.6	1.8	4.9	36	0.49
This	130 nm SiGe	x2 + Buffer	148	**46.2**	4.5	4	3.1	86	0.47

* output power 3-dB bandwidth　　# differential power　　† at peak P_{out}　　§ including pads

IV. CONCLUSION

This work presents a wideband doubler-amplifier chain with excellent agreement between simulations and measurements. Thanks to cascaded gain peaking and broadband balun and matching circuits, a peak output power of 4.5 dBm is demonstrated with a 3-dB bandwidth covering the entire D-band.

Table I compares the circuit performance to recently published silicon-based D-band doublers. Our proposed circuit exceeds the bandwidth demonstrated in prior work and occupies the smallest chip area among the buffered designs.

ACKNOWLEDGMENT

The authors would like to thank Andreas Lipp and Thorsten Fux from IHE for PCB manufacturing and assembly.

REFERENCES

[1] M. Möck, İ. K. Aksoyak, and A. Ç. Ulusoy, "A Ka-Band Colpitts–Clapp VCO With 30% Tuning Range and High Output Power," *IEEE Microwave and Wireless Technology Letters*, vol. 33, no. 4, pp. 439–442, 2023.

[2] C. Bredendiek, N. Pohl, K. Aufinger, and A. Bilgic, "Differential signal source chips at 150 GHz and 220 GHz in SiGe bipolar technologies based on Gilbert-Cell frequency doublers," in *2012 IEEE Bipolar/BiCMOS Circuits and Technology Meeting (BCTM)*, 2012, pp. 1–4.

[3] I. Lee, Y. Kim, and S. Jeon, "108–316- and 220–290-GHz Ultra-broadband Distributed Frequency Doublers," *IEEE Transactions on Microwave Theory and Techniques*, vol. 68, no. 3, pp. 1000–1011, 2020.

[4] C. Bohn, M. Kaynak, T. Zwick, and A. C. Ulusoy, "Ultra-Wideband Frequency Doubler with Differential Outputs in SiGe BiCMOS," in *2022 IEEE 22nd Topical Meeting on Silicon Monolithic Integrated Circuits in RF Systems (SiRF)*, 16.01.2022 - 19.01.2022, pp. 58–61.

[5] P.-H. Tsai, Y.-H. Lin, J.-L. Kuo, Z.-M. Tsai, and H. Wang, "Broadband Balanced Frequency Doublers With Fundamental Rejection Enhancement Using a Novel Compensated Marchand Balun," *IEEE Transactions on Microwave Theory and Techniques*, vol. 61, no. 5, pp. 1913–1923, 2013.

[6] Z. Yang, K. Ma, F. Meng, and B. Liu, "A 120-GHz Class-F Frequency Doubler With 7.8-dBm P_{OUT} in 55-nm Bulk CMOS," *IEEE Journal of Solid-State Circuits*, vol. 58, no. 8, pp. 2173–2188, 2023.

[7] M. Möck, İ. K. Aksoyak, and A. Ç. Ulusoy, "A High-Efficiency D-Band Frequency Doubler in 22-nm FDSOI CMOS," in *2022 17th European Microwave Integrated Circuits Conference (EuMIC)*, 2022, pp. 272–275.

[8] N. Oz and E. Cohen, "A compact 105–130 GHz push-push doubler, with 4dBm Psat and 18% efficiency in 28nm CMOS," in *2015 10th European Microwave Integrated Circuits Conference (EuMIC)*, 2015, pp. 101–104.

[9] S. Hao, Y. -W. Tang, X. Ding, L. Du, Y. Du, A. Tang, Q. J. Gu, and M. -C. F. Chang, "An 8.3% Efficiency 96–134 GHz CMOS Frequency Doubler Using Distributed Amplifier and Nonlinear Transmission Line," in *2020 IEEE Asian Solid-State Circuits Conference (A-SSCC)*, 2020, pp. 1–2.

[10] H.-C. Lin and G. M. Rebeiz, "A 135–160 GHz balanced frequency doubler in 45 nm CMOS with 3.5 dBm peak power," in *2014 IEEE MTT-S International Microwave Symposium (IMS2014)*, 01.06.2014 - 06.06.2014, pp. 1–4.

[11] I. Abdo, K. K. Tokgoz, T. Fujimura, K. Okada, and A. Matsuzawa, "A 100–123GHz CMOS frequency doubler with 5.5dBm output power and high fundamental rejection," in *2017 IEEE International Symposium on Radio-Frequency Integration Technology (RFIT)*, 2017, pp. 138–140.

[12] M. M. Pirbazari and A. Mazzanti, "High Gain 130-GHz Frequency Doubler With Colpitts Output Buffer Delivering P_{out} up to 8 dBm with 6% PAE in 55-nm SiGe BiCMOS," *IEEE Solid-State Circuits Letters*, vol. 4, pp. 36–39, 2021.

[13] R. B. Yishay and D. Elad, "A 17.8 dBm 110–130 GHz power amplifier and doubler chain in SiGe BiCMOS technology," in *2015 IEEE Radio Frequency Integrated Circuits Symposium (RFIC)*, 2015, pp. 391–394.

[14] A. Ergintav, F. Herzel, J. Borngraber, D. Kissinger, and H. J. Ng, "An integrated 122GHz differential frequency doubler with 37 GHz bandwidth in 130 nm SiGe BiCMOS technology," in *2017 IEEE MTT-S International Conference on Microwaves for Intelligent Mobility (ICMIM)*, 2017, pp. 53–56.

[15] C. Coen, S. Zeinolabedinzadeh, M. Kaynak, B. Tillack, and J. D. Cressler, "A highly-efficient 138–170 GHz SiGe HBT frequency doubler for power-constrained applications," in *2016 IEEE Radio Frequency Integrated Circuits Symposium (RFIC)*, 22.05.2016 - 24.05.2016, pp. 23–26.

[16] S. G. Rao, M. Frounchi, and J. D. Cressler, "A D-band SiGe Frequency Doubler with a Harmonic Reflector Embedded in a Triaxial Balun," in *Proceedings of the 2020 IEEE Radio Frequency Integrated Circuits Symposium*, 2020, pp. 255–258.

[17] R. B. Yishay and D. Elad, "A 14 dBm 110–130 GHz power amplifier and doubler chain in 90 nm SiGe BiCMOS technology," in *2016 IEEE 16th Topical Meeting on Silicon Monolithic Integrated Circuits in RF Systems (SiRF)*, 2016, pp. 120–122.

[18] İ. K. Aksoyak, M. Möck, M. Kaynak, and A. Ç. Ulusoy, "A D-Band Power Amplifier With Four-Way Combining in 0.13-μm SiGe," *IEEE Microwave and Wireless Components Letters*, vol. 32, no. 11, pp. 1343–1346, 2022.

A SiGe-Based Quadrature D-Band Up-Converter with High Output Power

İbrahim Kağan Aksoyak
Institute of Radio Frequency
Engineering and Electronics
Karlsruhe Institute of Technology
Karlsruhe, Germany
ibrahim.aksoyak@kit.edu

Matthias Möck
Institute of Radio Frequency
Engineering and Electronics
Karlsruhe Institute of Technology
Karlsruhe, Germany
matthias.moeck@kit.edu

Ahmet Çağrı Ulusoy
Institute of Radio Frequency
Engineering and Electronics
Karlsruhe Institute of Technology
Karlsruhe, Germany
cagri.ulusoy@kit.edu

Abstract—This paper presents a quadrature D-band up-converter, implemented in a 130-nm SiGe BiCMOS technology. The up-converter consists of a 90-degree hybrid coupler with broadband operation, two up-converting I/Q mixers employing double-balanced Gilbert cell architecture, and a differential power divider for LO distribution. The I/Q up-converter demonstrates remarkable performance, with a maximum conversion gain of 11.2 dB at the RF frequency of 145 GHz and IF frequency of 3 GHz when supplied with an LO power of 4.5 dBm. At the same frequency of operation, the output-referred 1-dB compression point is measured to be -5.5 dBm, and the maximum output power is -2.2 dBm. The proposed up-converter exhibits a wide RF bandwidth of 40 GHz. The chip consumes 22 mW from a 2.5 V supply, and the core area is only 0.15 mm^2.

Index Terms—Mixer, gilbert-cell, SiGe, millimeter-wave (mm-wave), D-band, up-conversion.

I. INTRODUCTION

The growing interest on developing radar systems that are highly integrated, low-cost and power-efficient has stimulated extensive research in the utilization of mm-wave frequencies. The combination of relatively low atmospheric attenuation below $2\,\mathrm{dB\,km^{-1}}$ and the availability of wide bandwidths (BW) makes the frequencies within the D-band ($110-170\,\mathrm{GHz}$) particularly enticing to address the requirements of achieving high-resolution and compact radar solutions.

Modern silicon technologies offer cost-effective solutions with exceptional integration capabilities and high-performance. Previous studies have employed SiGe BiCMOS processes to demonstrate the functionality of transmitter (TX) [1] and receiver [2] blocks operating at D-band. Also, numerous complete transceiver (TRX) systems [3] have been developed at D-band for radar and communication applications. In both applications, the linearity and power-handling capability of the up-converters in the transmit path are particularly important, as they directly influence the ability to drive the power-amplifier (PA) without necessitating an additional driver stage between the up-converting mixers and the PA.

II. CIRCUIT DESIGN

A. System Considerations

The proposed quadrature I/Q up-converter (QuC) is designed using a 130 nm SiGe BiCMOS technology with

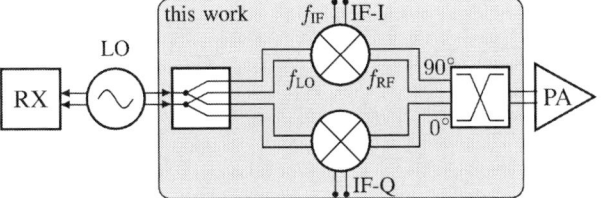

Fig. 1. Block diagram of the QuC in I/Q TRX system.

$f_\mathrm{T}/f_\mathrm{MAX}$ of $300/500\,\mathrm{GHz}$. It is to be employed in a broadband D-band radar TRX, where it drives the subsequent PA. The simplified block diagram of the envisioned TRX is represented in Fig. 1. The QuC architecture comprises two up-converting mixers operating in I/Q fashion, a differential power divider for LO distribution and a broadband 90° hybrid coupler facilitating the quadrature operation. The circuit is fully differential, and on-chip marchand baluns (not shown in Fig. 1) are incorporated at RF and LO ports for single-ended on-wafer probing.

B. Mixer Design

The up-converting I/Q mixers employ a double-balanced Gilbert cell core, and their circuit schematic together with the current mirror-based bias circuitry to generate the base bias voltages is shown in Fig. 2a. A layout view of the double-balanced Gilbert cell mixer core is shown in Fig. 2b. Extensive efforts have been made to preserve symmetry within the core layout, particularly at the high-frequency ports, RF and LO.

The impact of current density (J_c) on conversion gain (CG) is shown in Fig. 3a, where a trade-off exists between the CG and the power consumption (P_dc). In this work, a current density of $0.5\,\mathrm{mA}$ per finger is chosen for the transistors $Q_\mathrm{1-6}$ to keep the P_dc at a moderate level. To meet the specifications of the TRX, the target input power of the PA is set to be $-6\,\mathrm{dBm}$ and the size of the active devices is determined accordingly. Fig. 3b illustrates how the device size of the input stage transistors ($Q_{5,6}$) (and accordingly the $Q_\mathrm{1-4}$ to preserve the same current density) influence both CG as well as the maximum achieved output power ($P_\mathrm{out,max}$). Although increasing the device sizes favors both CG and

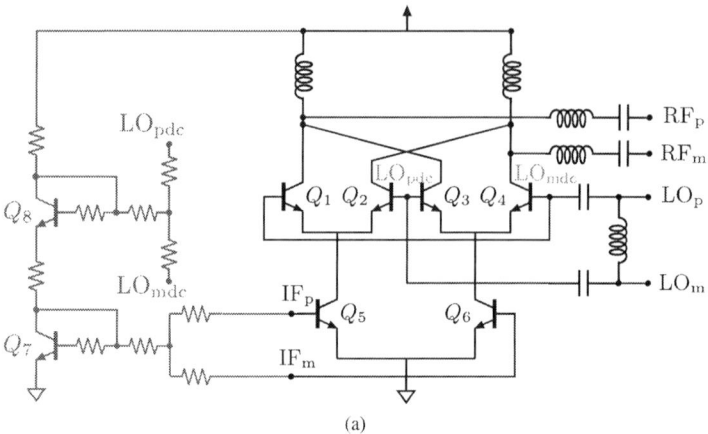 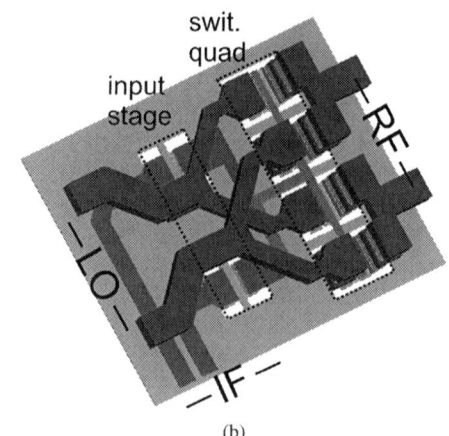

(a) (b)

Fig. 2. (a) Mixer schematic with bias circuitry (gray). (b) A layout image of the Gilbert cell core.

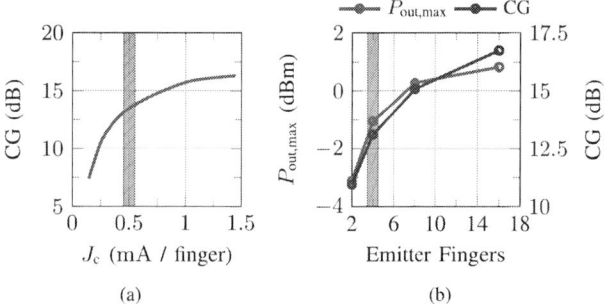

(a) (b)

Fig. 3. The gray bars indicate the chosen values. The impact of (a) current density (J_c) on CG. (b) transistor size (Q_{5-6}) on $P_{out,max}$ and CG.

$P_{out,max}$, a device size of $4 \times 0.07 \times 0.9\,\mu m^2$ for Q_{5-6}, and $2 \times 0.07 \times 0.9\,\mu m^2$ for Q_{1-4}, are chosen to relax the LO drive requirements and keep the P_{dc} low.

Lastly, the matching networks at the RF and LO ports are optimized based on load-pull and source-pull simulations. The inductors at the RF output of the Gilbert cell are carefully designed with ground cut-outs to reduce the losses over the output matching network, leading to an improved output power.

C. Coupler and Power Divider Design

The careful optimization of passives, specifically the coupler and power divider, are of crucial importance in this study, as they significantly impact the overall chip area. Their detailed optimization ensures area efficiency and compatibility with the layout geometry. The differential hybrid broadside coupler exhibits simulated differential excess losses of 0.7 dB and 0.2 dB, at 0° & 90° outputs, at the RF center frequency (f_{RF}) of 145 GHz. The amplitude and phase imbalances remain less than 1 dB and 3°, respectively, along the 3-dB RF BW while the isolation is higher than 24 dB. It presents wideband matching over the whole D-band. The hybrid coupler is placed at the RF path for its wideband-matching capabilities, superior isolation to prevent the I/Q cross-talk and lower excess losses. The differential power divider achieves an excess loss of

1.1 dB, with amplitude and phase imbalances lower than 0.1 dB and 1.6°, over the 3-dB RF BW, and is placed at the LO path. Both passives are designed using the thickest top-metal layers for minimized losses and are shown in Fig. 4.

III. EXPERIMENTAL RESULTS

The chip micrograph of the fabricated QuC is given in Fig. 4. The occupied chip area is $860 \times 630\,\mu m^2$ including the pads. Remarkably, the core of the chip is quite compact, measuring only $480 \times 315\,\mu m^2$. Additionally, the marchand balun, denoted as B_1, is separately taped-out and characterized in a back-to-back configuration. Since B_1 is not a part of the TRX in Fig. 1, the measured losses of the on-chip baluns are de-embedded to ensure accurate assessment of the QuC.

During the large-signal measurements, the differential I/Q IF signals (0°, 90°, 180°, 270°) are generated through an external coaxial coupler, and two off-chip coaxial baluns. The LO signal is applied through a VDI SGX-M D-band source module. A D-band sub-harmonic down-conversion mixer is connected at the RF output to down-convert the RF signal and examine the spectral content through a spectrum analyzer. All the losses due to external cables and equipment are de-embedded. The measurement setup is illustrated in Fig. 5.

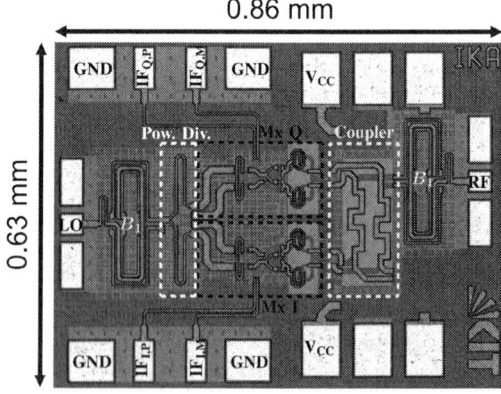

Fig. 4. Chip micrograph of the fabricated QuC.

TABLE I
STATE-OF-THE-ART SILICON-BASED D-BAND UP-CONVERTERS.

Ref.	Technology	Type#	f_{RF} (GHz)	RF BW$_{3dB}$* (GHz)	CG$_{max.}$ (dB)	OP_{1dB} (dBm)	$P_{out, max.}$ (dBm)	IF BW$_{3dB}$* (GHz)	P_{dc} (mW)	Area (mm^2)
[5]	40 nm CMOS	I-Q UC + PA	118	14	13.5	4.5	n.a	n.a	271	1.51
[6]	40 nm CMOS	UC	120	30	-4	-12.5	-11	n.a	9	0.22†
[7]	130 nm SiGe	I-Q UC	135	30	9.5	-8.5	-4.5	13§	52	0.3
[8]	130 nm SiGe	I-Q UC + PA	130	31	23	-4	0	8	240	1.54
This	130 nm SiGe	I-Q UC	145	40	11.2	-5.5	-2.2	>20	22$^\&$, 43‡	0.15†

UC: up-converter   * CG 3-dB bandwidth   § graphical estimation   † core area   & at small signal   ‡ at $P_{out, max}$

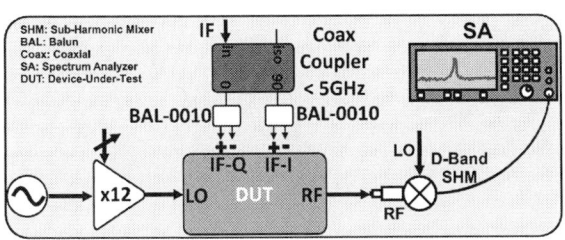

Fig. 5. Large-signal measurement setup.

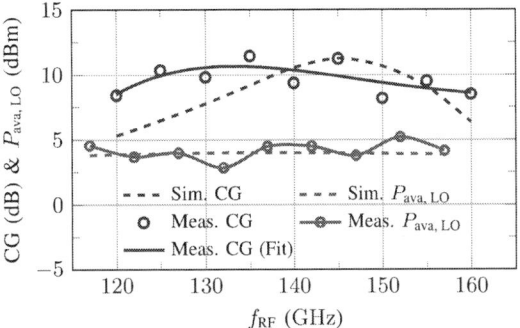

Fig. 7. CG vs. f_{RF} at $f_{IF} = 3$ GHz.

The simulated and measured upper side-band CG and output power (P_{out}) at f_{RF} of 145 GHz and f_{IF} of 3 GHz are shown in Fig. 6. CG is measured to be 11.2 dB while the output-referred $1 - $dB compression-point (O$P_{1dB}$) reaches as high as -5.5 dBm. The measured $P_{out,max}$ is -2.2 dBm when the QuC is supplied with an available LO power ($P_{ava, LO}$) of 4.5 dBm and an IF power of -5.7 dBm. The image rejection ratio (IRR) is measured to be better than 20 dBc, which points out to a maximum phase imbalance of 11°, or a maximum amplitude imbalance of 1.7 dB or a combination of both [4], while the simulated IRR is better than 25 dBc. This can be explained through the imperfect measurement setup, e.g. the unmatched cables, imbalances in the external hybrid and baluns, which complicates a precise differential I/Q feed at the IF input.

The 3-dB RF BW is measured while keeping the f_{IF} at a constant value of 3 GHz and sweeping the f_{LO}. The CG over the f_{RF} is given in Fig. 7. The 3-dB IF BW is measured at a fixed f_{RF} of 145 GHz. The external coaxial-hybrid has

a cut-off frequency of only 5 GHz, hence the measurement is performed up to an f_{IF} of 4 GHz. The measured and simulated CG with respect to varying f_{IF} are shown in Fig. 8a. Since the measurement results over the measured f_{IF} range shows a good agreement with the simulation results, it is safe to assume a 3-dB IF BW of more than 20 GHz.

The impact of $P_{ava, LO}$ on the CG and $P_{out,max}$ is shown in Fig. 8b. This measurement is performed at f_{RF} of 145 GHz, and f_{IF} of 3 GHz. It can be deducted from the measured plot that increasing the $P_{ava, LO}$ after a certain value results in only minor improvements in both parameters. The values reported in this work are based on an $P_{ava, LO}$ of 4.5 dBm. The $P_{ava, LO}$ can also be reduced to 1.7 dBm at the cost of 0.3 dB of CG and 1.2 dBm of $P_{out,max}$.

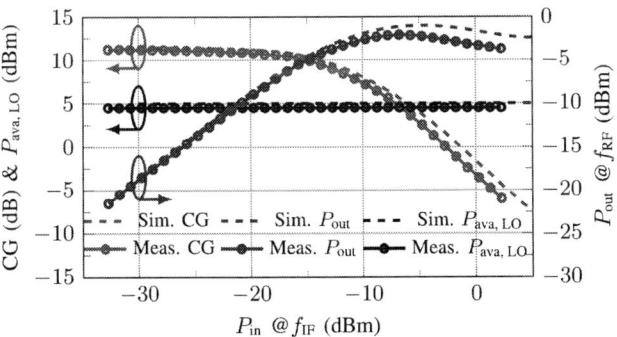

Fig. 6. CG and P_{out} vs. P_{in} at $f_{RF} = 145$ GHz, $f_{IF} = 3$ GHz.

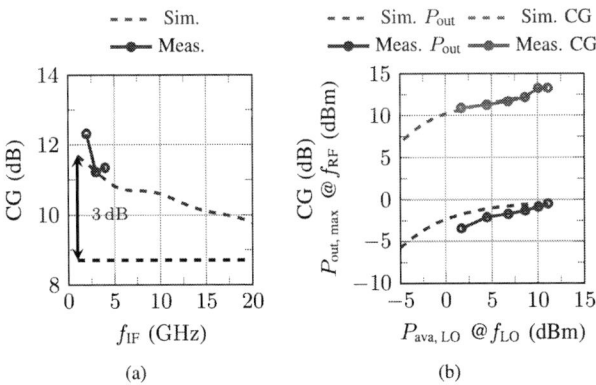

Fig. 8. (a) CG vs. f_{IF} at $f_{RF} = 145$ GHz. (b) CG & $P_{out, max}$ vs. $P_{ava, LO}$ at $f_{RF} = 145$ GHz, $f_{IF} = 3$ GHz.

IV. CONCLUSION

This paper has demonstrated a D-band I/Q up-converter fabricated in a 130 nm SiGe BiCMOS technology for a broadband radar application. The fabricated chip is fully characterized, and experimental results are presented. Table I compares the circuit performance to state-of-the-art silicon-based D-band up-converters and TX chains. Our proposed circuit exhibits the highest CG, OP_{1dB} and P_{out} among the designs that do not employ a PA, while presenting the highest RF and IF BWs with the smallest chip area.

ACKNOWLEDGMENT

The authors would like to thank Andreas Lipp and Thorsten Fux from IHE for PCB manufacturing and assembly.

REFERENCES

[1] İ. K. Aksoyak, M. Möck, M. Kaynak, and A. Ç. Ulusoy, "A D-Band Power Amplifier With Four-Way Combining in 0.13-um SiGe," *IEEE Microwave and Wireless Components Letters*, vol. 32, no. 11, pp. 1343–1346, 2022.

[2] İ. K. Aksoyak, M. Möck, and A. Ç. Ulusoy, "A Differential D-Band Low-Noise Amplifier in 0.13 um SiGe," *IEEE Microwave and Wireless Components Letters*, vol. 32, no. 8, pp. 979–982, 2022.

[3] D. Kissinger, G. Kahmen, and R. Weigel, "Millimeter-Wave and Terahertz Transceivers in SiGe BiCMOS Technologies," *IEEE Transactions on Microwave Theory and Techniques*, vol. 69, no. 10, pp. 4541–4560, 2021.

[4] B. Razavi, "Design considerations for direct-conversion receivers," *IEEE Transactions on Circuits and Systems II: Analog and Digital Signal Processing*, vol. 44, no. 6, pp. 428–435, 1997.

[5] C. J. Lee, S. H. Kim, H. S. Son, D. M. Kang, J. H. Kim, C. W. Byeon, and C. S. Park, "A 120 GHz I/Q Transmitter Front-end in a 40 nm CMOS for Wireless Chip to Chip Communication," in *2018 IEEE Radio Frequency Integrated Circuits Symposium (RFIC)*, 2018, pp. 192–195.

[6] C. J. Lee, J.-S. Kang, and C. S. Park, "A D-Band Low-Power Gain-Boosted Up-Conversion Mixer With Low LO Power in 40-nm CMOS Technology," *IEEE Microwave and Wireless Components Letters*, vol. 27, no. 12, pp. 1113–1115, 2017.

[7] S. Carpenter, Z. S. He, and H. Zirath, "A direct carrier I/Q modulator for high-speed communication at D-band using 130nm SiGe BiCMOS technology," in *2017 12th European Microwave Integrated Circuits Conference (EuMIC)*, 2017, pp. 265–268.

[8] S. Carpenter, H. Zirath, Z. S. He, and M. Bao, "A fully integrated D-band direct-conversion I/Q transmitter and receiver chipset in SiGe BiCMOS technology," *Journal of Communications and Networks*, vol. 23, no. 2, pp. 73–82, 2021.

A 300 GHz x9 Multiplier Chain With 9.6 dBm Output Power in 0.13-μm SiGe Technology

Arjith Chandra Prabhu
IHCT, University of Wuppertal
Wuppertal, Germany
chandra@uni-wuppertal.de

Janusz Grzyb
IHCT, University of Wuppertal
Wuppertal, Germany
grzyb@uni-wuppertal.de

Philipp Hillger
IHCT, University of Wuppertal
Wuppertal, Germany
hillger@uni-wuppertal.de

Thomas Bücher
IHCT, University of Wuppertal
Wuppertal, Germany
buecher@uni-wuppertal.de

Holger Rücker
*IHP-Leibniz-Institut für
innovative Mikroelektronik*
Frankfurt (Oder), Germany
ruecker@ihp-microelectronics.com

Ullrich Pfeiffer
IHCT, University of Wuppertal
Wuppertal, Germany
ullrich.pfeiffer@uni-wuppertal.de

Abstract—This paper presents a 300 GHz x9 multiplier chain in an advanced 0.13-μm SiGe HBT technology. It consists of 2 over-driven differential class-A cascode triplers (27-35 GHz and 81-105 GHz), a buffer amplifier, 2 bandpass mode-selective filters, and a 3-stage power amplifier (PA). The mode-selective filters were designed to allow broadband operation with high harmonic rejection. Band-selective matching transformers with high common mode (CM) suppression are implemented. The x9 multiplier chain has a peak 9.6 dBm output power at 270 GHz, fractional 3-dB bandwidth of 21% (244-300 GHz), 0.66 W DC power consumption, and 1.36% DC-RF efficiency. It has a wideband high harmonic suppression of >30 dBc/>40 dBc in the frequency range of 244-300 GHz/252-288 GHz for 7th, 8th, 10th, and 11th harmonics. The active area of the x9 multiplier chain is 0.92 mm^2.

Index Terms—SiGe HBT, mode-selective filter, multiplier chain, J-band, wideband harmonic suppression.

I. INTRODUCTION

The J-band (220-325 GHz) has gained increased interest in high-speed wireless communications due to its low atmospheric attenuation and large available bandwidth. The progress made in SiGe HBT technology over the past decade enables the design of fully integrated circuits in this frequency range. Local oscillator (LO) is a key component of communication systems. At such high frequencies, Voltage-controlled oscillators (VCO) suffer from low output power and degraded phase noise, which is not suitable for wireless communication systems. Another approach to generate power at such frequencies is through a frequency multiplier chain which can provide high output power and a wide tuning range. A major drawback of the multiplier chain is the presence of parasitic harmonics. The presence of parasitic harmonics in the LO, driving a mixer can distort the up/down converted signal and deteriorate the desired harmonic which degrades the performance of the transceiver [1]. With a wide tuning range and high multiplication factor, in-band harmonic rejection becomes a major challenge that requires suitable frequency planning of the cascaded circuit blocks.

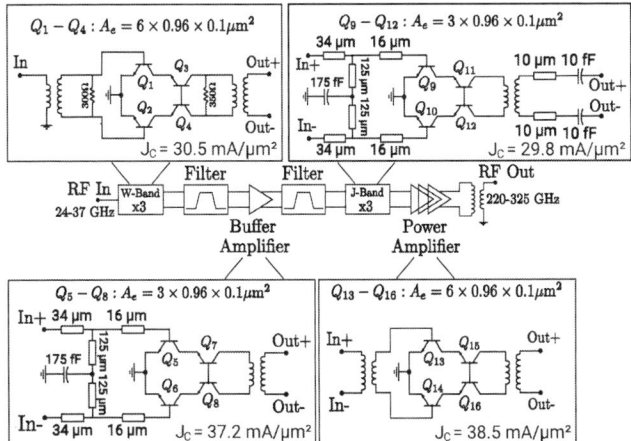

Fig. 1. Block diagram of the x9 multiplier chain with simplified schematic representation of triplers, buffer and power amplifier.

In this paper, we present a J-band wideband x9 multiplier chain consisting of 2 over-driven differential class-A cascode triplers, mode-selective filters [2], a W-band buffer amplifier, and a 3-stage 300 GHz PA [3]. High harmonic rejection is achieved in two ways. Firstly, compact mode-selective filters with broadband suppression of all CM harmonics (both in-band as well as out-of-band) with high suppression of the two strongest differential harmonics (x1 and x2). Secondly, a band-selective matching transformer with high suppression of CM harmonics by a suitable arrangement of decoupling capacitors at the center taps. A wideband pad integrated balun was implemented at the output to facilitate broadband on-chip measurements [4]. The x9 multiplier chain is fabricated in an early development lot for IHP's next generation 0.13-μm SiGe HBT BiCMOS technology SG13G3 with a target peak f_t/f_{max} values of 470/650 GHz.

II. CIRCUIT DESIGN

Fig. 1. show the block diagram of the x9 multiplier chain with a simplified schematic of the triplers, buffer, and PA. The layout of each block is full wave 3D EM simulated with parasitic interconnects in Ansys HFSS. The design detail of the above-mentioned blocks are discussed in the following section.

A. W-band Tripler, 100 GHz Filter, and Buffer Amplifier

The W-band tripler consists of a differential cascode core, input balun, and output transformer as described in [2]. The tripler/buffer operates at 2.4/2.5 V supply reaching a saturated output power of 4.2 dBm, with a 3-dB fractional bandwidth of 22.5% ranging from 84-106 GHz.

B. 300 GHz Mode-Selective Filter

The filter plays a crucial role in attenuating the unwanted harmonics and the design concept is described in [2]. The overall filter consists of two identical sections, as shown in Fig. 3. with the input/output (C_1/C_1) and inter-stage (C_2) series MIM capacitors controlling the coupling coefficients between them. The series capacitors (C_1/C_2) also contribute to the overall fundamental signal rejection independently from the operation mode (even/odd). Each filter section contains a tightly coupled asymmetrical 2-line system with a galvanic connection to the surrounding global ground through 2 shunt line sections (L_1) in its center. Shorter lengths of the shunt line sections (L_1) improve the common-mode rejection. The filter was synthesized and optimized with a 3-D EM field solver, wherein the MIM capacitors were simulated using their physical material constants. The filter is very compact in terms of wavelength at the operating frequency of 270 GHz ($\sim \lambda/4.5$). It provides very high fundamental suppression of >45 dB and all-stop-like rejection of CM harmonics at least to 500 GHz. Fig 4. show the simulated results of the J-band filter.

C. J-Band Tripler

The J-band tripler core consists of a differential cascode similar to the W-band tripler core and the transistors are sized at $Ae = 3 \times 0.96 \times 0.1 \mu m^2$. The input matching is implemented by series and shunt stubs in top metal 2 (TM2). The output transformer is implemented as an asymmetric broadside coupled line in top metal 1 (TM1) and metal 4 (M4) followed by a 10 μm series line and series 10 fF capacitor as shown in Fig. 2, transforming the broadband 100 Ω to the optimal output impedance of 20+80j at the 3rd harmonic, resulting in a peak output power of -1.2 dBm at 270 GHz driven at 4 dBm of input power. The collector of the cascode is biased through the transformer along with large decoupling capacitors. The other section of the transformer is left unconnected. Due to this, a global short established on one side is transformed into a global open on the other port, providing a broadband CM suppression. The J-band tripler is operated at 2.5V and consumes 96 mW.

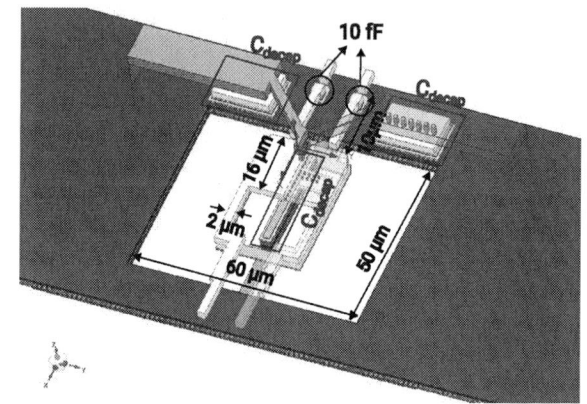

Fig. 2. Output Transformer implemented on top metal 1 (TM1) and metal 4 (M4) with 10 fF series capacitor and 10 μm series line.

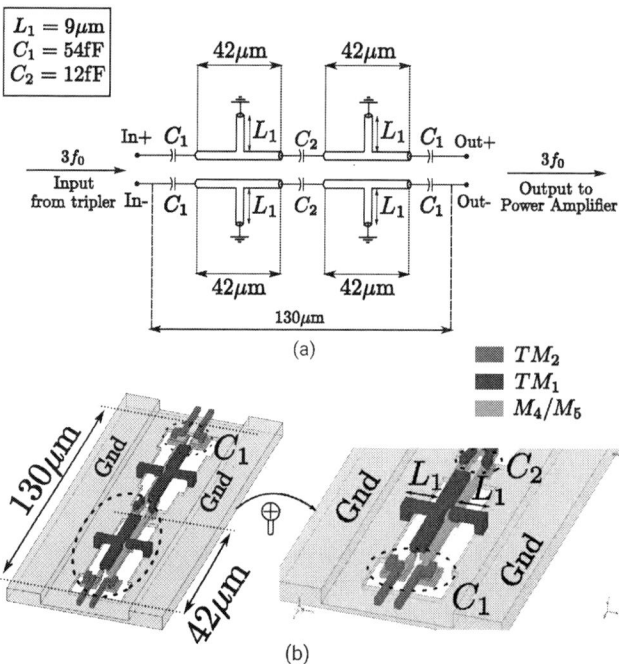

Fig. 3. Filter Structure: (a) Simplified representation of a two section bandpass filter; (b) 3-D model representing the coupled line structure.

D. 300 GHz Power Amplifier

The multiplier chain drives a 3-stage wideband 300 GHz power amplifier to boost its power. It is operated with a 3.5 V supply and shows a small-signal gain of 23 dB and a P_{sat}/OP_{1dB} of 9.7/6.7 dBm respectively with a 3-dB bandwidth of 63 GHz (239-302 GHz) in the small signal operation and 94 GHz (223-317 GHz) when saturated. To facilitate broadband measurements of the multiplier chain, a broadband Marchand-type balun with pad compensation was implemented at the output of the multiplier chain.

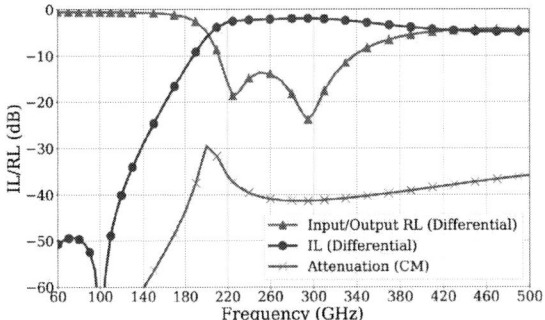

Fig. 4. Simulation results of differential insertion loss (IL), return loss (RL) and CM attenuation of the J-band filter.

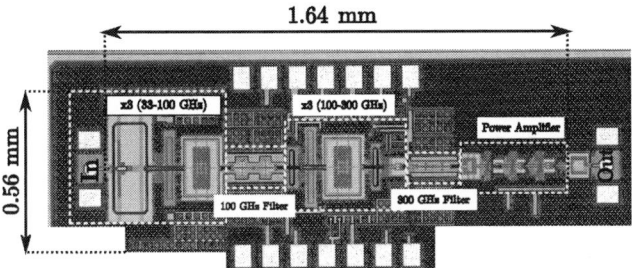

Fig. 5. Chip micrograph of the x9 multiplier chain with an active area of 0.92 mm² consisting of W and J-band tripler, filters, buffer amplifier and, PA.

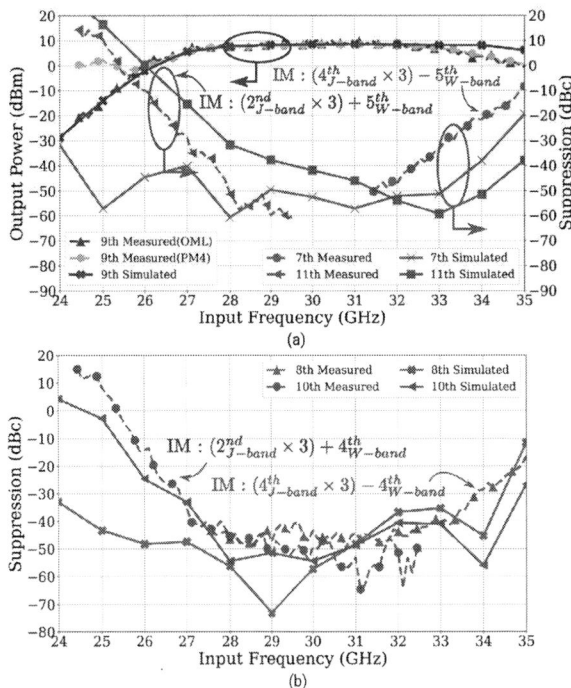

Fig. 6. Measured and simulated harmonic output powers at 2 dBm of input power: (a) 9th harmonic output power, 7th/11th harmonic suppression. The 7th/8th harmonics are produced due to the intermodulation product (IM) of J-band tripler 4th/2nd harmonics with W-band tripler 5th harmonic; (b) 8th/10th harmonic suppression. The 8th/10th harmonics are produced due to the IM of J-band tripler 4th/2nd harmonics with W-band tripler 4th harmonic

III. MEASUREMENT AND RESULTS

Fig 5. shows the chip micrograph. All measurements were performed on-wafer with OML and other frequency extension modules and PM4 power meter after initial calibration. All relevant setup losses, such as probes and cables were calibrated and de-embedded. The balun was measured separately showing 1.3 dB insertion loss and <-20 dB return loss and de-embedded. A peak 9th harmonic output power of 9.6 dBm at 270 GHz was measured consuming 0.66 W of DC power. The 3-dB operational fractional bandwidth is 21% in the range of 244-300 GHz. The measured output power at higher frequencies is limited by the PA bandwidth. The output power measured with PM4 increases at lower frequencies due to the influence of the 10th and 11th harmonics. In-band suppression of down to 40 dBc was measured. At band edges, the parasitic harmonics fall under the operation band of the PA and the suppression deteriorates quickly. The 7th, 8th,10th, and 11th parasitic harmonics can be attributed to the intermodulation products of the 4th/5th harmonics of the first tripler with the 2nd and 4th harmonics of the second tripler. Fig. 6 shows the harmonic measurement results and has a good correlation with the simulated results. Measurements were done across 283-325 GHz/248-325 GHz/220-293 GHz/220-266 GHz for 7th/8th/10th/11th harmonics due to out-of-band harmonic placement.

TABLE I
PERFORMANCE COMPARISON OF INTEGRATED MULTIPLIER CHAIN ABOVE 200 GHZ IN SIGE TECHNOLOGY

Ref	f_{out} (GHz)	MF	P_{out} (dBm)	3dB BW (%)	HR (dBc)	η (%)	Area mm²
[5]	235-265	x16	2.5	12.24	-	0.25	0.98
[6]	216-256	x8	0*	17	>25	0.4	1.2
[7]	220-261	x18	8	17.08	>25	1.47	0.28
[8]	223-350	x8	-2.3	44.32	>30	0.32	1
[9]	218-284	x18	-7.7	33.1	>28	0.07	0.86
This Work	244-300	x9	9.6	21	>30^a/>40^b	1.38	0.92

MF : Multiplication Factor, * Includes balun losses,
^{a} 244-300 GHz, ^{b} 252-288 GHz

IV. CONCLUSION

This paper presents a wideband 300 GHz x9 multiplier chain with a peak power of 9.6 dBm at 270 GHz and 21% 3-dB bandwidth ranging from 244-300 GHz. Compact mode-selective filters are implemented to suppress the parasitic harmonics and improve the performance of the transceiver by reducing the distortion in the up/down converted signal and of the desired harmonic. Parasitic harmonic rejection of >30 dBc/>40 dBc was measured in the frequency range of 244-300 GHz/252-288 GHz. The multiplier chain consumes 0.66 W of power while occupying an area of 0.92 mm².

979-8-3503-4331-1/24 $31.00 © 2024 IEEE

ACKNOWLEDGMENT

The research work presented in this paper was partially funded by the German Research Foundation ("Deutsche Forschungsgemeinschaft") (DFG) under project-ID 287022738 TRR 196 for project C04 and within the project DotSeven2IC (DFG PF 661/15-1).

REFERENCES

[1] Z. Zong and R. B. Staszewski, "Effects of Subharmonics in LO Generation on RF Transceivers," in *IEEE MTT-S Int. Microw. Symp. Dig.*, 2018, pp. 1–3.

[2] A. Chandra-Prabhu, J. Grzyb, P. Hillger, B. Heinemann, H. Rücker, and U. R. Pfeiffer, "A Wideband W-band Frequency Tripler With a Novel Mode-Selective Filter for High Harmonic Rejection," in *IEEE European Microw. Integr. Circ. Conf.*, (accepted).

[3] T. Bücher, J. Grzyb, P. Hillger, H. Rücker, B. Heinemann, and U. R. Pfeiffer, "A Broadband 300 GHz Power Amplifier in a 130 nm SiGe BiCMOS Technology for Communication Applications," *IEEE J. Solid-State Circuits*, vol. 57, no. 7, pp. 2024–2034, 2022.

[4] J. Grzyb, M. Andree, P. Hillger, T. Bücher, and U. R. Pfeiffer, "A Balun-Integrated On-Chip Differential Pad for Full/Multi-Band mmWave/THz Measurements," in *IEEE MTT-S Int. Microw. Symp. Dig.*, 2023.

[5] N. Sarmah, B. Heinemann, and U. R. Pfeiffer, "235–275 GHz (x16) frequency multiplier chains with up to 0 dBm peak output power and low DC power consumption," in *IEEE Radio Freq. Integr. Circuits Symp.*, 2014, pp. 181–184.

[6] M. H. Eissa, A. Malignaggi, M. Ko, K. Schmalz, J. Borngräber, A. C. Ulusoy, and D. Kissinger, "216–256 GHz fully differential frequency multiplier-by-8 chain with 0 dBm output power," in *IEEE Int. J. Microw. and Wireless Technol.*, 2017, pp. 216–219.

[7] E. Turkmen, I. K. Aksoyak, W. Debski, W. Winkler, and A. C. Ulusoy, "A 220–261 GHz Frequency Multiplier Chain (× 18) With 8-dBm Peak Output Power in 130-nm SiGe," *IEEE Microw. Wireless Compon. Lett.*, vol. 32, no. 7, pp. 895–898, 2022.

[8] A. Ali, J. Yun, M. Kucharski, H. J. Ng, D. Kissinger, and P. Colantonio, "220–360-GHz Broadband Frequency Multiplier Chains (x8) in 130-nm BiCMOS Technology," *IEEE Trans. Microw. Theory Techn.*, vol. 68, no. 7, pp. 2701–2715, 2020.

[9] T. Buecher, S. Malz, K. Aufinger, and U. R. Pfeiffer, "A 210–291-GHz (8×) Frequency Multiplier Chain With Low Power Consumption in 0.13-μm SiGe," *IEEE Microw. Wireless Compon. Lett.*, vol. 30, no. 5, pp. 512–515, 2020.

230 GHz Signal Generator for High-Bandwidth Data Links in 130 nm SiGe BiCMOS

Christian Hoyer
Chair for Circuit Design and Network Theory
Technische Universität Dresden
Germany
christian.hoyer1@tu-dresden.de

Luca Steinweg
Chair for Circuit Design and Network Theory
Technische Universität Dresden
Germany

Florian Protze
Chair for Circuit Design and Network Theory
Technische Universität Dresden
Germany

Franz Alwin Dürrwald
Chair for Circuit Design and Network Theory
Technische Universität Dresden
Germany

Tilo Meister
Chair for Circuit Design and Network Theory
Technische Universität Dresden
Germany
tilo.meister@tu-dresden.de

Frank Ellinger
Chair for Circuit Design and Network Theory
Centre for Tactile Internet with Human-in-the-Loop (CeTI)
Technische Universität Dresden
Germany

Abstract—**This study presents an analysis of an energy efficient signal generator designed to enable short-range wireless data links for innovative base station architectures. The focus of this research is to achieve efficient frequency generation at 230 GHz using 130 nm SiGe BiCMOS technology. The circuit consists of a 57.5 GHz fundamental oscillator followed by subsequent stages that quadruple the output frequency. The implemented design achieves an operating range from 208.7 GHz to 253.2 GHz with a maximum output power of 1.81 dBm. The DC-to-RF efficiency of the signal generator is measured to be 2.16 %, which is competitive with other signal generators in SiGe BiCMOS technology operating in this frequency range.**

Index Terms—**Oscillators, frequency synthesis, voltage-controlled oscillator, quadrupler, millimeter wave, Sub-THz, BiCMOS integrated circuits, SiGe**

I. INTRODUCTION

Research into the fifth (5G) and sixth (6G) generations of mobile data communications is increasingly addressing concrete problems and scenarios [1], [2]. In such ecosystems, edge cloud applications will play a critical role, enabling distributed computing and low latency applications [3], [4]. To support the dynamic nature of edge cloud services and to reduce overall energy consumption, baseband signal processing tasks can be efficiently completed by specialized compute nodes (CN) [2]. This enables highly energy-efficient, load-aware execution of the radio access network (RAN). As a result, the ability to dynamically switch high-speed data links between the CNs is of critical importance. In this context, a promising approach for distributing data can be the use of short-range, high-bandwidth wireless data links. An example of a system with four CNs is shown in Fig. 1. Since only one data link is assigned between CNs, each CN consists of a homodyne transceiver architecture and a dedicated baseband processing unit. The aim of this work is to analyze a possible power-efficient signal generator with a center frequency of about 230 GHz needed as a local oscillator (LO) building

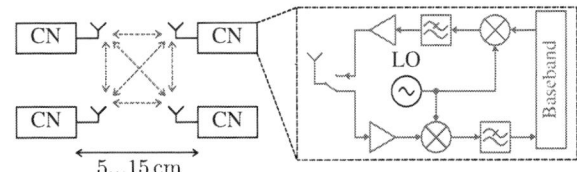

Fig. 1. Simplified overview of the system with four compute nodes (CNs). Each CN consists of a conventional homodyne transceiver architecture, where the presented local oscillator (LO) circuitry is highlighted.

block for such a data transmission link.

Designing oscillator circuits to generate fundamental frequencies above 200 GHz is a challenging task. It involves a complex trade-off between several key performance indicators (KPIs), including phase noise (PN), tuning range, DC-to-RF efficiency and center frequency. One way to address this challenge is to optimize each KPI with a dedicated building block. In the context of this paper, the most common approach is used. It consists of a fundamental voltage controlled oscillator (VCO) at a frequency of 57.5 GHz, which is then multiplied by a factor of four using a phase controlled push-push (PCPP) frequency quadrupler. A buffered injection-locked oscillator (ILO) improves the spectral purity of the output signal and provides sufficient output power for subsequent mixers, modulators, and power dividers. In addition, the tuning range of the signal generator should be as reconfigurable as possible to allow channel selection in larger base-stations with multiple compute nodes by assigning data channels to widely separated carrier frequencies. In a further step, the presented frequency generator is supposed to be integrated into a transmitter system including a phase locked loop (PLL).

II. SIGNAL GENERATOR DESIGN

To prove the concept, the monolithic integration is done in a 130 nm SiGe BiCMOS technology with heterojunction bipolar

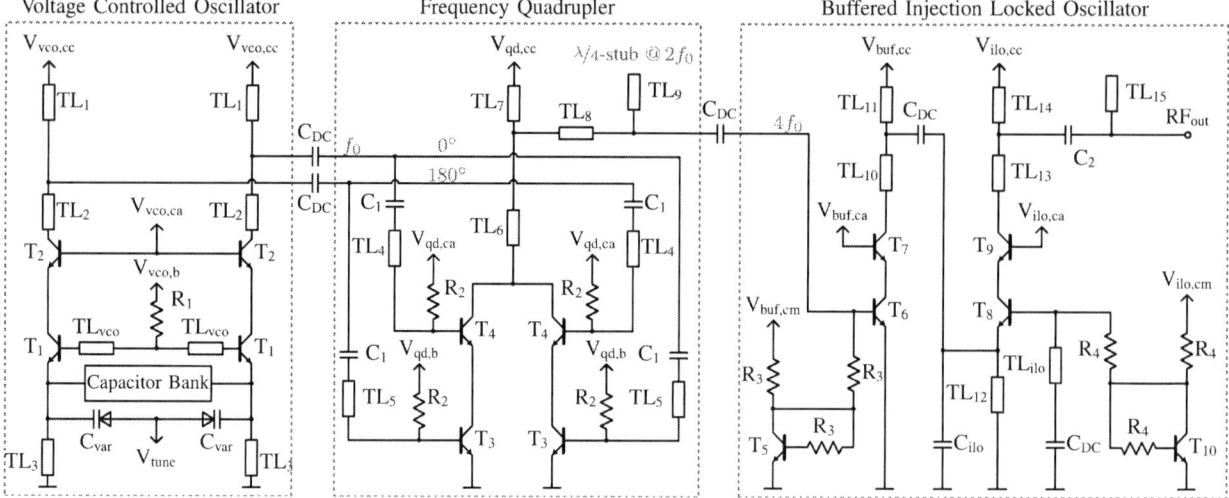

Fig. 2. Simplified circuit diagram of the proposed signal generator consisting of an oscillator, frequency quadrupler and injection locked oscillator.

transistors (HBTs) having a maximum oscillation frequency f_{max} of $450\,\text{GHz}$. The high-frequency transmission lines (TL) are based on the concept of grounded coplanar waveguides, using the top metalization as the signal trace and one of the lower metal layers as ground. Each block is DC powered using a zero-ohm line approach [5]. This minimizes cross-talk and provides a low-ohmic high-capacitance DC trace as close to the transistors as possible. A schematic of the proposed signal generator is shown in Fig. 2.

The circuit topology of the fundamental VCO is based on a fully differential Colpitts oscillator architecture [6], [7], [8]. The resonator consists of the inductive line TL_{VCO} and a binary-weighted capacitor bank, which is complemented by additional metal-oxide-semiconductor (MOS) varactors C_{var} to ensure a wide overall operating range. This provides an analog continuous frequency tuning range (FTR) in addition to the digital configuration. For the resonator TL_{VCO}, the grounded coplanar TL have been modified with a patterned ground plane. This minimizes losses and increases the quality factor by forcing the return current into the sidewalls of the TL. A common-base buffer-stage with HBT T_2 is used to increase the load impedance seen by the oscillator core. The transistor T_1 of the oscillator core is biased through TL_{VCO} by the high impedance R_1. An interstage matching consisting of TL_1, TL_2 ensures optimal power matching with the subsequent frequency quadrupler. This frequency quadrupler is used to up-convert the VCO output signal in frequency by a factor of four [9], [10]. To keep the power consumption and the total area of the chip as small as possible, this circuit consists of only two stacked HBTs T_3 and T_4 per path. The lower ones are class AB biased and the upper ones are class C biased. Building on the well established push-push principle for frequency multiplication-by-2, the single-stage frequency quadrupler employs the phase-controlled push-push architecture which allows a frequency multiplication-by-4 [10]. In this approach,

the collector current in one of the differential paths is clipped at $90°$ at f_0, causing doubling the frequency. The subsequent push-push concept then leads to a quadrupling of the input frequency. Input power matching of the transistors with different operating points is realized by TL_4, TL_5 and C_1. On the output side, the matching as well as the DC supply of this block is done by TL_6 and TL_7. A second harmonic suppression is realized by the $\lambda/4$ open stub line TL_9. To achieve higher output power and spectral purity, a buffered ILO is used [10]. The first buffer stage is realized in cascode topology consisting of T_6 and T_7 to amplify the quadrupled signal. This signal is then fed to the emitter node of the common-collector Colpitts oscillator modified as an ILO, which consists of the HBT T_8 and the resonant circuits TL_{ilo} and C_{ilo}. The free-running frequency of the ILO is $209\,\text{GHz}$. The quality factor of the resonator was intentionally chosen to be lower, as this results in a larger lock-in range of the ILO. Compared to a conventional oscillator, this would result in higher phase noise. With the ILO, however, the phase noise is only source-dependent [10]. Similar to the VCO circuitry with T_2, the HBT T_9 is used as an output buffer. The output matching of the overall signal generator was realized with the elements TL_{13}, TL_{14} and C_2 as well as TL_{15}.

III. MEASUREMENTS

The signal generator has been characterized in a lab environment. As shown in Fig. 3, the chip is glued and bonded to a carrier board. This additional layer of wiring allows precise filtering of all supply voltages using decoupling capacitors, which is essential for characterizing individual oscillators without using a phase-locked loop. The proposed circuit has been characterized using two different measurement setups attached to a spectrum analyzer. The first setup covers the frequency range of $140\,\text{GHz}$ to $220\,\text{GHz}$ and the second setup covers $220\,\text{GHz}$ to $330\,\text{GHz}$. Power measurements are made with a power meter. All supply voltages were kept constant throughout the measurements, resulting in a total

979-8-3503-4331-1/24 $31.00 © 2024 IEEE

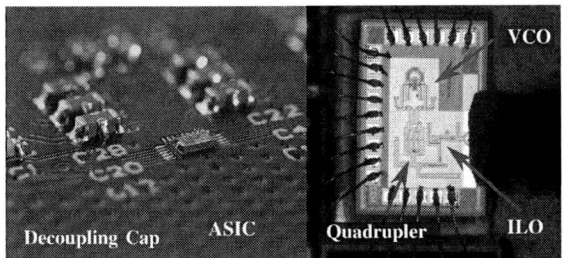

Fig. 3. Photographs of the chip glued and bonded to a carrier printed circuit board for RF measurements on a probe station while the DC power is supplied via the carrier board. The chip size is $810\,\mu\text{m} \times 1210\,\mu\text{m}$.

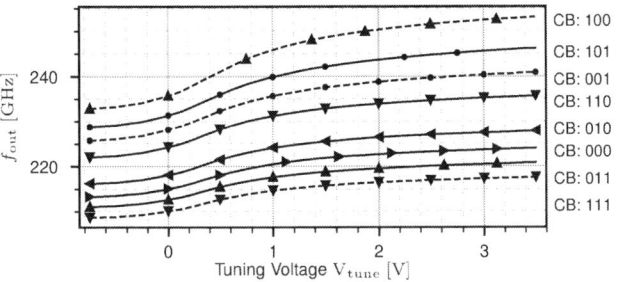

Fig. 4. Measured quadrupled output frequency f_{out} as a function of tuning voltage V_{tune} for different capacitor bank configurations (CB).

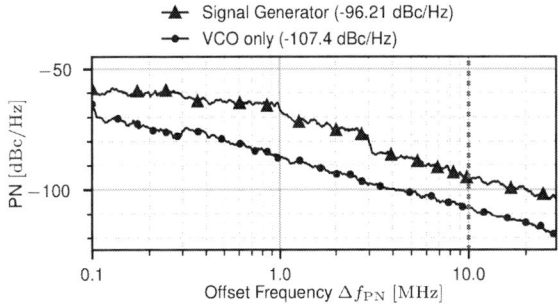

Fig. 5. Measured phase noise (PN) at an offset frequency of 10 MHz and as a function of the offset frequency for a tuning voltage V_{tune} of 0 V and for CB=000 and for the entire signal generator as well as for the VCO only.

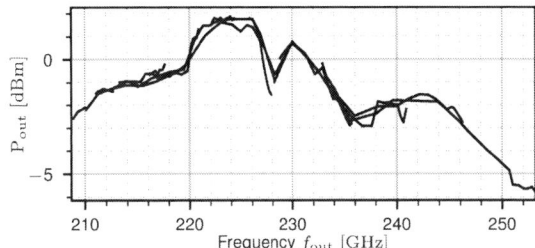

Fig. 6. Measured output power P_{out} as a function of the entire quadrupled output frequency f_{out} range for different CB and V_{tune}.

power consumption of $71.6\,\text{mW}$. The VCO consumed the most power with $52.8\,\text{mW}$.

Adjusting the tuning voltage V_{tune} from $-0.5\,\text{V}$ to $3.5\,\text{V}$ and the binary-weighted capacitor bank results in the operating ranges shown in Fig. 4. Since the tuning characteristics overlap, it is also possible to use two different configurations for a given frequency, resulting in a possible switchable, segmented total operating frequency range (TR) of $208.7\,\text{GHz}$ to $253.2\,\text{GHz}$. The corresponding continuous frequency tuning range varies between $4.15\,\%$ to $8.24\,\%$, depending on the selected capacitor bank configuration (CB), denoted by the 3 bits on the right. As shown in Fig. 5, the measured phase noise at an offset frequency of $10\,\text{MHz}$ is $-96.21\,\text{dBc/Hz}$ for a configuration of CB=000 and a tuning voltage V_{tune} of $0\,\text{V}$. Compared to a measurement of a single VCO under the same operating conditions [7], the phase noise increases by $12\,\text{dB}$, as expected from the theory. The measured output power of the signal generator as a function of the output frequency is shown in Fig. 6. The maximum output power is $1.81\,\text{dBm}$ at a frequency of $222.4\,\text{GHz}$. The losses caused by the measurement setup are calibrated. The resulting DC-to-RF efficiency is $2.16\,\%$.

Tab. I summarizes the performance of the proposed signal generator and compares it to the state of the art. The presented circuit concept provides competitive KPIs for signal generators in SiGe technology in the $200\,\text{GHz}$ range. This is shown by the figure of merit FOM_T, a large negative value means an improvement in performance. A comparison of the FOM_T as a function of the center frequency f_0 is shown in Fig. 7. Compared to [11], the presented work has a

higher DC-to-RF efficiency at higher output power, while the phase noise and center frequency are comparable. In [13], a similar FOM_T for the oscillator is achieved using a frequency doubler and an oscillator at $120\,\text{GHz}$, with a comparable operating range to the present work. However, further use of this signal generator in a phase-locked loop is difficult to achieve because the oscillator core operates at a higher frequency, making frequency division difficult. [14] achieved a comparable DC-to-RF efficiency using a single-core harmonic oscillator. However, in contrast to the results presented in this work, the overall operating range is more limited. A fundamental oscillator, as presented in [15], may offer better efficiency. However, it is difficult to use in a phase-locked loop because of the frequency division required. [16] presents a highly efficient oscillator with a DC-to-RF efficiency of $15.19\,\%$. However, the limited tuning range prevents the use of multiple parallel data links as required in the proposed base

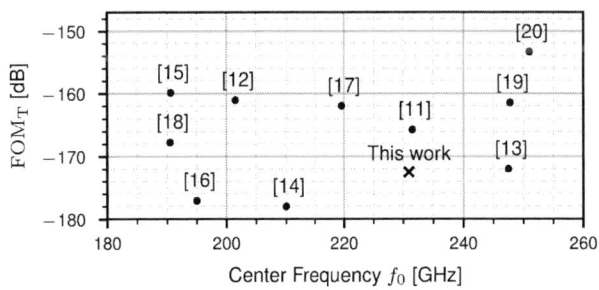

Fig. 7. Comparison of the figure of merit FOM_T of state-of-the-art signal generators between $180\,\text{GHz}$ and $260\,\text{GHz}$ in SiGe BiCMOS as a function of the center frequency f_0.

979-8-3503-4331-1/24 $31.00 © 2024 IEEE

TABLE I

COMPARISON TO THE STATE-OF-THE-ART SIGNAL GENERATORS BETWEEN 180 GHz AND 260 GHz IN BiCMOS TECHNOLOGY.

Ref.	f_0 [GHz]	PN [dBc/Hz]	P_{out} [dBm]	DC-to-RF [%]	FOM_T [dB]
[11]	231.5	$-98^\ddagger$	-3.6	0.81	-165.7
[12]	201.5	$-87^\dagger$	-7.2	0.63	-161.0
[13]	247.5	$-82^\dagger$	7.2	1.3	-171.9
[14]	210	$-87.5^\dagger$	1.4	2.26	-178.0
[15]	190.6	$-91^\ddagger$	-0.8	5.1	-159.8
[16]	195	$-98.6^\dagger$	6.5	15.19	-177.0
[17]	219.6	$-68.9^\dagger$	-3.7	0.52	-161.9
[18]	190.5	$-102.6^\ddagger$	-2.1	0.21	-167.7
[19]	247.8	$-81^\dagger$	-2.6	1.3	-161.4
[20]	251.0	$-86^\dagger$	-18	0.01	-153.3
This	230.92	$-96.2^\ddagger$	1.81	2.16	-172.5

Phase noise at 1 MHz ($\dagger$) and 10 MHz ($\ddagger$) offset Δf_{PN} to the carrier f_0.
$FOM_T = PN - 20\log\left(f_0/\Delta f_{PN} \cdot TR/10\right) + 10\log\left(P_{DC}/1\,mW\right) - P_{out}$,
where TR is the total operating frequency range

station. In [17] and [18], a wide tuning range is reported, but in contrast to the presented work, the DC-to-RF efficiency is rather low, making it unsuitable for the proposed system.

IV. CONCLUSION

This study analyzes an efficient signal generator for short-range data links to be applied in new base-station architectures. The goal was to provide efficient and reconfigurable frequency generation at 230 GHz using 130 nm SiGe BiCMOS technology to enable high-speed data distribution between specialized compute nodes within a base-station. The circuit consists of a 57.5 GHz fundamental oscillator, of which the output frequency is quadrupled by a subsequent stage. This approach allows further use of the presented signal generator in a phase-locked loop, since frequency division can be easily implemented. A maximum output power of 1.81 dBm is achieved over a segmented operating range of 208.7 GHz to 253.2 GHz. The DC-to-RF efficiency of 2.16 % is competitive with other oscillators in this frequency range. The presented feasibility study indicates that the proposed concept is well suited for efficient and high performance data links for edge-cloud applications.

ACKNOWLEDGMENT

This research project was funded in part by the German Research Foundation (DFG) project BAUDOT with project ID 458074433, project MORE with project ID 434434888 and by the German Federal Ministry of Education and Research (BMBF) project E4C under grant 16ME0426K.

REFERENCES

[1] R. Wittig, A. Goens, C. Menard, E. Matus, G. P. Fettweis, and J. Castrillon, "Modem Design in the Era of 5G and Beyond: The Need for a Formal Approach," in *2020 27th International Conference on Telecommunications (ICT)*. IEEE, Oct. 2020.

[2] M. A. Habibi, M. Nasimi, B. Han, and H. D. Schotten, "A Comprehensive Survey of RAN Architectures Toward 5G Mobile Communication System," *IEEE Access*, vol. 7, May 2019.

[3] K. Cao, Y. Liu, G. Meng, and Q. Sun, "An overview on edge computing research," *IEEE Access*, vol. 8, pp. 85 714–85 728, May 2020.

[4] J. Robledo and J. Castrillon, "Parameterizable mobile workloads for adaptable base station optimizations," in *2022 IEEE 15th International Symposium on Embedded Multicore/Many-core Systems-on-Chip (MCSoC)*. IEEE, Dec. 2022.

[5] G. Tretter, D. Fritsche, J. Leufker, C. Carta, and F. Ellinger, "Zero-Ohm transmission lines for millimetre-wave circuits in 28 nm digital CMOS," *Electronics Letters*, vol. 51, no. 11, pp. 845–847, May 2015.

[6] C. Hoyer, J. Wagner, and F. Ellinger, "A 60 GHz VCO with 654 MHz direct Frequency Modulation Bandwidth in 0.13-μm SiGe BiCMOS," in *2021 Int. Conference on Electrical, Computer, Communications and Mechatronics Engineering (ICECCME)*. IEEE, Oct. 2021.

[7] C. Hoyer, F. Protze, J. Wagner, and F. Ellinger, "A Reconfigurable 60-GHz VCO with -103.2 dBc/Hz Phase Noise in a 0.13-μm SiGe BiCMOS Technology," in *2022 Asia-Pacific Microwave Conference (APMC)*. IEEE, Nov. 2022.

[8] L. Pantoli, H. Bello, G. Leuzzi, H. J. Ng, and D. Kissinger, "Design considerations on the realization of signal sources at mm-waves," in *2020 15th European Microwave Integrated Circuits Conference (EuMIC)*. EuMA, Jan. 2021, pp. 129–132.

[9] L. Steinweg, V. Rieß, P. Stärke, P. V. Testa, C. Carta, and F. Ellinger, "A Low-Power 255-GHz Single-Stage Frequency Quadrupler in 130-nm SiGe BiCMOS," *IEEE Microwave and Wireless Components Letters*, vol. 30, no. 11, pp. 1101–1104, Sep. 2020.

[10] L. Steinweg, P. V. Testa, C. Carta, and F. Ellinger, "A 213 GHz 2 dBm Output-Power Frequency Quadrupler with 45 dB Harmonic Suppression in 130 nm SiGe BiCMOS," in *ESSCIRC 2021 - IEEE 47th European Solid State Circuits Conference (ESSCIRC)*, Sep. 2021, pp. 447–450.

[11] S. P. Voinigescu, A. Tomkins, E. Dacquay, P. Chevalier, J. Hasch, A. Chantre, and B. Sautreuil, "A Study of SiGe HBT Signal Sources in the 220–330-GHz Range," *IEEE J. Solid-State Circuits*, vol. 48, no. 9, pp. 2011–2021, Sep. 2013.

[12] P.-Y. Chiang, O. Momeni, and P. Heydari, "A 200-GHz Inductively Tuned VCO With -7-dBm Output Power in 130-nm SiGe BiCMOS," *IEEE Trans. Microw. Theory Techn.*, vol. 61, no. 10, Oct. 2013.

[13] S. Shopov, A. Balteanu, J. Hasch, P. Chevalier, A. Cathelin, and S. P. Voinigescu, "A 234–261-GHz 55-nm SiGe BiCMOS Signal Source with 5.4–7.2 dBm Output Power, 1.3% DC-to-RF Efficiency, and 1-GHz Divided-Down Output," *IEEE J. Solid-State Circuits*, vol. 51, no. 9, pp. 2054–2065, Sep. 2016.

[14] C. Jiang, A. Cathelin, and E. Afshari, "An efficient 210 GHz compact harmonic oscillator with 1.4dBm peak output power and 10.6% tuning range in 130nm BiCMOS," in *2016 IEEE Radio Frequency Integrated Circuits Symposium (RFIC)*. IEEE, May 2016, pp. 194–197.

[15] D. Fritsche, S. Li, N. Joram, C. Carta, and F. Ellinger, "Design and characterization of a 190-GHz voltage-controlled oscillator," in *2016 46th European Microwave Conference (EuMC)*. London, UK: IEEE, Oct. 2016, pp. 493–496.

[16] H. Khatibi, S. Khiyabani, A. Cathelin, and E. Afshari, "A 195 GHz single-transistor fundamental VCO with 15.3% DC-to-RF efficiency, 4.5 mW output power, phase noise FoM of -197 dBc/hz and 1.1% tuning range in a 55 nm SiGe process," in *2017 IEEE Radio Frequency Integrated Circuits Symposium (RFIC)*. Honolulu, HI, USA: IEEE, Jun. 2017, pp. 152–155.

[17] A. Mostajeran and E. Afshari, "An ultra-wideband harmonic radiator with a tuning range of 62GHz (28.3%) at 220GHz," in *2017 IEEE Radio Frequency Integrated Circuits Symposium (RFIC)*. Honolulu, HI, USA: IEEE, Jun. 2017, pp. 164–167.

[18] R. Kananizadeh and O. Momeni, "A 190-GHz VCO With 20.7% Tuning Range Employing an Active Mode Switching Block in a 130 nm SiGe BiCMOS," *IEEE J. Solid-State Circuits*, vol. 52, no. 8, Aug. 2017.

[19] S. Khiyabani, H. Khatibi, and E. Afshari, "A Compact 275 GHz Harmonic VCO with -2.6 dBm Output Power in a 130 nm SiGe Process," in *2019 IEEE Custom Integrated Circuits Conference (CICC)*. IEEE, Apr. 2019, pp. 1–4.

[20] Y. Mao, E. Shiju, K. Schmalz, J. Borngraber, and J. C. Scheytt, "A novel 245 GHz 4th index push-push VCO," in *2019 IEEE International Symposium on Radio-Frequency Integration Technology (RFIT)*. IEEE, Aug. 2019, pp. 1–3.

The Chip-Level in-Plane Stress Distribution over BiCMOS Wafers

Zhibo Cao
IHP – Leibniz-Institut für innovative Mikroelektronik
Frankfurt (Oder), Germany
cao@ihp-microelectronics.com

Thomas Voss
IHP – Leibniz-Institut für innovative Mikroelektronik
Frankfurt (Oder), Germany
voss@ihp-microelectronics.com

Matthias Wietstruck
IHP – Leibniz-Institut für innovative Mikroelektronik
Frankfurt (Oder), Germany
wietstruck@ihp-microelectronics.com

Corrado Carta
Technische Universität Berlin
IHP – Leibniz-Institut für innovative Mikroelektronik
Frankfurt (Oder), Germany
carta@ihp-microelectronics.com

Mehmet Kaynak
Kilby Lab, Texas Instrument
Dallas, Texas, USA
kaynakmehmet@icloud.com

Abstract—In this study, the in-plane stress distribution over BiCMOS wafers is investigated. Both analytical and finite element models are exploited to characterize the global and local stresses. By analyzing the single dies' thicknesses and curvatures, the corresponding stress values at specific positions can be extracted. It is found that although the layout design of the die results in a few hundred N/m stresses and is the major contributor to the die stress, the process on-wafer variation, which can bring a variation of ± 50-60 N/m to the total stress, is also not negligible.

Keywords—BiCMOS process, in-plane stress, on-wafer distribution, ultra-thin dies

I. INTRODUCTION

With the advancement of "More than Moore", the chips get thinner to prepare for the 3D heterogeneous integration. However, the thin substrate poses great challenges to the integration yield and stability due to its excessive bow. There are already intensive studies about wafer deformations, both regarding the front- and back-end process steps [1] and regarding the post-processing [2-4]. These studies are dedicated to investigating the impacts on the wafer-level processes (e.g. lithography, wafer chucking, bonding, etc.). However, the shapes of dies over the wafer, which are important for packaging and 3D integration, are rarely characterized. The warpages of dies are especially critical for both chip-level flip-chip assembly and embedded packaging approaches (e.g. fan-in and fan-out), which demand very high coplanarity to achieve good yields. Large deflections on single chips may lead to non-uniform contacts over the chip region during flip-chip bonding or chip-wafer 3D integration. On one hand, the bump size for chip-level integration is within tens of micrometers. To ensure an acceptable packaging yield, the coplanarity requirement is usually a few micrometers. On the other hand, the chip-bow for thin chips dedicated to 3D stacking is much larger compared with 700 μm chips. Considering these two factors, a better understanding of the chip in-plane stress and bow can be helpful to achieve better yield in the chip-level hetero-integration.

In this study, a fully processed BiCMOS wafer is thinned down to 35 μm and the stresses associated with chips located at different positions on the wafer are extracted. Through such an approach, an overview of chip-level in-plane stresses over the

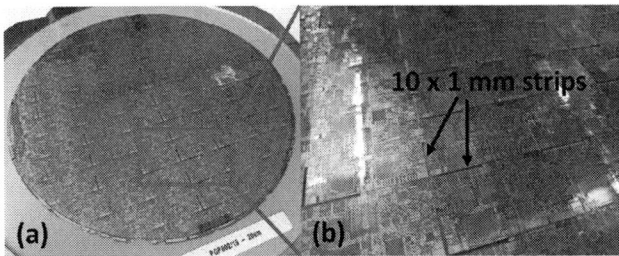

Fig. 1 10 x 1 mm thin strips are taken from a fully processed BiCMOS wafer for profilometry. (a) is an overview of the wafer, showing that the strips are taken from the same position within each reticle size. (b) shows the zoomed-in view of the picked strip, one horizontal and one verticle are taken from the same test field for comparison.

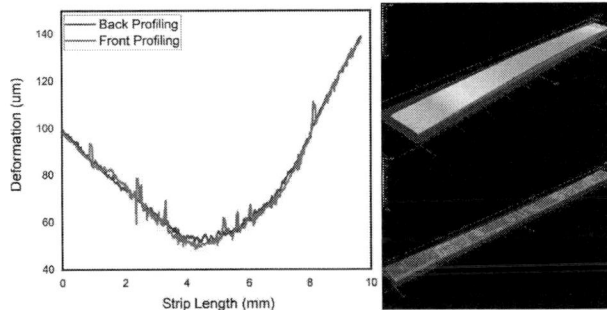

Fig. 2 Front and back laser profiling of the strip. The strip is placed face-up and face-down for curvature comparison.

BiCMOS wafer is obtained, facilitating a deeper understanding of the chip-level deformation and paving ways for better integration yields and stability.

II. METHODOLOGY

To characterize the in-plane residual stress distributions over the wafer, a detailed view of the wafer processing steps is needed. A multi-project wafer (MPW) first goes through the standard BiCMOS process provided by IHP (SG13G2 technology), and then through a post-grinding process. Since thinner wafers can lead to larger deformations, making the characterization easier. The wafers are thinned and polished down to 35 μm (15 μm back end and 20 μm silicon substrate) using a #6000 poly grind wheel. Worth mentioning that the

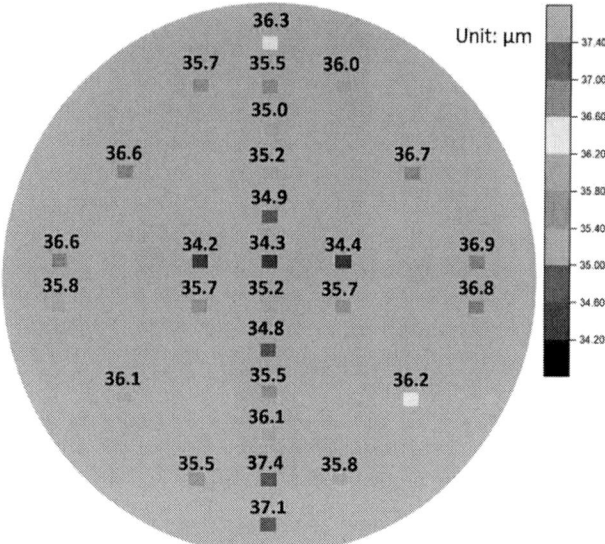

Fig. 3 The thickness wafer map. The target thickness value is 35 μm. 28 pieces of 0.5 mm x 0.5 mm squares are picked from different locations of the wafer for thickness measurement. And the colors indicate the corresponding thickness values.

grinding and dicing for a fully processed wafer down beyond 50 μm is not always straightforward due to their excessive bow and fragility. Therefore, a consecutive from coarse to fine dicing-before-grinding (DBG) process is exploited. With such a process, the back-end manufacturing and grinding contribute the most to the wafers' final shapes, as is explained in detail in [1].

According to the above processing information, the in-plane stress is assumed to be evenly distributed all over the wafer. However, there are some stress variations on the chip level. To extract and characterize the local stresses, both horizontally and vertically positioned 1 x 10 mm thin strips are picked within each reticle size (it is referred to as test field here) for profilometry, as is illustrated in Fig. 1.

The stresses associated with each picked strip are extracted using the stress identification method. Stress identification, similar to the load identification method, is a process to extract the stress information based on the measured component shapes.

The measurement is carried out using Keyence laser confocal microscope VK-X1000. The large magnification (20x objective) is used together with the stitching function to minimize the measurement error and handle the large aspect ratio of the samples. Worth mentioning that the profilometry is done from both the front side and the back side of the samples to characterize the influences of gravity. As illustrated in Fig. 2, the front and back profiles of the silicon strip overlap with each other, indicating that the influences of gravity can be neglected for the 10 x 1 x 0.035 mm samples.

Regarding the stresses accumulated in the silicon strips, they are lumped into two categories – the global in-plane stress that is uniform over the strip and the local stress that results in a spatial wavelength smaller than the strip length. The former can be roughly described using the famous Stoney equation.

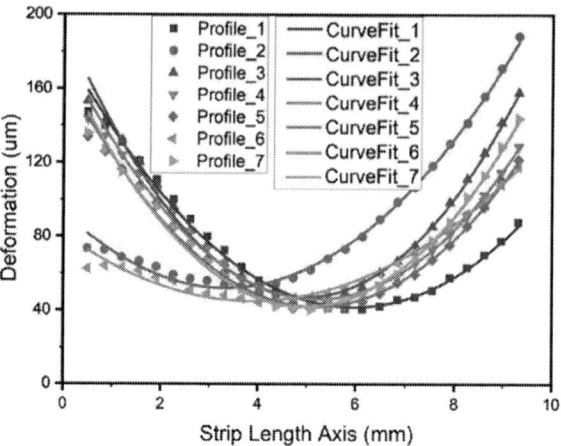

Fig. 4 The measured profiles of the thin strips picked from different positions over the BiCMOS wafer, together with a circular-arc fitting to extract the global in-plane stresses.

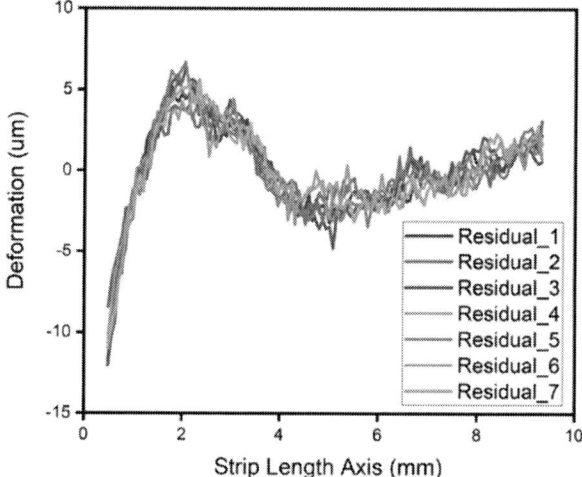

Fig. 5 The profile residuals of the picked thin strips over the BiCMOS wafer after the circular-arc curve fitting is applied.

$$\sigma_f t = \frac{E_s h^2}{6(1-v_s)}\left(\frac{1}{\rho}\right) \qquad (1)$$

Where σ_f is the film in-plane stress, ρ is the strip curvature radius, t and h are film and substrate thicknesses, E_s and v_s are Young's modulus and Poisson ratio of the substrate material respectively. In this paper, since the thickness of stressed layer t is unknown, $\sigma_f t$, the product of stress and stressed layer thickness (PST) with a unit of N/m is used as the figure of merit to characterize the in-plane stresses within each dielet. For the PST extraction, the left unknowns are the substrate thicknesses, as well as the lengthwise curvature of the strips. As can be seen in Fig. 3, the thickness values over the wafer can be directly measured by laser profiling to be 35.8 ± 0.8 μm. The curvatures are measured using the aforementioned laser microscope, with the corresponding results shown in Fig. 4. It is noticed that the shapes of the strips are not purely circular-arc, which is described by equation (1). This can be attributed to the local out-of-plane stresses. In this case, a circular-arc fitting can

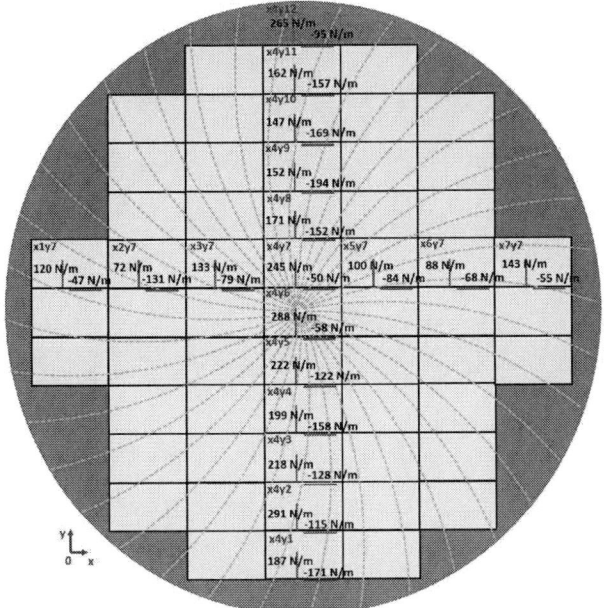

Fig. 6 Stress wafer map. The strip positions and grind marks are also illustrated in this figure. Tensile stress has a positive value and compressive stress has a negative value.

effectively get rid of the noises and extract the needed curvature information. Interestingly, the residuals after the fitting shown in Fig 5 are identical regarding different strips picked from different positions of the wafer. This means the local out-of-plane stresses, which are reflected by the residuals, are only dictated by the layout designs on the chips.

Besides, a finite element model is also developed in ANSYS 18.2 to assist the equation (1) for stress extractions. Due to the fact that silicon dioxide and aluminum take almost half of the strips due to the ultra-thin silicon substrate, the Stoney equation is not able to precisely describe the chip-level deformation given such a mixture of materials. Therefore, a more sophisticated model is necessary to ensure the accuracy of the extracted stresses. Specifically, shell elements SHELL181 with a layered section are used to model the strips. In this way, the complexity of the model is minimized for modelling object bending behaviors.

III. Stress Distribution and Analysis

Using the above finite element model, the PST values of the strips over the wafer are summarized in Fig. 6. Since all the dies from the same position of the test field are with the same layout, the PST difference can only come from the process variations and anisotropic characteristic of the grinding marks [5]. The PST on-wafer variation is observed to be around $150 - 200$ N/m. Specifically, for the horizontally oriented die, the extracted stress values vary from -47 N/m to -194 N/m (average: -109.6 ± 47.0 N/m), and for the vertically oriented die, the extracted stress values vary from 72 N/m to 291 N/m (average: 181.5 ± 64.6 N/m). Two different designs make the resultant global stresses on the horizontal and vertical strips -109.6 N/m and 181.5 N/m respectively, leading to a difference of almost 300 N/m. Considering the results obtained in [1], the variation of metal density in the back-end-of-line can lead to a PST variation of $-340 - 910$ N/m, implying that the stress variation associated with layout design is still much larger than the on-wafer variation. However, the stress variations of the chips with the same design and manufactured in the same lot are not negligible either – a PST variation of $\pm$ 50-60 N/m is observed. With these numbers, a more concrete understanding of chip shapes is obtained, paving the way for further development of 3D chip-level integration technologies.

IV. Conclusion

Through such an analysis, an overview of the stress distribution regarding a BiCMOS wafer is made. The local stresses that induce deformations with spatial wavelengths below 10 mm are mostly associated with the layout design, whereas the global stress induced deformations always follow a circular-arc shape and are subjected to process on-wafer variations. It is found that despite the fact that the layout design contributes the most to the shape of a single dielet, the process on-wafer variation can also bring about $\pm$ 50-60 N/m stress deviations over an 8-inch wafer.

Acknowledgment

The authors would like to show their gratitude for the FLEXCOM and Bend-IT projects, which are funded by the European Union's Horizon 2020 Programme and Deutsche Forschungsgemeinschaft (DFG) under grant agreement 101004233 and Project number 356463945 respectively.

References

[1] Z. Cao and A. Göritz and M. Stocchi and M. Wietstruck and C. Hoyer and L. D. Steinweg and C. Carta and F. Ellinger and B. Tillack and M. Kaynak, "An Advanced Finite Element Model for BiCMOS Process Oriented Ultra-Thin Wafer Deformation," in IEEE Transactions on Semiconductor Manufacturing, 2022, vol. 35(1), pp. 2-10.

[2] L. Di Cioccio and I. Radu and P. Gueguen and M. Sadaka, "Direct bonding for wafer level 3D integration," in Proceedings IEEE International Conference on Integrated Circuit Design and Technology, 2010, pp. 110-113.

[3] K. T. Turner and S. M. Spearing and W. A. Baylies and M. Robinson and R. Smythe, "Effect of nanotopography in direct wafer bonding: modeling and measurements," in IEEE transactions on semiconductor manufacturing, 2005, vol. 18(2), pp. 289-296.

[4] K. T. Turner and S. Veeraraghavan and J. K. Sinha, "Predicting distortions and overlay errors due to wafer deformation during chucking on lithography scanners," in Journal of Micro/Nanolithography, MEMS and MOEMS, 2009, vol. 8(4), pp. 043015.

[5] P. Zhou and Y. Ying and H. Ning and Z. Wang and R. Kang and D. Guo, "Residual stress distribution in silicon wafers machined by rotational grinding," in Journal of Manufacturing Science and Engineering, 2017, vol. 139(8), pp. 08

Characterization of Silicon Substrates for sub-THz Electronics, Benefit of The Beatty Resonator Test-Structure

Luca Lucci
CEA-Leti
Univ. Grenobles-Alpes
38054 Grenoble, France
luca.lucci@cea.fr

Olivier Valorge
CEA-Leti
Univ. Grenobles-Alpes
38054 Grenoble, France
olivier.valorge@cea.fr

Alexandre Oliviera
CEA-Leti
Univ. Grenobles-Alpes
38054 Grenoble, France
alexandre.oliviera@cea.fr

Herve Boutry
CEA-Leti
Univ. Grenobles-Alpes
38054 Grenoble, France
herve.boutry@cea.fr

Christophe Dubarry
CEA-Leti
Univ. Grenobles-Alpes
38054 Grenoble, France
christophe.dubarry@cea.fr

Fred Gaillard
CEA-Leti
Univ. Grenobles-Alpes
38054 Grenoble, France
fred.gaillard@cea.fr

Blandine Duriez
CEA-Leti
Univ. Grenobles-Alpes
38054 Grenoble, France
blandine.duriez@cea.fr

Abstract—The Beatty resonator is a classical test structure with significant utility in characterizing PCB substrate materials. This structure is widely acknowledged within the signal-integrity community and is even included in network-analyzer cal-kits.

As integrated RF circuits now target frequencies exceeding 100 GHz, the Beatty resonator's overall size has reduced to just a few millimeters, making it compatible with modern integrated circuit (IC) processes.

This contribution explores the Beatty resonator's application on a typical metallization stack for IC and highlights its advantages, particularly in complementing the substrate effective resistivity extraction.

Index Terms—Beatty, CMOS, substrate resistivity, mmW.

I. INTRODUCTION

The pursuit of RF and CMOS integration has led to a multitude of manufacturing techniques proposed as cost-effective solutions for combining RF-mmW and cutting-edge digital circuitry. Simultaneously, there is a growing focus on designs and developments, such as 5G/6G, SATCOM, and antenna metasurfaces, that require even higher operational frequencies. To achieve this, advanced techniques like hybrid-bonding, die-to-wafer hybridization, and local epitaxial re-growth of exogenous materials are actively being explored, especially for mmW technologies.

Consequently, there is a pressing need to find practical ways to assess and fairly compare the frequency-dependent loss properties of the various proposed substrates and back-end-of-line (BEOL) combinations.

Among all alternatives, the high-resisitive (HR) and trap-rich (TR) Silicon subtrates [1] are among the most promising options since are fully compatible with CMOS manufacturing and guarantee state-of-the-art RF performances. The RF characterization of a CMOS substrate is generally based on the common assessment of the losses due to the

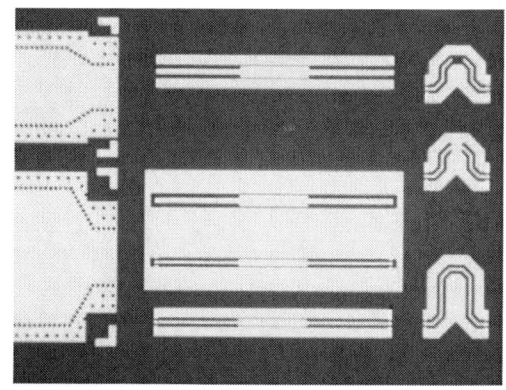

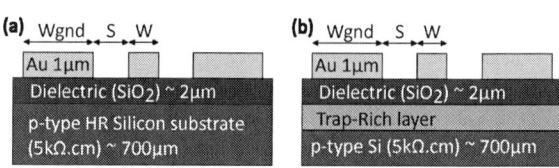

Fig. 1. (top) Die micro-photograph of the implemented Beatty resonator structures embedded in 100 and 50 μm GSG pads. The resonator is 700 μm long and the complete test-structure, with pads and accesses, is less than 2.4×0.3 mm². (Bottom) cross section of the two silicon substrate investigated, not to scale, (a) high-resistivity (HR), (b) trap-rich (TR).

BEOL and the substrate by the extraction of an *effective resistance* parameter (ρ_{eff}) that summarizes all the loss contributions of all the dielectric and semiconductor layers [2].

The effective resistance extraction is delicate, especially above 50/60 GHz where it presents some pitfalls [3]. It is valuable and recommended to use additional structures to double-check and validate the extracted results against expected physical effects.

979-8-3503-4331-1/24 $31.00 © 2024 IEEE
48
SiRF 2024

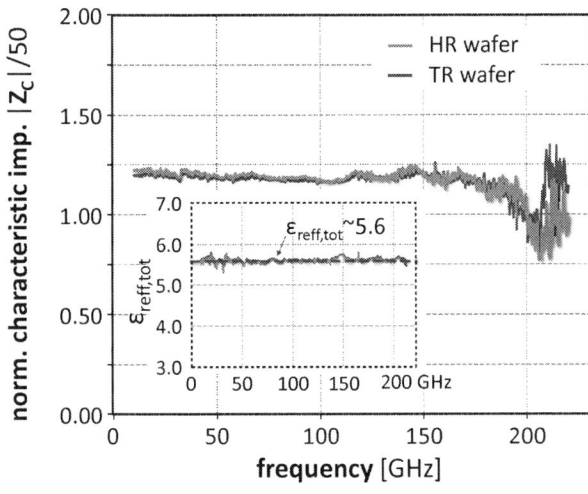
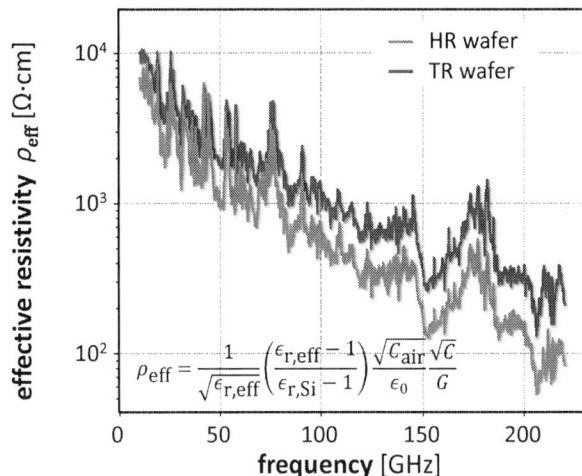

Fig. 2. Extracted characteristic impedance and, in the inset, effective dielectric constant. Both extracted parameters on HR and TR results in indistinguishable values, underlying the similarity of the two starting wafers. Extraction procedure follows [2].

Fig. 3. The extracted effective resistivity has a consistently slightly higher value for the TR wafer. For completeness the used formula for the extraction is also reported [2] where ϵ_0 and ϵ_{Si} are the vacuum permittivity and Silicon dielectric constant respectively, $\epsilon_{r,eff}$ is reported in Fig. 2, C_{air} is obtained by simulations, C and G are obtained extracting the *RLCG* parameters of a reference transmission line.

The validation test-structures are even more important if are not the ones used for the extraction itself to avoid falling into the trap of circular "self-consistency," ensuring more robust and reliable verification of the results.

One structure that fits all these requirements is the Beatty resonator [4]. The Beatty resonator utilizes a simple change in trace width to create a measurable and predictable change in impedance. The change in the impedance across the structure introduces reflections and standing waves that can be used to identify the correct values for *e.g.* the effective dielectric permittivity of the transmission lines used in the series resonator structure [5], [6]. The drop-off in the insertion loss characteristic can be used to separate the losses due to the tangent loss of the substrate and the losses due to the resistivity of the conductors [7].

In this study, we explore the application of a Beatty resonator to enhance our understanding and complement the extracted effective properties of the substrate. We conduct an effective resistance extraction on two test cases, namely a TR and a HR substrate, which are prepared and processed in parallel, thereby sharing the same back-end-of-line (BEOL).

II. WAFER AND TEST-STRUCTURE PREPARATION

Starting wafer are p-type 700 μm thick 200 mm wafers with nominal resistivity ρ_{DC}=5kΩ·cm. Both a high-resistivity (HR) version and a trap-rich (TR) version with a un-doped polysilicon interface layer were processed. The absence of a front-end processing (no MOS or junctions) reduces the overall thermal budget and helps in achieving very good RF performances in both cases. The back-end-of-line consists of a 2 μm thick thermal oxide (SiO_2) followed by a 1 μm thick gold metal layer for metallization and finally a 0.5 μm SiN capping (see cross-section in Fig. 1 where SiN capping is omitted).

Preliminar HFSS simulations were carried out (not shown) to define the target dimensions the Beatty structure and of the necessary transmission lines for the *effective resistivity* extraction. As shown in Fig. 1, two different CPWG transmission line flavours are implemented: (A) a version with ground width of 100 μm on left and right sides and (B) a version with a common ground plane shared among neighboring structures and in any case wider than 230 μm. For each of the (A) and (B) version, Beattys and transmission lines are replicated on both 100 μm and 50 μm GSG pads, to allow maximum flexibility in the choice of measurement probes. Only results from set (A) are reported, set (B) resulting in slight degradation of performances.

The Beatty line has a target length of 700 μm for its $\lambda/4$-section, giving an expected resonance every 75/85 GHz depending on the actual effective substrate dielectric constant.

TABLE I
GEOMETRICAL SPECIFICATIONS [μm]

Beatty line	signal width [W]	spacing [S]	Ground [Wgnd]
CPWG-50 μm type (A)	34	25	100
CPWG-50 μm type (B)	34	25	>230

Transmission lines: 450, 800, 1000, 1200, 1400, 2200, 2500 μm

III. MEASUREMENTS

The Beatty resonator an the transmission lines were measured on-wafer using a broadband 220 GHz setup on a manual probe station in a temperature controlled environment. The S-parameters were collected on an Anritsu ME7838G4 VNA with MA25400A extenders to 220 GHz. Probes were

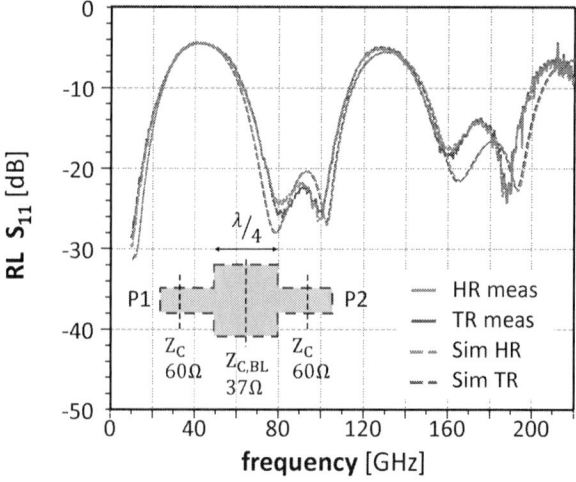
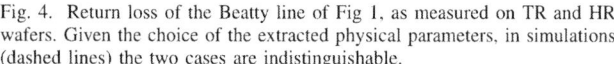
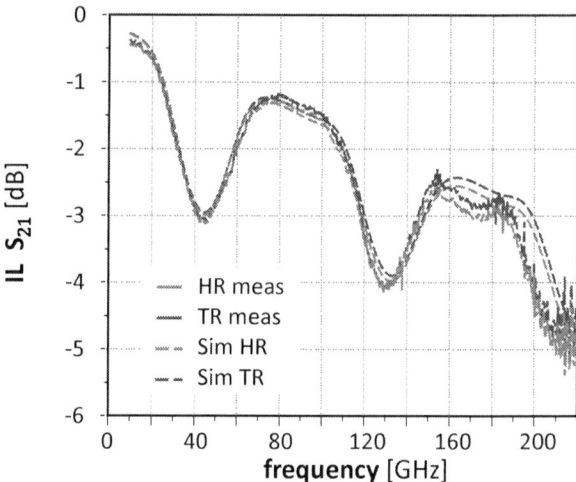

Fig. 4. Return loss of the Beatty line of Fig 1, as measured on TR and HR wafers. Given the choice of the extracted physical parameters, in simulations (dashed lines) the two cases are indistinguishable.

Fig. 5. Insertion loss of the Beatty lines. Measurements (continuous lines) vs. ADS simulations (dashed lines).

MPI T220, and a standard ISS used for calibration to probe tips.

The measurement setup was calibrated on a impedance substrate ISS using an enhanced LRRM calibration [8]. Pad and accesses were subtracted from the measurement with the ABCD-matrix approach detailed in [9].

IV. RESULTS AND DISCUSSION

The transmission lines on the two wafers are characterized and the effective resistivity of the two substrates is extracted following [2], [3]. Given the similarity of the starting wafers and the common BEOL, the characteristic impedance Z_C and the effective dielectric constant $\epsilon_{r,eff}$, reported in Fig. 2, are very close in value. The characteristic impedance of the lines turned out to be higher that the targeted, around 60 Ω. This is due to over-etching of the gold layer that unfortunately yielded thinner signal and wider spacing than the expected values in Tab. I. Fortunately the over-etching is very similar between the two wafers. Overall the extraction is satisfactory at least up to 150 GHz and it degrades above 200 GHz, probably the limits for usage of the LRRM calibration.

As seen in Fig. 3 there is a significant difference in the extracted value for effective resistivity ρ_{eff}, the TR wafer having higher, better, values. Regarding the excellent values for the HR substrate, the low-temperature processing may have helped and also the fact that all measurements are carried out without any bias applied.

To perform effective resistance ρ_{eff} extraction, we obtain an RLCG model of the transmission line. This data can then be used in any spice simulator to simulate the actual lines on which we performed the extraction. By doing so, we can check the model's correlation with the hardware.

This validation step becomes more useful if a Beatty standard is available. It provides immediate feedback on the quality and accuracy of the extraction by comparing measurements and simulations on a simple structure that was not part of the initial extraction process.

In Fig. 4 the resonances in the return losses are dependent on the $\epsilon_{r,eff}$ and the ratio of the characteristic impedances Z_C, $Z_{C,BL}$ between the different sections of the Beatty line, all values are indeed extracted or known during the effective resistivity procedure, e.g. reported in Fig. 2.

Depending on the transmission line model used, the effective dielectric constant $\epsilon_{r,eff}$ may have to be converted back to the effective substrate dielectric constant $\epsilon_{rsub,eff}$, with the equation $\epsilon_{r,eff} = 1 + q(\epsilon_{rsub,eff} - 1)$, were q is the form-factor extracted from metal geometry [2].

Simulating the Beatty insertion loss and comparing to the measures in Fig. 5 it is immediate to realize that, as expected, the conductor losses estimated by R, have a negligible contribution at all frequencies. The only possible tuning of the model is the repartion of the losses between the substrate dielectric losses, modeled with the $\tan \delta$, whose starting value can be chosen as $\tan \delta = 1/(\omega \epsilon_r \epsilon_0 \rho_{eff})$ and the conductive substrate losses. Substrate conductive and dielectric losses have different frequency dependence, so a correct redistribution can be found.

V. CONCLUSION

A simple resonator structure that is widely used for the characterization of PCB material, the Beatty line, has been implemented in a typical BEOL for integrated circuits. As the frequency targeted by CMOS electronics is extending well above 100 GHz, the length of the structure, roughly three times the $\lambda/4$ at the target frequency, becomes manageable. This useful structure was so far ignored in the substrate characterization flow for CMOS since prohibitively long for integrated technologies. We show that it is a valuable addition

979-8-3503-4331-1/24 $31.00 © 2024 IEEE

to the device modeling test-set, in particular for modeling and characterization of advanced RF silicon substrates.

ACKNOWLEDGMENT

The experiments in this contribution were supported by the French authorities within the framework of the ANR-Carnot institute funding. Authors also acknowledge the assistance of Loic Vincent (CIME/INPG) for the measurements.

REFERENCES

[1] K. Ben Ali, C. Roda Neve, A. Gharsallah, and J.-P. Raskin, "RF SOI CMOS technology on commercial trap-rich high resistivity SOI wafer," in *2012 IEEE Int. SOI Conference*, 2012.

[2] D. Lederer and J.-P. Raskin, "Effective resistivity of fully-processed SOI substrates," *Solid-State Electronics*, vol. 49, no. 3, pp. 491–496, 2005.

[3] L. Nyssens, "A systematic analysis of Silicon substrates toward improvements of integrated FD-SOI circuit performance at mm-wave frequencies," Ph.D. dissertation, Université Catholique de Louvain, 2023.

[4] R. Beatty, "2-port $\lambda_g/4$ waveguide standard of voltage standing-wave ratio," *Electronics Letters*, vol. 2, no. 9, pp. 24–26, 1973.

[5] H. Barnes, R. Schaefer, and J. Moreira, "Analysis of test coupon structures for the extraction of high frequency PCB material properties," in *17th IEEE Workshop on Signal and Power Integrity*, 2013.

[6] Y. Shlepnev, A. Neves, T. Dagostino, and S. McMorrow, "Measurement-assisted electromagnetic extraction of interconnect parameters on low-cost FR-4 boards for 6-20 Gb/sec applications," *Proc. DesignCon 2009*.

[7] H. Barnes, J. Moreira, and M. Walz, "Non-destructive analysis and EM model tuning of PCB signal traces using the Beatty standard," in *DesignCon*, 2017.

[8] L. Hayden, "An enhanced Line-Reflect-Reflect-Match calibration," in *67th IEEE ARFTG Conf.*, 2006, pp. 143–149.

[9] A. Mangan, S. Voinigescu, M.-T. Yang, and M. Tazlauanu, "De-embedding transmission line measurements for accurate modeling of IC designs," *IEEE Transactions on Electron Devices*, vol. 53, no. 2, pp. 235–241, 2006.

f_T extraction of HEMT transistors at mm-waves through EM-simulated de-embedding devices

Mohammed Medbouhi
CEA-Leti, Grenoble-Alpes University
Grenoble, France
mohammed.medbouhi@cea.fr

Jose Lugo-Alvarez
CEA-Leti, Grenoble-Alpes University
Grenoble, France
jose.lugo@cea.fr

Philippe Ferrari
TIMA laboratory, Grenoble-Alpes University
Grenoble, France
philippe.ferrari@univ-grenoble-alpes.fr

Erwan Morvan
CEA-Leti, Grenoble-Alpes University
Grenoble, France
erwan.morvan@cea.fr

Abstract— **De-embedding structures (transmission lines, open, short) are costly, whether in terms of silicon area or measurement time. In addition, they lead to measurement inaccuracies increasing with frequency in the millimeter-wave bands. In this paper, we show the possibility to perform de-embedding of HEMT transistors measurements through EM simulations in order to extract the unity-gain cut-off frequency (f_T). Only an open must be measured to calibrate the EM simulation. Several de-embedding techniques are studied and compared up to 67 GHz, based on measurements or EM simulations. The agreement for the extraction of f_T between measured or EM-simulated de-embedding structures validates the proposed approach.**

Keywords— *millimeter-waves, CMOS, cut-off frequency (f_T), de-embedding, EM simulations.*

I. INTRODUCTION

The unity-gain cut-off frequency f_T of transistors is a key figure of merit (FoM) describing the performance limits of circuits such as amplifiers and VCOs. With the increasing demand of millimeter-wave (mm-wave) systems, the required f_T today exceeds 300 GHz, whether in CMOS or BiCMOS technologies [1]. For this reason, it would be necessary to be able to extract f_T around 400 GHz, facing the challenges of very high frequency measurements due to the imprecision of both the calibration and the de-embedding procedures. This inaccuracy is linked to several aspects: (i) the precision of the VNAs decreases due to a lower measurement dynamic, significantly above 100 GHz, (ii) the quality of the calibration standards decreases, whether for an off-wafer calibration for which it is very complicated to carry out precise loads, or an on-wafer calibration of the TRL type where one must know the characteristic impedance of the LINE standard which serves as a reference for the S-parameters, (iii) the quality of the tip-wafer contacts is degraded, and (iv) parasitic couplings between probes and probes to wafer [2].

In this context, the measurements are rather carried out low in frequency, often below 100 GHz, and an extrapolation is realized to extract f_T. An off-wafer calibration is often previously performed, followed by a de-embedding procedure [3]. Several more or less complex de-embedding procedures have been developed in recent years in order to define reference planes as close as possible to the transistor under test [4-8]. All

these de-embedding methods require the design of on-wafer de-embedding standards, e.g. open, short and transmission lines type. These devices generate on the one hand a consumption of silicon surface impacting the cost of the test, and on the other hand additional measurements impacting the accuracy of the extraction due to problems of contact quality and measurement noise.

In order to solve problems related to de-embedding, one solution could consist relying on simulated de-embedding standards. This is the subject of this article, in which we show that it is possible to fully de-embed using simulated standards. We show that it is only necessary to measure an "open" type standard in order to calibrate the electromagnetic simulation, the other de-embedding standards being fully simulated.

The paper is organized as follows. The device under test, i.e. a GaN transistor, is described in section II. The proposed de-embedding approach is presented in section III, and results are shown in section IV, before a brief conclusion in section V.

II. DEVICE UNDER TEST DESCRIPTION

In this paper, a Gallium Nitride (GaN) based High Electron Mobility Transistor (HEMT) is employed to validate the proposed approach. GaN transistors, owing to their remarkable characteristics of high breakdown field and high carrier density (Ns), have exhibited significant performance in Radio-Frequency (RF) and mm-waves. Consequently, it has been widely used in diverse applications including but not limited to switching, power amplification and radars within space, aeronautic and automotive industries [9-11].

III. PROPOSED METHOD

In order to obtain the intrinsic frequency response of the transistor and be able to extract f_T, one must de-embed pads and access lines from the initial raw measurements (Fig. 1).

This classical process involves measuring one or several de-embedding standards such as through, open, short, etc. and apply a de-embedding method.

The method proposed in this paper consists of two steps: At first, only one standard is measured, namely the "open" standard, because it is the easiest to fabricate. This measurement is used to calibrate the EM simulations. The second step then consists

979-8-3503-4331-1/24 $31.00 © 2024 IEEE

SiRF 2024

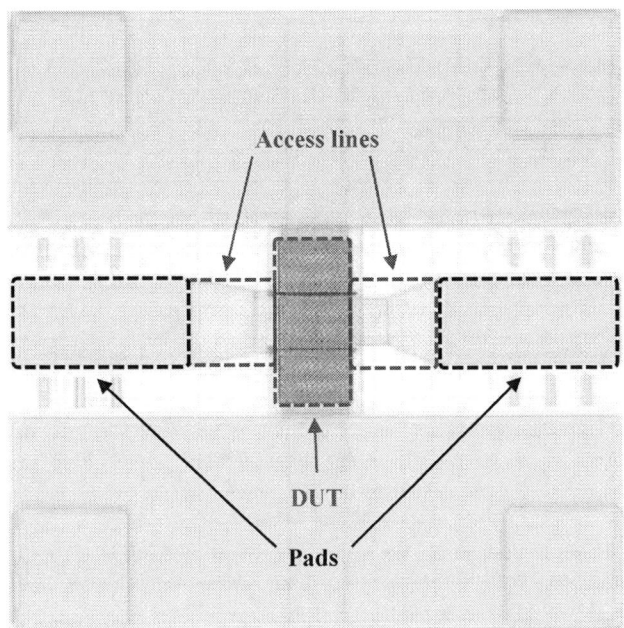

Fig. 1. GaN transistor structure layout

of simulating all the other standards, regardless of the de-embedding method used, and applying the de-embedding method using the simulated standards. These two steps are described in this section.

A. Open standard verification

Fig. 2 illustrates the open standard structure, having dimensions of 420 x 400 μm^2. S-parameters were measured using 100 μm pitch GSG probes and a N5227B PNA from Keysight, operated on a semi-automatic prober from MPI. The frequency range is 100 MHz to 67 GHz. To conduct EM simulations, Ansys HFSS software was employed with the same frequency range. The excitation of the structure was achieved by applying an RF signal through a wave port terminal. The assumed width and height of this latter is three times the width of the signal strip.

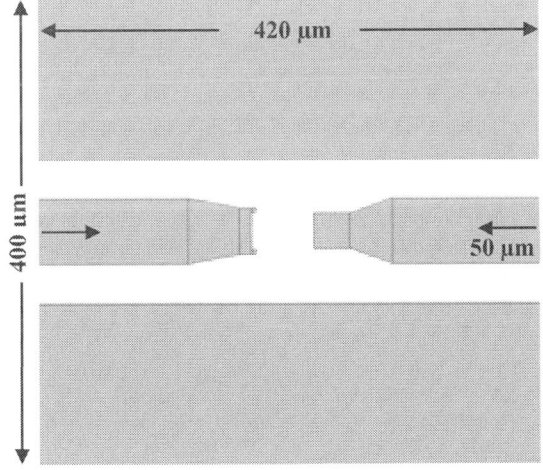

Fig. 2. Open standard HFSS layout.

To ensure an accurate comparison, a 50 μm distance was de-embedded from both the input and the output of the structures, considering that the probe measurements are not placed at the edges of the pads (see Fig. 2).

This is seen in Fig. 3 where good agreement in S_{11} phase is presented between measurements and the EM simulation after considering the displacement of 50 μm.

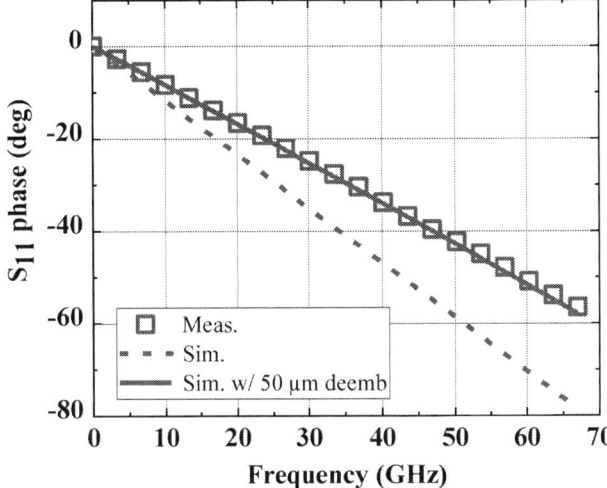

Fig. 3. S_{11} EM simulation with and without de-embedded 50μm pads compared to measurements.

Fig. 4 presents the magnitude of S_{11} and S_{21}. Here again a very good agreement is obtained between measurement and simulation results. As a result, the EM simulation is effectively validated, thus enabling the simulation of other de-embedding standards.

B. De-embedding methods

In order to accurately get the S-parameters of the device under test (DUT), a de-embedding procedure is necessary. At mm-waves, distributed element-based de-embedding techniques tend to perform better compared to methods based on

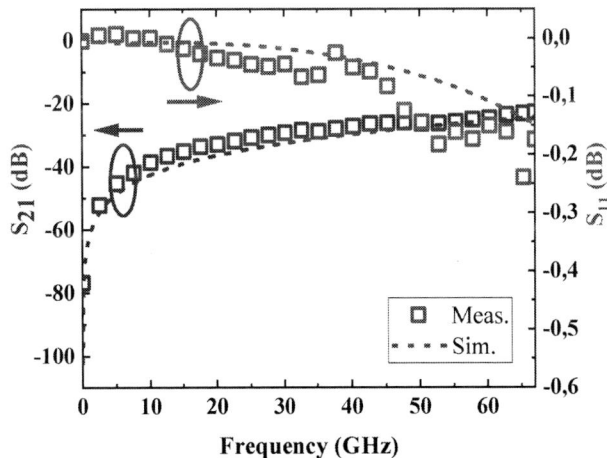

Fig. 4. Comparison of S-parameters results between measurements and EM simulation.

979-8-3503-4331-1/24 $31.00 © 2024 IEEE

lumped elements. In this work, five distinct de-embedding techniques were selected and applied to analyze the DUT:

- Through-only [4]: implies only the through (or thru) standard, by de-embedding each side of the raw structure (i.e. input and output) with the half standard.

- Open-Short [5]: the DUT is characterized by the effects of open and short circuit discontinuities, removing the impact of the parallel admittance of pads and the series impedance of interconnects.

- Vandamme [6]: similar to Open-Short technique, this method takes into account the coupling between the input and the output of the DUT. It uses open, short and through de-embedding standards.

- Cascade [7]: this method involves decomposing the raw structure into smaller cascaded sub-networks. Through calculations, the impact of each sub-network is removed from the raw structure.

- Multiline TRL [8]: by using through, open or short, and Tlines with different lengths, the TRL can be used for de-embedding but also for first-tier calibration.

IV. Results and Discussion

As explained above, in this paper, the proposed simulation-based de-embedding is applied to the extraction of a GaN transistor f_T. However, the proposed approach is totally general and can be applied to any kind of DUT, the goal being to de-embed the S-parameters.

In RF and microwave domains, the Figure of Merit (FoM) for a transistor is characterized by two essential parameters: transition frequency (f_T) and maximum oscillation frequency (F_{max}). We will focus on f_T in this work. This parameter is determined when the short-circuit current gain ($H_{21,dB}$) is equal to unity (0 dB). The computation of this frequency is achieved through the use of the following equation:

$$H_{21} = \frac{Y_{21}}{Y_{11}} \qquad (1)$$

The aforementioned 2-fingers GaN transistor was biased at $V_{DS} = 10$ V with drain current I_D of 350 mA/mm. Small-signal measurements were then performed using the same setup and frequency range as the de-embedding standards.

In the next section, we first show the comparison between EM-simulation-based and measurements-based extraction of H_{21} for the Open-Short de-embedding technique (section IV.A). Next, we show the extraction of H_{21} for the five well-known de-embedding techniques briefly described in section III by using only EM simulations (section IV.B).

A. Open-short de-embedding

Fig. 5 shows the H_{21} extracted from raw and Open-Short de-embedding for EM-simulated and measured de-embedding standards. The extraction based on raw measurements displays a slope change above 30 GHz, leading to an inaccurate extraction of f_T. Consequently, as well known, a de-embedding

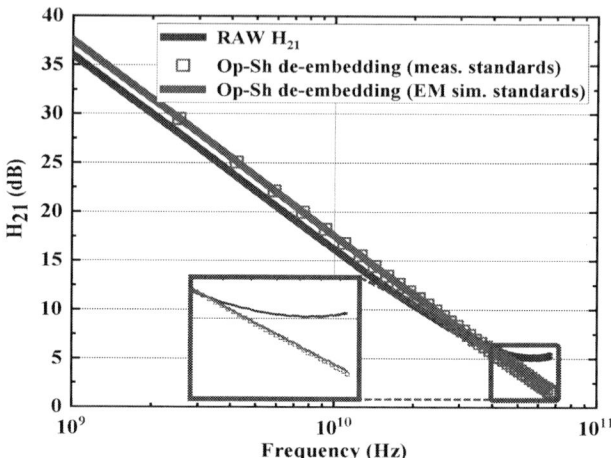

Fig. 5. Raw vs de-embedded H_{21} results with measured and EM simulated standards.

process is essential. Next, the agreement of the H_{21} extraction between measurements-based and EM-simulations-based de-embedding is very good.

B. Other de-embedding techniques

Five distinct de-embedding techniques were applied to extract H_{21}, using only EM-simulated de-embedding standards. Results are depicted in Fig. 6.

All of the techniques effectively corrected the issue of slope change problem above 30 GHz compared to raw measurements. Notably, the Open-Short and Vandamme methods yielded similar results, while the cascade and multiline TRL showcased superior results, as anticipated, exhibiting greater H_{21} in comparison to the other approaches. This distinction can be attributed to the distributed element-based principle of the latter two methods, enabling more accuracy at mm-waves as compared to the lumped element-based techniques.

Since the de-embedding curves extend above 0 dB within the maximum frequency range of measurements, an extrapolation of these curves was performed to derive the f_T value. Table I provides a summary of the corresponding f_T values obtained through the application of the five de-embedding techniques.

TABLE I. f_T Value Of Different De-embedding Techniques

De-embedding	Through	Op-Sh.	Vand.	Casc.	mTRL
f_T (GHz)	88.3	82.8	83.1	88.5	97.4

The most important conclusion is to note that the extraction of f_T after a de-embedding based only on EM simulations seems perfectly validated because the results of the various de-embedding techniques used are in good agreement with the measurement results for an "Open-Short" de-embedding.

V. Conclusion

We demonstrated, by a very simple approach to implement, that the extraction of the cut-off frequency f_T of transistors could be fully carried out using only EM simulations-based de-embedding methods. The only measurement of an "Open" type

979-8-3503-4331-1/24 $31.00 © 2024 IEEE

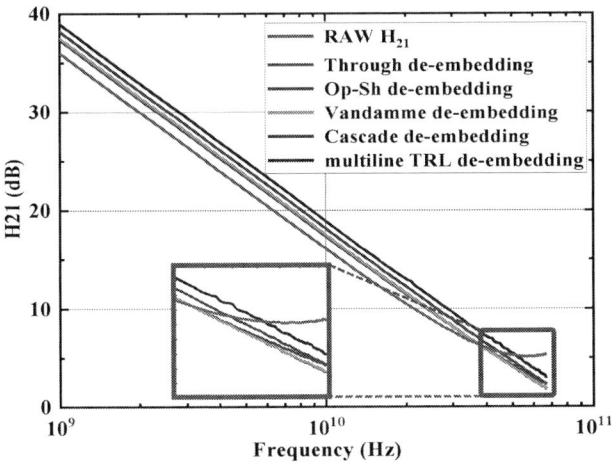

Fig. 6. H_{21} extraction results with different de-embedding techniques based on EM-simulated standards.

standard is necessary in order to calibrate the EM simulation with certainty. This approach provides a double advantage: saving significant silicon surface because the de-embedding standards are expensive in terms of surface, due in particular to the size of the associated pads; and on the other hand saving in measurement time and again in extraction precision due to the inaccuracy of the measurements when several standards are to be measured at mm-waves.

REFERENCES

[1] D. Manger *et al.*, "Integration of SiGe HBT with ft=305 GHZ, fmax=357 GHz in 130nm and 90nm CMOS," *2018 IEEE BiCMOS and Compound Semiconductor Integrated Circuits and Technology Symposium (BCICTS)*, San Diego, CA, USA, 2018.

[2] S.Panda *et al.*, " TCAD and EM co-simulation method to verify SiGe HBT measurements up to 500 GHz," Solid-State Electronics, Vol. 174, 2020.

[3] S. Fregonese *et al.*, "Comparison of On-Wafer TRL Calibration to ISS SOLT Calibration With Open-Short De-Embedding up to 500 GHz," in *IEEE Transactions on Terahertz Science and Technology*, vol. 9, no. 1, pp. 89-97, Jan. 2019.

[4] H. Ito *et al.*, *"A simple through-only de-embedding method for on-wafer S-parameter measurements up to 110 GHz,"* 2008 IEEE International Microwave Symposium, *Atlanta, GA, USA, 2008.*

[5] M. C. A. M. Koolen, "On-wafer high-frequency device characterization," *ESSDERC '92: 22nd European Solid State Device Research conference*, Leuven, Belgium, 1992, pp. 679-686, doi: 10.1016/0167-9317(92)90521-R.

[6] E. P. Vandamme, *et al.*, "Improved three-step de-embedding method to accurately account for the influence of pad parasitics in silicon on-wafer RF test-structures," in *IEEE Trans. on Electron Devices*, vol. 48, no. 4, pp. 737-742, April 2001.

[7] *Ming-Hsiang Cho, et al., "A novel cascade-based de-embedding method for on-wafer microwave characterization and automatic measurement,"* 2004 IEEE International Microwave Symposium, *Fort Worth, TX, USA, 2004.*

[8] D. C. DeGroot, *et al.*, "Multiline TRL revealed," 60th ARFTG Conference Digest, Washington, DC, USA, 2002.

[9] M. Zerarka *et al.*, "Cosmic ray immunity of COTS Normally-Off Power GaN FETs for space, aeronautic and automotive applications," 20th European Conference RADECS, Toulouse, France, 2020.

[10] G. Brocero, *et al.*, "Determination of AlGaN/GaN power transistor junction temperature for radar applications," 21st International Conference MIKON, Krakow, Poland, 2016.

[11] R. Kabouche et al., "Comparison of C-Doped AlN/GaN HEMTs and AlN/GaN/AlGaN Double Heterostructure for mmW Applications," 13th European Microwave Integrated Circuits Conference (EuMIC), Madrid, Spain, 2018.

979-8-3503-4331-1/24 $31.00 © 2024 IEEE

CMOS LNA and VGA for 5G NR Using Gain-Linearity-Boosting and Body Floating Techniques

Jin-Fa Chang
Department of Electronic Engineering
National Changhua University of Education
Changhua, Taiwan
jfchang@cc.ncue.edu.tw

Yo-Sheng Lin
Department of Electrical Engineering
National Chi Nan University
Puli, Taiwan
stephenlin@ncnu.edu.tw

Abstract— We report a 9.5-mW 21.3-27.9-GHz CMOS low-noise amplifier (LNA) with auxiliary-gain-linearity-enhancement (AGLE) stage. It adopts body-floating and coupled-transmission-line (CTL)-based gain-boosting techniques. The LNA constitutes a common-source (CS) input stage, followed by CS gain and output stages. The bias current of the output stage is reused by the gain stage for low power dissipation (P_{dc}). The CTL in conjunction with a coupling capacitance (C_{ctl}) contributes an in-phase gain at the output of the input stage. Over 21.3-27.9 GHz, 0.75-4.28 dB boosting in S_{21} and 0.25-0.46 dB reduction in noise figure (NF) are achieved. Moreover, based on the LNA topology, we report a 13.2-mW 21-28-GHz CMOS variable-gain amplifier (VGA). Analog switch transistor M_4 is in parallel with the output stage to tune its overdrive and drain-source voltage (V_{DS}) for fine tuning of S_{21}. Digital switch transistor M_5 is in parallel with the gain stage to control its AC V_{DS} for coarse tuning of S_{21}. The VGA achieves S_{21} of 19.5 ±1.5 dB for 21-28 GHz (i.e., 3-dB bandwidth f_{3dB}=7 GHz), S_{21} tuning range of 35.6 dB (21~ -14.6 dB), minimum NF of 1.99 dB at 24 GHz and average NF (NF_{avg}) of 2.19 dB for 21-28 GHz, and figure-of-merit (FOM_2) of 61 $nm \cdot GHz^{2/3}/mW^{1/3}$. The NF_{avg} and FOM_2 are one of the best results ever reported for VGAs/LNAs with f_{3dB} greater than 5 GHz and P_{dc} lower than 15 mW.

Keywords— CMOS, LNA, VGA, coupled TL, gain boosting

I. INTRODUCTION

Nowadays, CMOS ICs for the 5G n257 (26.5-29.5 GHz) and n260 (37-40 GHz) new radio (NR) bands have become popular due to its capability of providing a high throughput and low latency experience. For LNA and VGA in an RF transceiver, in addition to low power dissipation (P_{dc}), the basic requirements include good input/output matching, high gain, low noise figure (NF), wideband, and good phase/power linearity. Up to now, a few 28 GHz LNAs/VGAs have been reported [1]-[7]. However, the performance still has room for improvement. For instance, in [2], a 28 GHz CMOS VGA with 3-dB gain bandwidth (f_{3dB}) of 25.3 GHz (18.7-44 GHz) is reported. Though remarkable gain tuning range of 16 dB and minimum NF (NF_{min}) of 2.4 dB are achieved, its maximum S_{21} ($S_{21,max}$) of 10.5 dB and P_{dc} of 25 mW are not good enough. In [6], a 28 GHz CMOS VGA with f_{3dB} of 3.5 GHz and P_{dc} of 8 mW is demonstrated. Though decent $S_{21,max}$ of 21.2 dB is achieved, its gain tuning range of 5 dB is not satisfactory. To demonstrate that low P_{dc}, low NF, high S_{21}, and wide f_{3dB} can be achieved for CMOS LNA and VGA for 28 GHz 5G systems,

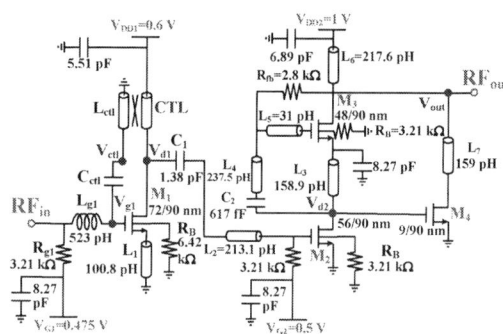

Fig. 1. Schematic diagram of the LNA.

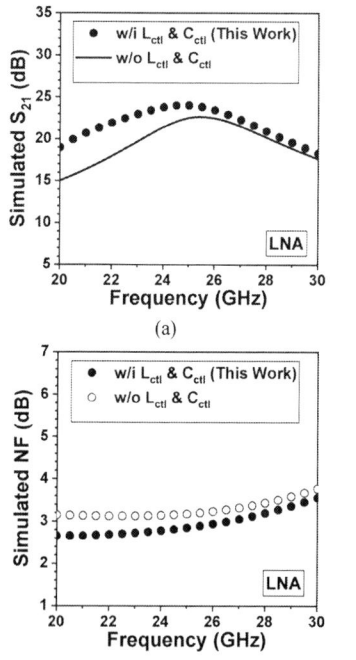

Fig. 2. Simulated (a) S_{21}, and (b) NF of the coupled-TL-based LNA.

we demonstrate a 9.5-mW 21.3-27.9-GHz CMOS LNA (with S_{21} of 21.1-24.1 dB and NF of 2.66-3.18 dB) using coupled-transmission-line (CTL)-based gain-enhancement and body-floating techniques. Moreover, based on the LNA topology, we

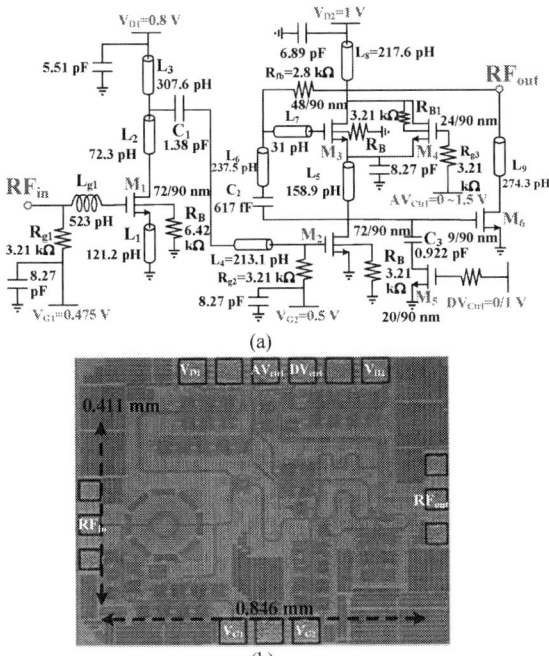

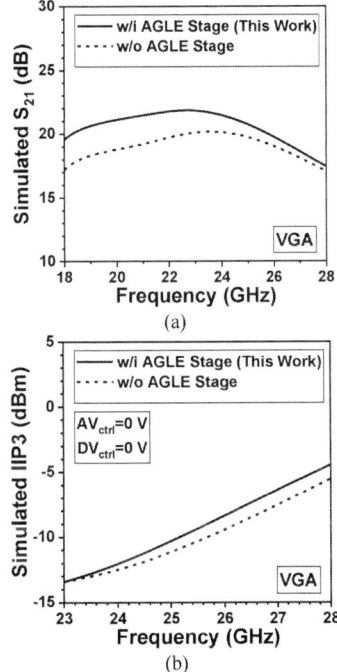

Fig. 3. (a) Schematic diagram and (b) chip photo of the VGA.

Fig. 4. Simulated (a) S_{21}, and (b) IIP3 of the VGA using AGLE stage.

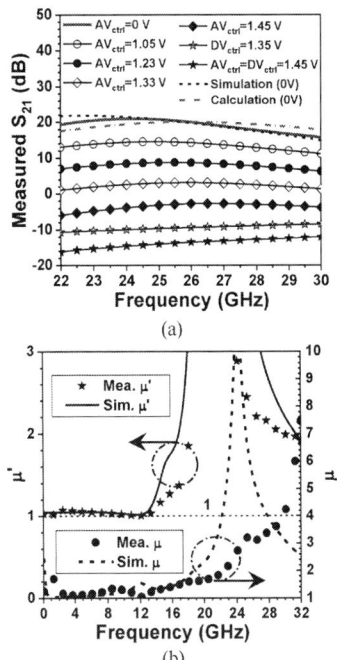

Fig. 5. Measured (a) S_{21}, and (b) μ and μ' of the VGA.

report a 13.2-mW 21-28-GHz CMOS VGA with S_{21} tuning range of 35.6 dB (21~ -14.6 dB) and NF_{avg} of 2.19 dB. Gain-boosting is attained by the inductive peaking of L_{eff} (equal to $L_6 + L_7 - 1/(\omega^2 C_2)$, refer to Fig. 2(a)) in the output stage and the additional gain-boosting path constituting transistor M_6 and the peaking inductor L_9. Body floating of the transistors (for NF enhancement) is achieved by B-to-GND with R (R_B in Fig. 2(a)).

II. LNA AND VGA DESIGN

The LNA and VGA using auxiliary-gain-and-linearity enhancement (AGLE) technique were designed by a 1P9M 90 nm CMOS process with substrate resistivity of 10 Ω·cm. The TL inductors and spiral inductor L_{g1} (provided by the foundry) are placed on the 3.4-μm-thick top metal (MT_9). The bottom metal (MT_1) is used as the ground plane. The distance (D) between MT_9 and MT_1 is 6.06 μm. The space between the TLs is at least five times of D (30.3 μm) to control the mutual coupling and parasitics. Fig. 1 shows the schematic diagram of the LNA with power consumption (P_{dc}) of 9.5 mW. The LNA consists of a common-source (CS) input stage, and current-reused CS gain and output stages. A resistive-feedback via R_{fb} of 2.8 kΩ (close to an open circuit in AC) is used for self-bias of the output stage. The bias current of the output stage is reused by the gain stage for low P_{dc}. A large body floating resistance R_B (connected between the body node and ground (B-to-GND)) of 6.42 kΩ and 3.21 kΩ are used in transistors M_1 and M_2/M_3, respectively, for NF and gain enhancement. The proposed AGLE technique is achieved by including an auxiliary G_m-boosting stage (P_{dc}= 0.9 mW) consisting of transistor M_4 (with B-to-S) and inductor L_7 in parallel with the output stage. An R_B (negligible to overall NF) is not included in M_4. Gain-boosting is achieved since it contributes an in-phase gain of $-g_{m4}Z_{load}/(1-\omega^2 L_7 C_{d4})$ at output. Z_{load} is the load impedance at the output node RF_{out}. C_{d4} is the capacitance at drain of M_4. The inclusion of L_7 leads to 5% gain boosting. Linearity boosting is achieved since V_{DS4} (1 V) is about twice of V_{DS3} (0.53 V).

Moreover, the input stage of the LNA adopts CTL-based gain-boosting and body-floating techniques. CTL in conjunction with C_{ctl} contributes an in-phase gain at the output

TABLE I
SUMMARY OF THE CMOS VGA AND RECENT REPORTED STATE-OF-THE-ART CMOS LNAS/VGAS WITH SIMILAR OPERATION FREQUENCY

	Gain Tuning Range (dB)	S_{11} (dB)	S_{21} (dB)	f_{3dB} (GHz)	NF_{min}/ NF_{avg} (dB)	IIP3 (dBm)	P_{dc} (mW)	Area (mm²)	FOM_1 (GHz)	FOM_2 (nm·GHz$^{2/3}$/mW$^{1/3}$)	CMOS Process
This Work-VGA	35.6 (-14.6~ 21)	< -10	19.5±1.5	21-28	1.99/ 2.19	-8.7	13.2	0.348	1.03	61	90 nm
[1], 2022 MWCL	23.4 (-5.4~ 18)	< -10	10.4±1.5	3.1-23.7	2.53/ 3.97	-6.5	5.85	0.364	1.75	32.4	90 nm
[2], 2020 TMTT	16 (-5.5~ 10.5)	< -9.5	9±1.5	18.7-44	2.4/ 4.05	0*	25	0.49	1.85	11.9	45 nm
[3], 2020 IMS	35.4 (-5.5~ 29.9)	< -5	28.4±1.5	22.3-25.7	5.63/ 8.75	-28.9*			0.001	0.32	
	37.9 (-5.5~ 32.4)	< -10	30.9±1.5	31.5-34.6	4.55/ 7.46	-30.9*	25.6	0.671	0.001	0.47	28 nm
	33.7 (-11.5~ 22.2)	< -10	20.7±1.5	42.6-57.4	5.96/ 10.5	-22.9*			0.003	0.46	
[4], 2017 MWCL	N/A	< -10	9.2±1.5	7.6-29	4.5/ 5.05	-6	12.1	0.3	0.58	10	65 nm
[5], 2022 TMTT	N/A	< -9.5	15.1±1.5	7.2-27.3	3.3/ 3.51	-6	13.2	0.14	1.75	27.4	65 nm
[6], 2019 MWCL	5 (16.2~21.2)	< -10	19.7±1.5	3.5	N/A	-12.4	8	0.17	N/A	N/A	65 nm

*Estimation

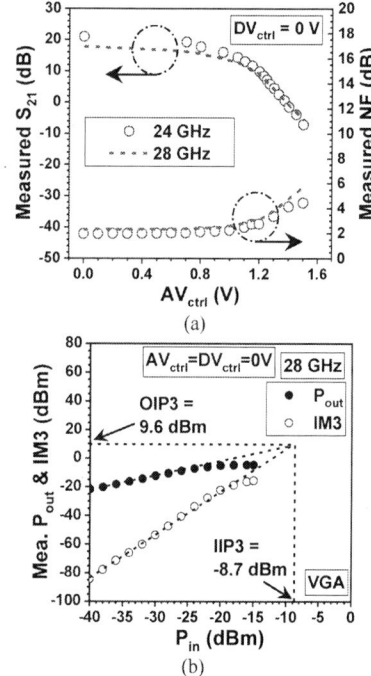

Fig. 6. Measured (a) S_{21} and NF, and (b) IIP3 of the VGA.

of the input stage. Fig. 2(a) shows the simulated S_{21} versus frequency characteristics of the LNA. Fig. 2(b) show the simulated NF versus frequency characteristics of the LNA. Over 21.3-27.9 GHz, the inclusion of L_{ctl} (in the CTL) and C_{ctl} leads to 0.75-4.28 dB S_{21} increase and 0.25-0.46 dB NF reduction (mainly due to gain boosting). The reason of gain boosting can be analyzed as follows. The feedback factor β of the input stage over 21.3-27.9 GHz is given by

$$\beta = \frac{v_{g1}}{v_{d1}} = -CC \frac{1-\omega^2 L_{ctl}C_{ctl}}{\omega^2 L_{ctl}C_{ctl}} \approx -0.102 \sim -0.184 \quad (1)$$

in which CC is the net coupling factor of the CTL (~0.0048), L_{ctl} is equal to 201.5 pH, and C_{ctl} is equal to 6.67 fF. Suppose the load impedance of the input stage is Z_{load1} and the voltage gain without consideration of the feedback is A (~6 dB), then

the (closed-loop) voltage gain A_f of the input stage over 21.3-27.9 GHz is given by

$$|A_f| = \left| \frac{A}{1-A\beta} \right| = \left| \frac{-g_{m1}Z_{load1}}{1-g_{m1}Z_{load1}CC \frac{1-\omega^2 L_{ctl}C_{ctl}}{\omega^2 L_{ctl}C_{ctl}}} \right| = 8.2 \sim 10.3 \text{ dB}$$

$$(2)$$

That is, theoretically, the inclusion of L_{ctl} and C_{ctl} leads to 2.2-4.3 dB boosting in S_{21}, consistent with the simulated result (0.75 -4.28 dB).

Based on the LNA topology, a 13.2-mW 21-28-GHz CMOS VGA is designed and implemented. Figs. 3(a) and 3(b) show the circuit diagram and chip photo, respectively, of the VGA. The analog switch transistor M_4 is in parallel with the output-stage transistor M_3 to tune its overdrive voltage (V_{ov}) and drain-source voltage (V_{DS}) for fine S_{21} tuning. The digital switch transistor M_5 is in parallel with the gain-stage to control its AC V_{DS} for coarse S_{21} tuning. An AGLE stage (P_{dc}=0.8 mW) consists of M_6 and L_9 is included in parallel with the output stage. Gain-boosting is achieved since it contributes an in-phase gain of $-g_{m6}Z_{load1}/(1-\omega^2 L_9 C_{d6})$ at output, in which Z_{load1} is the load impedance at the output node RF_{out}, and C_{d6} is the capacitance at drain of M_6. The inclusion of L_9 leads to 5% gain boosting. Linearity boosting is achieved since V_{DS6} (1 V) is about twice of V_{DS3} (0.54 V). Fig. 4(a) shows the simulated S_{21} versus frequency characteristics of the VGA. Over 21-28 GHz, the inclusion of the AGLE stage leads to 0.4-2.2 dB boosting in S_{21}. Fig. 4(b) shows the simulated input third-order intercept point (IIP3) versus frequency characteristics of the VGA. Over 23-28 GHz, the inclusion of the AGLE stage leads to 0.44-1.2 dB enhancement in IIP3.

III. MEASUREMENT RESULTS AND DISCUSSIONS

The VGA consumes 13.2 mW since the input and gain/ output stage drain 7.5 mA and 7.2 mA current from 0.8 V and 1 V power supply, respectively. Fig. 5(a) shows the measured S_{21} versus frequency characteristics of the VGA. In the

979-8-3503-4331-1/24 $31.00 © 2024 IEEE

condition of the analog control voltage (AV_{ctrl}) and the digital control voltage (DV_{ctrl}) both equal to 0 V, the VGA achieves measured $S_{21,max}$ of 21 dB (at 24 GHz) and f_{3dB} of 7 GHz (21-28 GHz), close to the simulated one ($S_{21,max}$ of 21.8 dB (at 22.7 GHz) and f_{3dB} of 9.1 GHz (17.7-26.8 GHz)) and the calculated one ($S_{21,max}$ of 20.1 dB (at 26 GHz) and f_{3dB} of 9.4 GHz (21.6-31 GHz)). For AV_{ctrl} of 0-1.45 V and fixed DV_{ctrl} of 0 V, the VGA achieves wide S_{21} tuning range of 24.8 dB (21~ -3.8 dB) at 24 GHz, and 20.7 dB (17.8~ -2.9 dB) at 28 GHz. Moreover, the inclusion of the results at DV_{ctrl} of 1.35-1.45 V and fixed AV_{ctrl} of 1.45 V leads to a wider S_{21} tuning range of 35.6 dB (21~ -14.6 dB) at 24 GHz, and 30.6 dB (17.8~ -12.8 dB) at 28 GHz. Fig. 5(b) shows the measured and simulated stability factors μ and μ' versus frequency characteristics of the VGA. The VGA is unconditionally stable since μ and μ' are larger than one over 0.1-32 GHz.

Fig. 6(a) shows the measured S_{21} and NF versus AV_{ctrl} characteristics of the VGA at 24 and 28 GHz. DV_{ctrl} is fixed at 0 V. The VGA achieves S_{21} of 17.8-21 dB and NF of 1.99-2.33 dB at AV_{ctrl} of 0 V, and NF_{avg} of 2.28-2.67 dB for AV_{ctrl} of 0-1.5 V. In the condition of the VGA being turned off (i.e., AV_{ctrl} equal to 1.5 V, and S_{21} equal to -5.4~ -7 dB), decent NF of 4.48-5.72 dB is achieved. Fig. 6(b) shows the measured fundamental output power (P_{out}) and third-order inter-modulation output power (IM3) versus input power (P_{in}) characteristics of the VGA at 28 GHz. The VGA achieves IIP3 of -8.7 dBm and OIP3 of 9.6 dBm. Two figure-of-merits (FOMs) suitable for evaluation of wideband, high gain, low noise, and decent P_{dc} and linearity LNAs/VGAs are given by [7]-[8]

$$FOM_1[GHz] = \frac{BW[GHz] \cdot OIP3[mW]}{(F-1)[1] \cdot P_{dc}[mW]} \qquad (3)$$

$$FOM_2[\frac{nm \cdot GHz^{2/3}}{mW^{1/3}}] = \frac{(S_{21}[1]-1)^{4/3} \cdot L}{S_{21}[1] \cdot (F-1)[1]} \left(\frac{IP_{1dB} \cdot f_0 \cdot BW}{P_{dc}^2} \right)^{1/3} \qquad (4)$$

Table I is a summary of the implemented VGA, and recently reported state-of-the-art CMOS LNAs and VGAs. As can be seen, our VGA occupies small area, consumes low P_{dc}, and achieves decent S_{21}, f_{3dB}, IIP3, and FOM_1, and the best NF_{min}, NF_{avg}, and FOM_2.

IV. CONCLUSION

We report an LNA and a VGA for 28 GHz 5G NR. Bias current of the output stage is reused by the gain stage for low P_{dc}. An AGLE stage (in parallel with the output stage) is included for gain and linearity boosting. The proposed CTL-based gain-boosting and body-floating techniques are useful for S_{21} and NF enhancement of the LNA. The combination of an analog current-steering switch and a digital AC-ground switch is useful for S_{21} tuning range enhancement of the VGA.

REFERENCES

[1] J. F. Chang et al., "5.85 mW 3.1-23.7 GHz Two-Stage CMOS VGA with 2.53 dB NF_{min} Using Concurrent Current Steering," *IEEE Microw. Wireless Compon. Lett.*, vol. 32, no. 11, pp. 1331-1334, Nov. 2022.

[2] Li Gao et al., "A 22–44-GHz phased-array receive beamformer in 45-nm CMOS SOI for 5G applications with 3–3.6-dB NF," *IEEE Trans. Microw. Theory Techn.*, vol. 68, no. 11, pp. 4765-4774, Nov. 2020.

[3] C. J. Liang et al., "A Tri (K/Ka/V)-Band Monolithic CMOS Low Noise Amplifier with Shared Signal Path and Variable Gains," *IEEE MTT-S International Microwave Symposium (IMS)*, 2020, pp. 333-336.

[4] P. Qin et al., "Design of wideband LNA employing cascaded complimentary common gate and common source stages," *IEEE Microw. Wireless Compon. Lett.*, vol. 27, no. 6, pp. 587-589, Jun. 2017

[5] H. Chen et al., "A 7.2–27.3 GHz CMOS LNA With 3.51±0.21 dB Noise Figure Using Multistage Noise Matching Technique," *IEEE Trans. Microw. Theory Techn.*, vol. 70, no. 1, pp. 74-84, Jan. 2022.

[6] C. W. Byeon et al., "A Ka-Band Variable-Gain Amplifier With Low OP1dB Variation for 5G Applications," *IEEE Microwave and Wireless Components Letters*, vol. 29, no. 11, pp. 722-724, Nov. 2019.

[7] J. F. Chang et al., "A 13.7-mW 21-29 GHz CMOS LNA with 21.6 dB Gain and 2.74 dB NF for 28 GHz 5G Systems," *IEEE Microwave and Wireless Components Letters*, vol. 32, no. 2, pp. 137-140, Feb. 2022.

[8] L. Belostotski et al., "Figures of merit for CMOS low-noise amplifiers and estimates for their theoretical limits," *IEEE Trans. Circuits Syst. II, Exp. Briefs*, vol. 69, no. 3, pp. 734-738, Mar. 2022.

Parametric-Oscillation-Free Efficient SiGe:C Power Amplifier Design for *K*u-/*K*a-Band SATCOM

Tsung-Ching Tsai
IHE, Karlsruhe Institute of Technology
Karlsruhe, Germany
tsung-ching.tsai@kit.edu

Václav Valenta
ESTEC, European Space Agency
Noordwijk, The Netherlands
vaclav.valenta@esa.int

Ahmet Çağrı Ulusoy
IHE, Karlsruhe Institute of Technology
Karlsruhe, Germany
cagri.ulusoy@kit.edu

Abstract—**For *K*u- and *K*a-Band satellite communications (SATCOM), two SiGe:C transformer-based current-combined power amplifiers (PAs) with different damping capabilities against parametric oscillation are presented and compared. Node analysis is conducted and proves its effectiveness of detecting oscillation by the measurement. At the targeted 18.8GHz, the parametric-oscillation-free PA achieves 22.8dBm saturated output power (P_{sat}) with 37.7% power-added-efficiency (PAE).**

Index Terms—**Ku-band, Ka-band, SATCOM, power amplifier, SiGe:C, transformer-based, parametric oscillation**

I. INTRODUCTION

Deployment of various SATCOM direct radiating arrays (DRAs) typically requires power levels per element in the order of a few watts, which is supported by III-V semiconductor technologies [1]. Now it is apparent that larger deployable active antennas with thousand elements will require much lower power per radiating element with DRA. As a result, alternative technologies such as SiGe are being considered for tapered DRAs or extremely large deployable DRAs, where the required power levels might be in the range of 15-25 dBm. The possibility of multi-function integration and high efficiency makes this technology very promising for emerging *K*u-/*K*a-band DRA applications. In this paper, SiGe power amplifier (PA) designs suitable for the next generation of large-scale DRAs are presented. The targeted frequency is mainly 17.3-20.3 GHz and additionally 20.2-21.2 GHz for the upcoming IRIS² satellite constellation [2].

Parametric oscillations in PA designs have been reported in many III-V semiconductor technologies. Different bifurcations lead to distinct anomalous behaviors and oscillations [3]. In some cases, due to signal modulation on variable parasitic capacitances, half-frequency oscillation or even chaotic spectrum sustain in the steady state [3]–[5]. In an example transformer-based VDMOS PA, an RF hysteresis effect is further observed [3]. This paper presents a new case of parametric oscillation in SiGe integrated PA design.

A variety of methods of stability check in simulation have been proposed over the past few decades [6]. In this paper, harmonic-balance-based node analysis [7], [8] is adopted. Two PAs are designed and oscillation start-up conditions are examined at critical nodes. The simulation shows its effectiveness of predicting parametric oscillations in measurement.

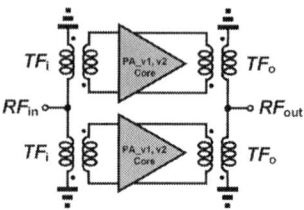

(a) Overall circuit diagram

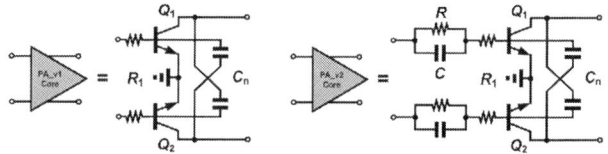

(b) Core of PA_v1 (original) and PA_v2 (modified)

Fig. 1. Schematic of the proposed PAs

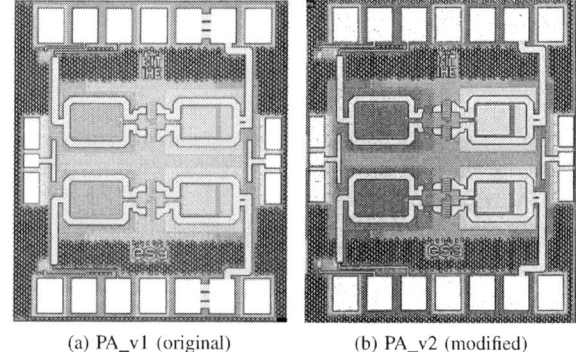

(a) PA_v1 (original) (b) PA_v2 (modified)

Fig. 2. Chip micrographs

II. CIRCUIT DESIGN

The prototype PAs are designed in IHP 130-nm SiGe:C BiCMOS process where the npn-HBT has $f_{\text{T}}/f_{\text{max}}$ of 250/340 GHz. The complete schematic is shown in Fig. 1.

A. Transistor Configuration

Cascode configuration is widely adopted in *K*u-/*K*a-band SiGe power amplifiers [9], [10]. However, in this technology and the targeted frequency, large parasitic collector-emitter capacitance of the common-base device causes negative impedance at the output port. The issue has been reported in CMOS technology and can be solved by

---

979-8-3503-4331-1/24 $31.00 © 2024 IEEE

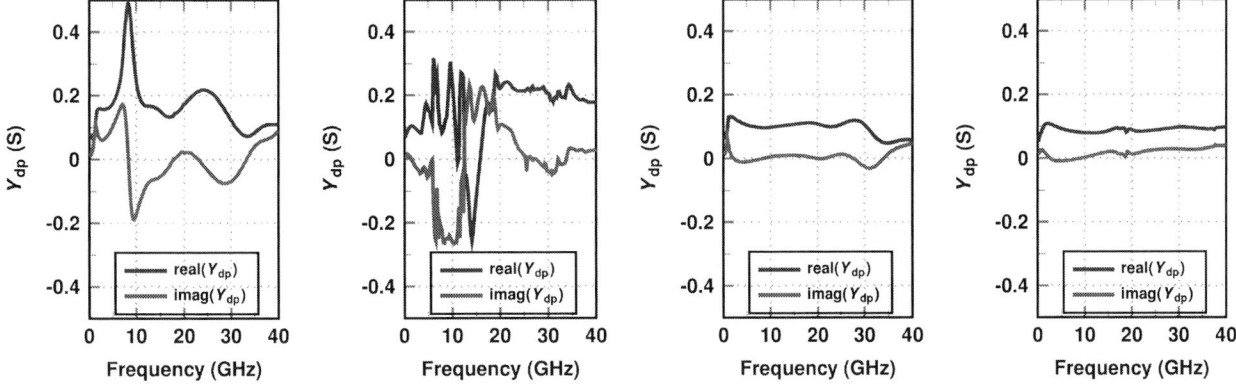

(a) PA_v1 (@small-signal region) (b) PA_v1 (@6-dB gain compression) (c) PA_v2 (@small-signal region) (d) PA_v2 (@6-dB gain compression)

Fig. 3. Simulated driving-point admittance (Y_{dp}) from input node analyses with driven signals at 18.8 GHz.

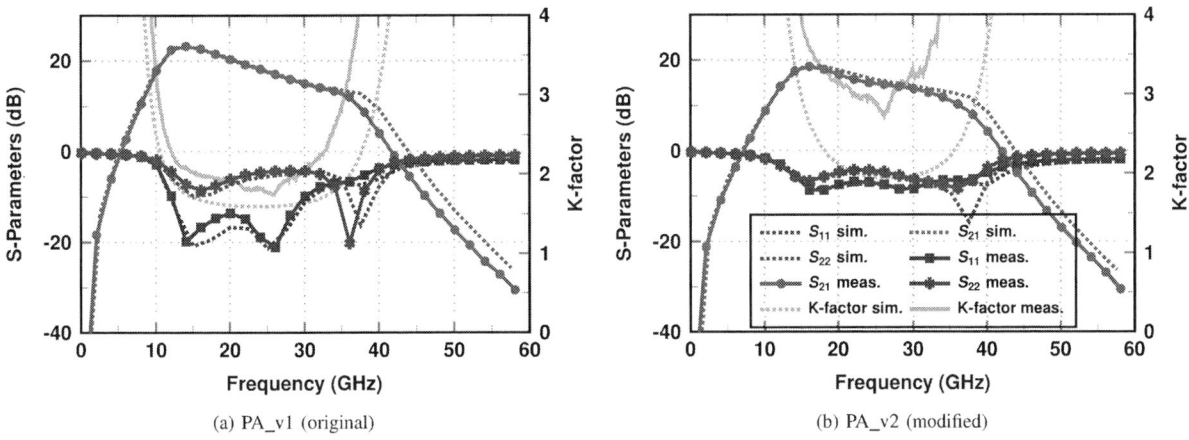

(a) PA_v1 (original)

(b) PA_v2 (modified)

Fig. 4. Simulated and measured S-parameters and K-factor of the original and modified PAs.

neutralization for common-base device [11] at the cost of degradation of power-added-efficiency (PAE). To avoid this, capacitive-neutralized common-emitter configuration is used.

B. Transformer-Based Two-Way Current-Combining

To develop a compact single-ended SiGe prototype PA for satellite application, transformer-based two-way current combining instead of voltage-combining is chosen. The circuit diagram is shown in Fig. 1(a). $Q_{1,2}$ consists of 48 multipliers and the equivalent turn ratio of the output transformer (TF_o) is 1.7 (Fig. 1). The supply voltage is 1.5 V.

C. Parametric-Oscillation-Free Design

To eliminate parametric oscillation, a compact RC network is proposed for MMIC design [12]. However, the empirical solution does not provide a clear guideline on selecting the size of the RC network. Excessive resistance leads to unnecessary considerable degradation of gain. Meanwhile, sizing of the capacitor has an impact on the filtering of lower frequency and thus on the effectiveness of oscillation elimination. By examination of oscillation start-up conditions at critical nodes, sizing of the RC network has become certain.

As shown in Fig. 1(b), two PAs are designed: an original PA without RC network (named PA_v1) and a modified PA with RC network (named PA_v2). With 18.8-GHz driven signals, node analyses are made in simulation for both PAs at the interface of input matching network and base terminal of HBTs. Resulted driving-point admittances (Y_{dp}) at small-signal region and fully-driven region (at 6 dB gain compression) are shown in Fig. 3. In small-signal regions, both PAs present no negative conductance, whereas in fully-driven region as shown in Fig. 3(b), PA_v1 presents negative conductance at 11-16 GHz. Kurokawa condition [6] is even fulfilled at 11-12 GHz. In PA_v2, RC network is designed with $R = 40\,\Omega$ and $C = 750\,\text{fF}$ such that oscillation start-up conditions are removed in fully-driven region (Fig. 3(d)).

III. EXPERIMENTAL RESULTS

The chip micrographs are demonstrated in Fig. 2. The core area of PA_v1 and PA_v2 measure 0.524mm² and 0.547mm² respectively. Both PAs are characterized with identical bias voltage (V_{bias}) of 880 mV.

979-8-3503-4331-1/24 $31.00 © 2024 IEEE   61

TABLE I
COMPARISON WITH STATE-OF-THE-ART 16-20GHz SiGe PAs [13]

	This Work	ISSCC 2018	TMTT 2022	EuMIC 2022	TCAS-I 2023
Supply Voltage (V)	1.5	1.9	3.6	3.3	4.2
Frequency (GHz)	18.8	20	16	18	18
Gain (dB)	17.5	20	10	13.7	21.4
P_{sat} **(dBm)**	22.8	17.4	20.5	25.2	30
PAE_{max} **(%)**	37.7	37	18	27.2	23.5
Core Area (mm²)	0.55	0.29	0.77	1.2	1.1
Power Density* (mW/mm²)	348.5	189.5	145.7	275.9	885

*P_{sat}/Core Area

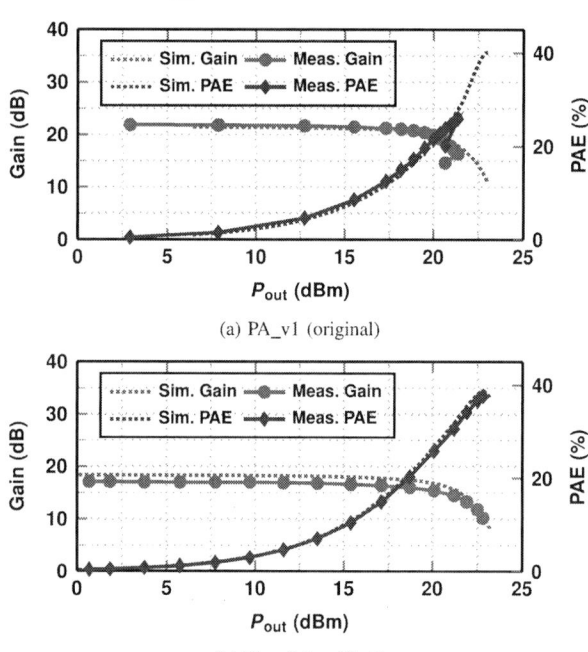

(a) PA_v1 (original)

(b) PA_v2 (modified)

Fig. 5. Simulated and measured 18.8 GHz large-signal sweep of the original and modified PAs.

A. S-parameters

S-parameters are characterized with a 125-GHz 4-port vector signal analyzer (VNA). The simulated and measured S-parameters of the PA_v1 and PA_v2 up to 60 GHz are shown in Fig. 4(a) and Fig. 4(b) respectively. In measurement, PA_v2 achieves gain of 17.5 dB at 18.8 GHz with BW_{-3dB} of 9.7 GHz centered at 17.65 GHz. By contrast, PA_v1 provides 20.1 dB gain at 18.8 GHz.

B. Continuous-Wave Large-Signal Results

18.8-GHz large-signal simulated and measured results are shown in Fig. 5. In measuring PA_v1, bifurcation occurs in the fully-driven region and it fails to provide more than 21.4 dBm output power. With the designed RC network, PA_v2 is free from parametric oscillation, achieving 22.8 dBm P_{sat} with 37.7 % PAE (Fig. 5(b)).

IV. CONCLUSION

In this paper, two transformer-based SiGe PAs with different damping characteristics are designed for *Ku-/Ka*-band SATCOM, examined with node analyses and measured. The parametric oscillation taking place in measurement verifies the effectiveness of the adopted node analyses in detecting oscillations. The parametric-oscillation-free PA achieves 22.8 dBm P_{sat} with 37.7 % PAE at 18.8 GHz. Performance summary and comparison with state-of-the-art SiGe PAs are shown in TABLE I.

REFERENCES

[1] TEC, European Space Agency. Going GaN: Novel Chips Powering Space Missions. [Online]. Available: https://technology.esa.int/news/going-gan-novel-chips-powering-space-missions

[2] IRIS²: the new EU Secure Satellite Constellation. [Online]. Available: https://defence-industry-space.ec.europa.eu/eu-space-policy/iris2_en

[3] A. Suarez *et al.*, "Stability Analysis and Stabilization of Power Amplifiers," *IEEE Microw. Mag.*, vol. 7, no. 5, pp. 51–65, 2006.

[4] L. Pantoli *et al.*, "Stability Analysis and Design Criteria of Paralleled-Device Power Amplifiers Under Large-Signal Regime," *IEEE Trans. Microw. Theory Techn.*, vol. 64, no. 5, pp. 1442–1455, 2016.

[5] A. Piacibello *et al.*, "A Simple Method to Identify Parametric Oscillations in Power Amplifiers Using Harmonic Balance Solvers," *IEEE Microw. Wireless Compon. Lett.*, vol. 31, no. 3, pp. 269–271, 2021.

[6] A. Suarez, "Check the Stability: Stability Analysis Methods for Microwave Circuits," *IEEE Microw. Mag.*, vol. 16, no. 5, pp. 69–90, 2015.

[7] Bode, H.W., *Network Analysis and Feedback Amplifier Design.* D. Van Nostrand Company, 1945.

[8] N. Ayllon *et al.*, "Systematic Approach to the Stabilization of Multitransistor Circuits," *IEEE Trans. Microw. Theory Techn.*, vol. 59, no. 8, pp. 2073–2082, 2011.

[9] I. Ju *et al.*, "An Efficient, Broadband SiGe HBT Non-Uniform Distributed Power Amplifier Leveraging a Compact, Two-Section λ/4 Output Impedance Transformer," *IEEE Trans. Microw. Theory Techn.*, vol. 70, no. 7, pp. 3524–3533, 2022.

[10] T.-C. Tsai *et al.*, "A Linear and Efficient Power Amplifier Supporting Wideband 64-QAM for 5G Applications from 26 to 30 GHz in SiGe:C BiCMOS," in *2021 IEEE Radio Frequency Integrated Circuits Symposium (RFIC)*, 2021, pp. 127–130.

[11] S. V. Thyagarajan *et al.*, "A 60 GHz Linear Wideband Power Amplifier using Cascode Neutralization in 28 nm CMOS," in *Proceedings of the IEEE 2013 Custom Integrated Circuits Conference*, 2013, pp. 1–4.

[12] D. Teeter *et al.*, "A Compact Network for Eliminating Parametric Oscillations in High Power MMIC Amplifiers," in *1999 IEEE MTT-S International Microwave Symposium Digest (Cat. No.99CH36282)*, vol. 3, 1999, pp. 967–970 vol.3.

[13] H. Wang *et al.* Power Amplifiers Performance Survey 2000-Present. [Online]. Available: https://ideas.ethz.ch/Surveys/pa-survey.html

979-8-3503-4331-1/24 $31.00 © 2024 IEEE

A 94 GHz Bandwidth Transimpedance Amplifier in 55nm SiGe BiCMOS for High Speed Optical Receivers

Lachlan Cuskelly
Electrical and Computer Engineering
University of California, Los Angeles
Los Angeles, California
lcuskelly@g.ucla.edu

Christopher Falt
ASIC Design
Ciena Corporation
Ottawa, Canada
cfalt@ciena.com

Peter Schvan
ASIC Design
Ciena Corporation
Ottawa, Canada
pschvan@ciena.com

Abstract—A transimpedance amplifier with a bandwidth of 94 GHz is presented and on chip measurements are discussed. The measured gain of the amplifier at low frequencies is 55 dBΩ, peaking to 61 dBΩ at 78 GHz. The circuit was designed in a cutting edge 55nm SiGe BiCMOS technology, occupies a pad limited area of 0.02 mm^2, and consumes 14 mA of current from a 3.3 V supply. The circuit achieves an average input referred current noise of 11 pA/$\sqrt{Hz}$, making it suitable for future high speed optical receivers.

Index Terms—broadband amplifiers, heterojunction bipolar transistors, MIMICs, optical communication, optical receivers

I. INTRODUCTION

With the advent of coherent optics and an ever increasing need for high data rates, transimpedance amplifiers (TIAs) have become a critical part of broadband fiber optic receivers. Data rates are now exceeding 100 Gbit/s per channel, demanding TIAs with bandwidth into the millimeter-wave range with low noise amplification. Such TIAs can be generally divided into lumped and distributed topologies [1]. Lumped designs include shunt feedback [2], common base [3], and noise cancelling [4]. Similarities can be drawn between broadband low noise amplifier (LNA) design and TIA design. Firstly, this paper demonstrates enhanced performance of a TIA in an improved BiCMOS process node; and secondly, demonstrates the application of a cross coupled cascode topology.

II. LITERATURE REVIEW

A comparison of various TIA topologies was performed in simulation using a 55nm SiGe BiCMOS technology. This technology has a 9 metal layer BEOL and achieves a peak f_t of 345 GHz at a nominal emitter length of 5.6 μm and at a current density of 1.7 mA/μm. At 100 GHz, a current density of 0.5mA/μm is required to achieve the lowest minimum noise figure. Fig. 1 and 2 outlines transimpedance gain and input referred current noise for some selected topologies.

From this comparison, the shunt feedback topology is the preferred choice for lowest noise operation with low power consumption. It has limited bandwidth compared to common base topology but shunt feedback features lower noise than

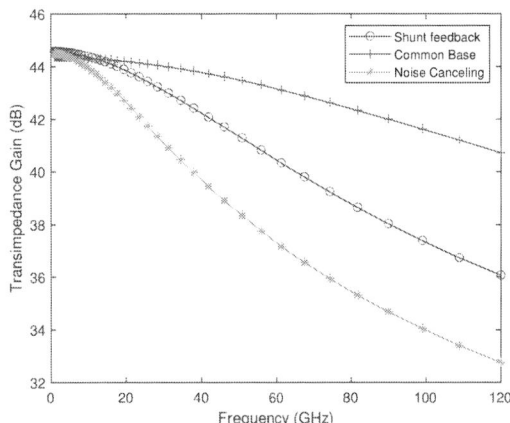

Fig. 1. Comparison of transimpedance gain of various TIA topologies.

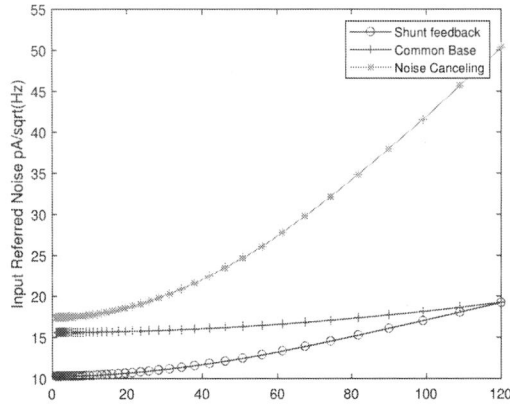

Fig. 2. Comparison of input referred noise of various TIA topologies.

both common base and noise canceling, especially at high frequencies. A noise canceling TIA will also consume more

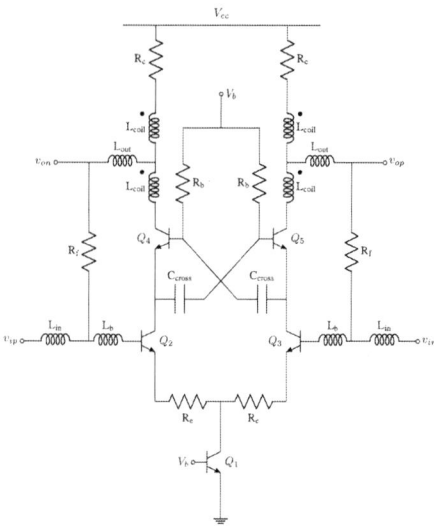

Fig. 3. Circuit schematic of the TIA input stage.

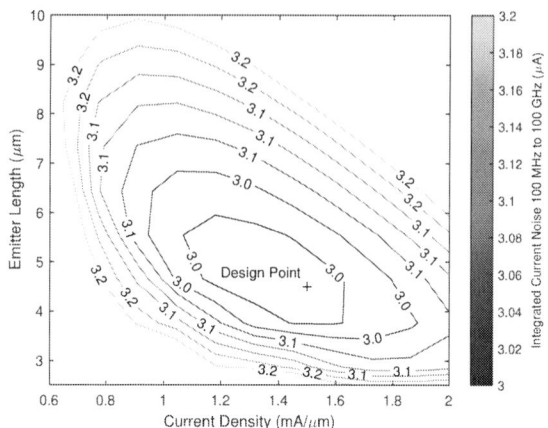

Fig. 4. Integrated current noise (μA) as function of emitter length and current density.

power since more transistors are needed to facilitate the noise cancellation effect. With the addition of appropriate bandwidth extension techniques, shunt feedback has been shown to have bandwidths greater than 90 GHz [5].

While cross coupled pairs are usually reserved for VCOs and mixers, many works have demonstrated their merit as a method for boosting the g_m in an LNA, thus lowering its noise figure and extending bandwidth. As shown in [6], a differential common base LNA is capacitively cross coupled to reduce the noise figure. The disadvantage of crossed coupled common base LNAs is reduced reverse isolation leading to instability. Common emitter LNAs tend to have lower noise and higher output resistance, which combined with a cascode device widens bandwidth. A capacitor cross coupled cascode LNA thus achieves the g_m-boosting effect of a cross coupled common base LNA while maintaining high reverse isolation, wide bandwidth and low noise [7].

Given the similarities between the cross coupled cascode LNA and a shunt feedback TIA, the proposed topology consists of the former in a shunt feedback configuration. A cross coupled cascode extends the bandwidth as compared to a conventional shunt feedback TIA while also consuming the same amount of power, since the common emitter and cascode devices share approximately the same collector current. In addition to bandwidth extension provided by the cross coupled cascode, several peaking inductors are used along with a symmetrical t-coil at the collectors of Q_4 and Q_5. The input inductors L_{in} tune out the capacitance at the base of the input devices while the output inductors L_{out} assist in peaking the gain at high frequencies. An output buffer is used to drive the pad capacitance and provide a 50Ω output match.

III. DESIGN METHODOLOGY

The design process as is follows:

1) Begin with the proposed topology in Fig. 3 while omitting the peaking inductors and t-coil.
2) Select the collector resistor and emitter degeneration resistors for a moderate amount of open loop gain.
3) Select the feedback resistor, keeping in mind larger values decrease noise at the expense of bandwidth.
4) Simultaneously sweep emitter width and current density, while keeping track of input referred current noise, bandwidth and linearity. Select a design point which gives the lowest noise while providing wide bandwidth and high linearity.
5) Select the peaking inductors and t-coil inductance to give the desired bandwidth and frequency response.

Resistors R_c and R_e were selected to be 75Ω and 20Ω respectively to provide enough open loop gain without comprising the bandwidth or linearity. R_e acts to provide emitter degeneration which improves linearity at the expense of gain and noise. The target open loop bandwidth was approximately 30 GHz (one third of the final bandwidth) and the open loop voltage gain is around 10 to 15 dB. The target transimpedance gain is 50 dBΩ at low frequencies and therefore $R_f = 280\Omega$ was chosen to satisfy this specification. The feedback resistor is DC-coupled to the input, as any sufficiently large decoupling cap would create too much parasitic shunt capacitance. Emitter width (the same for all HBT devices) and current density were simultaneously swept from 0.2 μm to 10 μm and 0.5 mA/μm to 2.0 mA/μm respectively. The contour plot in Fig. 4 shows the resulting integrated input referred current noise as a function of emitter length and current density (Note: Only 2.5-10 μm and 0.6 - 2.0 mA/μm are shown for clarity). The design with an emitter width of 4.5 μm at a current density of 1.5 mA/μm has excellent bandwidth and linearity while having a slightly compromised integrated input referred noise current of 3.0 μA (integrated from DC to 100 GHz). By sweeping the detuned circuit first, the effect of the inductors are removed and only the RC roll off is considered. Finally, a 100 GHz

979-8-3503-4331-1/24 $31.00 © 2024 IEEE

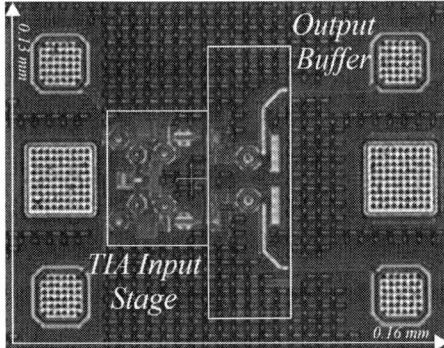

Fig. 5. Micrograph of fabricated TIA breakout.

bandwidth is required with approximately +8dB peaking at high frequencies. The intended application of this peaking is high frequency compensation for a lossy interconnect between the BiCMOS die and an ADC on a separate CMOS die. The resulting inductors have values of L_{in} = 180 pH, L_b = 112 pH, L_{out} = 55 pH, and the t-coil has a single ended inductance of L_{coil} = 200 pH which are chosen based upon the capacitance of the circuit in Fig 3. Spiral inductors were chosen due to their compact area and ability to create a t-coil structure. An additional shunt capacitor at the input of the TIA helps to tune the peaking frequency.

All components were laid out with 100 GHz operation in mind, thus the interconnect length within the TIA core must be minimized as much as possible. The feedback path requires special attention as it is sensitive to parasitics. Therefore the feedback resistor, decoupling capacitor and inductors are placed in such a way to minimize overall interconnect length. All inductors and the t-coil were designed to ensure a self resonant frequency well above 100 GHz.

IV. MEASUREMENTS

The micrograph in Fig. 5 shows the final fabricated TIA and output buffer, with the pad limited area on the die being 0.02 mm^2. The core and buffer consumes 65 mA, with the core itself consuming 14 mA from a 3.3 V supply.

Measurements were conducted on a sawn die which was glued onto a 4 inch silicon wafer. The wafer was placed on a 8 inch temperature controlled chuck and probed using GGB Picoprobe 110H dual probes and calibration was made with a GGB CS-2-100 calibration substrate. Power and bias was provided with two 3-probe wedges. A Keysight N5222B VNA with four N5295A frequency extenders was used for 4-port S-parameter measuerments without the use of a balun. A Keysight N9042B UXA spectrum analyzer with a V3050A frequency extender was used to measure noise on one terminal, with the other terminated to 50Ω while all other RF inputs were left open circuit.

At low frequencies the observed transimpedance gain is 55 dBΩ, rising to 61 dBΩ at 78 GHz, yielding a +7 dBΩ peak and a 3dB baseband bandwidth of 94 GHz. The integrated input referred current noise is 3.3 μA, resulting in an input referred

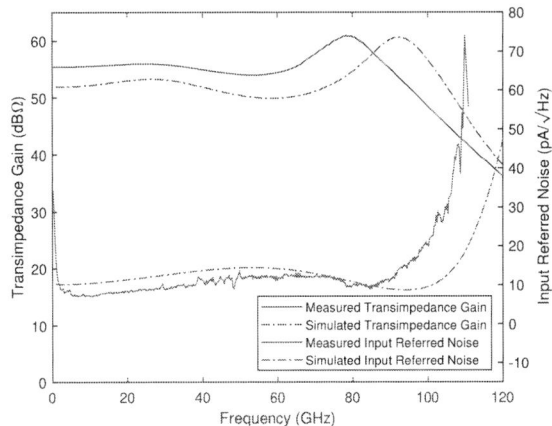

Fig. 6. Measured and simulated gain and noise of proposed topology.

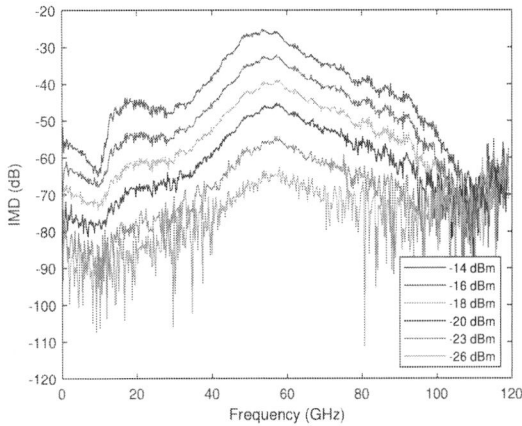

Fig. 7. Measured two tone IMD of proposed topology.

current noise of 11 pA/$\sqrt{Hz}$ when averaged over the same 94 GHz bandwidth. The noise includes the input stage as well as the output buffer. Assuming a photo diode with a differential signal current of 250 μA_{rms}, the SNR is approximately 38 dB.

A two tone test signal (spaced 250 MHz apart) was applied to the input of the TIA and swept up to 120 GHz and at power levels ranging -26 dBm to -14 dBm. The high frequency emphasis significantly reduces the linearity around 60 GHz.

Table 1 compares this design to other broadband TIAs with similar bandwidth. Considering the +8dB peaking requirement, this work maintains comparable or better power consumption, wide bandwidth and low noise operation.

V. CONCLUSION

The TIA design presented in 55nm SiGe BiCMOS demonstrates the viability of a cross coupled cascode topology originally designed for LNAs. It is suitable for greater than 100 Gbit/s channel operation, while having 46 mW of TIA core power consumption and only 3.3 μA of integrated input

979-8-3503-4331-1/24 $31.00 © 2024 IEEE

TABLE I
COMPARISON TO PREVIOUS WORKS

Parameter	[2]	[8]	[9]	[10]	**This Work**
Process	55nm SiGe BiCMOS	32nm CMOS SOI	65nm CMOS	500nm InP HBT	**55nm SiGe BiCMOS**
BW (GHz)	92	72	70	107	**94^+**
Zt (dBΩ)	47	38	40	55	**55**
Supply (V)	3.3	1.5	1.2	5.2	**3.3**
Power (mW)	48	28	24	365	**46***
Area (mm^2)	0.14	0.09	0.16	0.02	**0.02**
Noise (pA/$\sqrt{Hz}$)	18**	84	31	44	**11**
Topology	SE	SE/Diff	Diff	Diff	**Diff**

*Simulated result excluding buffer. Full power: 215 mW **Calculated from reported noise figure. $^+$Bandwidth as measured relative to baseband gain.

referred noise current. These performance characteristics make this design a possible candidate for future high speed optical receivers.

ACKNOWLEDGMENT

The authors would like to acknowledge Ciena for providing access to the design kit, simulations tools, services for fabrication, and measurement facilities.

REFERENCES

[1] R. Liu and H. Wang, "DC-to-15- and DC-to-30-GHz CMOS distributed transimpedance amplifiers," *2004 IEEE Radio Frequency Integrated Circuits (RFIC) Systems. Digest of Papers*, 2004, pp. 535-538

[2] K. Vasilakopoulos, S. P. Voinigescu, P. Schvan, P. Chevalier and A. Cathelin, "A 92GHz bandwidth SiGe BiCMOS HBT TIA with less than 6dB noise figure," *2015 IEEE Bipolar/BiCMOS Circuits and Technology Meeting - BCTM*, 2015, pp. 168-171

[3] S. Giannakopoulos, Z. S. He, I. Darwazeh and H. Zirath, "Differential common base TIA with 56 dB Ohm gain and 45 GHz bandwidth in 130 nm SiGe," *2017 IEEE Asia Pacific Microwave Conference (APMC)*, 2017, pp. 1107-1110

[4] Y. -H. Chien, K. -L. Fu and S. -I. Liu, "A 3–25 Gb/s Four-Channel Receiver With Noise-Canceling TIA and Power-Scalable LA," in *IEEE Transactions on Circuits and Systems II: Express Briefs*, vol. 61, no. 11, pp. 845-849, Nov. 2014

[5] E. Bloch, H. -c. Park, Z. Griffith, M. Urteaga, D. Ritter and M. J. W. Rodwell, "A 107 GHz 55 dB-Ohm InP Broadband Transimpedance Amplifier IC for High-Speed Optical Communication Links," *2013 IEEE Compound Semiconductor Integrated Circuit Symposium (CSICS)*, 2013, pp. 1-4

[6] W. Zhuo et al., "A capacitor cross-coupled common-gate low-noise amplifier," in *IEEE Transactions on Circuits and Systems II: Express Briefs*, vol. 52, no. 12, pp. 875-879, Dec. 2005

[7] M. S. Khalili and M. Jalali, "A capacitor cross-coupled differential cascade low-noise amplifier," *2012 IEEE International Conference on Electronics Design, Systems and Applications (ICEDSA)*, 2012, pp. 212-215

[8] J. Chong and D. S. Ha, "A 100 Gb/s transimpedance amplifier with diode-connecting input-resistance-reduction in 32 nm CMOS technology," *2015 IEEE 58th International Midwest Symposium on Circuits and Systems (MWSCAS)*, 2015, pp. 1-4

[9] M. N. Ahmed, J. Chong and D. S. Ha, "A 100 Gb/s transimpedance amplifier in 65 nm CMOS technology for optical communications," *2014 IEEE International Symposium on Circuits and Systems (ISCAS)*, 2014, pp. 1885-1888

[10] E. Bloch, H. -c. Park, Z. Griffith, M. Urteaga, D. Ritter and M. J. W. Rodwell, "A 107 GHz 55 dB-Ohm InP Broadband Transimpedance Amplifier IC for High-Speed Optical Communication Links," *2013 IEEE Compound Semiconductor Integrated Circuit Symposium (CSICS)*, 2013

A 200-325 GHz Gain-Boosted J-Band Low-Noise Amplifier in a 130 nm SiGe BiCMOS Technology

Manuel Koch
Institute for Electronics Engineering
Friedrich-Alexander-Universität Erlangen-Nürnberg
Erlangen, Germany
manuel.koch@fau.de

Sascha Breun
Institute for Electronics Engineering
Friedrich-Alexander-Universität Erlangen-Nürnberg
Erlangen, Germany
sascha.breun@fau.de

Robert Weigel
Institute for Electronics Engineering
Friedrich-Alexander-Universität Erlangen-Nürnberg
Erlangen, Germany
robert.weigel@fau.de

Abstract—This paper presents a wideband low-noise amplifier covering the complete J-Band up to the band edge of 325 GHz. A peak gain of 17.4 dB is achieved by a four-stage cascode-based prototype using inductive and capacitive gain boosting techniques. It is manufactured in a 130 nm SiGe BiCMOS technology with f_t/f_{max} of 350 GHz/450 GHz, respectively. Zero-Ohm lines are applied to bias the amplifier and low-loss Marchand baluns facilitate single-ended measurements. At both edges of the measured frequency range, a gain of at least 17 dB is shown, while a minimum gain of 12.1 dB is reported. Simulations predict a noise figure of 13.1 dB to 17.2 dB and an input-referred compression point better than -23 dBm, making the amplifier suitable for sub-terahertz radar and wireless communication within IEEE 802.15.3d frequency bands. A core chip area of $250 \times 230 \, \mu m^2$ and a DC power of 162 mW are required.

Index Terms—BiCMOS integrated circuits, Millimeter wave integrated circuits, Broadband amplifiers, Low-noise amplifiers, Baluns.

I. INTRODUCTION

The increasing performance of high-speed silicon-based semiconductor technologies leads to higher data rates requiring huge bandwidths that are available in millimeter and submillimeter wave bands. The beginning standardization of sub-terahertz communication by the IEEE standard 802.15.3d shows the interest in the J-Band for 6G and beyond communication. Due to the wide bandwidths it is also very attractive for high precision radar systems. State of the art, production grade semiconductor processes offer a f_t/f_{max} performance of approximately 300/500 GHz, which enables systems operating at millimeter wave frequencies [1]. Capabilities of radio frequency (RF) systems, especially at very high frequencies, are vastly determined by the performance of power amplifier (PA) and low-noise amplifier (LNA) building blocks. Therefore, a wideband LNA, close to the processes upper frequency limit, is investigated by this work. The design is based on IHPs SG13G2 process offering f_t/f_{max} of 350/450 GHz.

Despite the excellent RF capabilities, additional gain boosting techniques are used to further improve the performance above 300 GHz. Accordingly, Section II describes gain enhancement methods of the differential cascode core cell, the design of critical layouts and components, followed by measurement results in Section III and a conclusion.

II. CIRCUIT DESIGN

A. Amplifier Core

Due to the targeted frequency range, a cascode topology is chosen and the transistors are biased at a high current density of 2.5 mA per unit element to achieve sufficient gain and suppress the noise of following stages. A small multiplier of $2 \times 0.07 \times 0.9 \, \mu m^2$ keeps parasitic capacitances as low as possible. To further increase the gain, an inductance is added to the base of the common base stage. As a big inductance can lead to an output reflection magnitude greater than unity, care must be taken when sizing the inductor to prevent instabilities [2]. For this design, an inductance of 10.6 pH, with a parasitic resistance of $2 \, \Omega$, was chosen. Capacitive neutralization is considered as a second gain boosting technique. Because of the small transistor size, a capacitance of 2 fF is sufficient. The modified cascode is depicted in Fig. 1a. All combinations of the two techniques are compared by means of the maximum available gain of an idealized core cell, which is shown in Fig. 2. It can be seen that the individual techniques increase the gain only moderately, but by combining them a gain increase of almost 5 dB at 300 GHz is achieved.

To avoid additional parasitics in the layout, the common emitter and common base transistors are placed at an angle of 90°, which simplifies capacitive cross coupling. A 3D view of the layout, with hidden ground plane, is provided in Fig. 1b. The input signal is fed into the circuit through a via stack that integrates the DC blocking metal-insulator-metal (MIM) capacitor C_{C1} and connects with the base of T_1. The collector of transistor T_1 is cross-coupled with the base of its counterpart T_2 by the capacitor C_1, which is

SiRF 2024

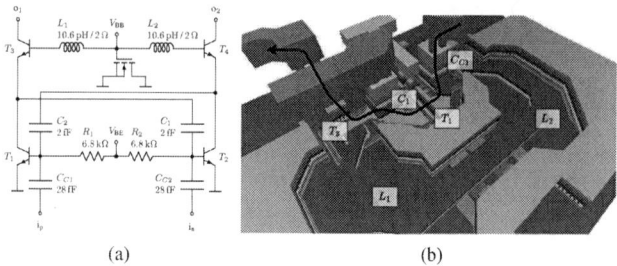

(a)   (b)

Fig. 1. (a) Circuit representation of the core cell with base inductance and capacitive neutralization boosting techniques. For the final amplifier, bias resistors and DC coupling capacitors are omitted for the first stage. (b) Rendering of the core cell with hidden ground layer and signal flow passing the components C_{C1}, T_1, C_1 and T_3.

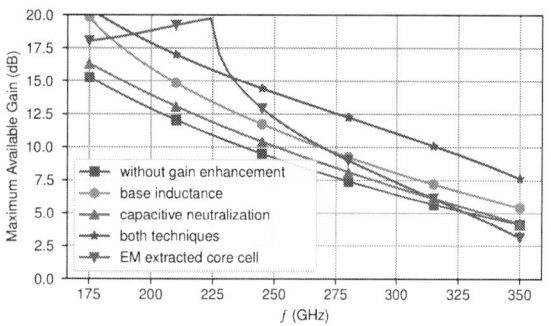

Fig. 2. Simulated maximum available gain of the schematic-level core cell with different gain enhancement techniques and the EM simulated core cell with both techniques.

realized by the overlapping areas of Metal 2 (gray) and Metal 3 (red) underneath the other via stack. The gain-boosting base inductance L_1 of transistor T_3 is realized by a $30\,\mu m$ long transmission line on stacked layers Metal 2 and Metal 3, thereby avoiding additional vias. The maximum available gain, resulting from an EM simulation of the entire core cell, is plotted together with the results of the idealized circuit in Fig. 2.

B. Interstage Matching

As a single core cell does not provide enough gain, an interstage matching network is required for cascading multiple core cells. To achieve an overall wideband gain characteristic, a T-type transmission line network is used. Due to the high frequency, lengths of matching structures become a sensitive parameter and precise lengths are required that are commonly achieved by using virtual grounds in the symmetry plane of circuits. However, this technique can not be applied here because of limited space due to the base inductances L_1 and L_2. As a solution, zero-Ohm lines are chosen to function as RF shorts and to apply the supply voltage [3]. To prevent impedance peaking at multiples of $\lambda/2$, a DC feeding zero-Ohm line was complemented by two open-ended zero-Ohm stubs. The structure, shown in Fig. 3a, uses a MIM layer directly between the thick metal and the ground layer. The

input impedance is determined by EM simulation for different combinations of stubs. Figure 3b shows that the stubs are able to further reduce the input impedance for a specific frequency range. The stub lengths were chosen to provide a very low impedance, even for harmonic frequencies.

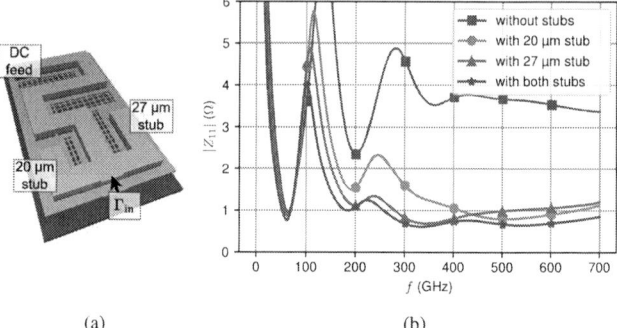

(a)   (b)

Fig. 3. (a) Rendering of the zero-Ohm line based RF short with dual open-ended stubs and (b) EM simulated input impedance.

C. Baluns

As test and measurement interfaces are commonly single-ended, Marchand baluns are used for interfacing with the differential LNA. Due to the short wavelength, a broadside coupled architecture without bends is chosen. The physical implementation has a length of $257\,\mu m$ and is shown in Fig. 4. The balun was optimized for minimum insertion loss since the LNAs noise figure (NF) is directly affected by it. Insertion loss data, plotted in Fig. 6, has been extracted from a back-to-back test structure used for measurement and simulation. Comparing insertion loss, simulated phase imbalance and amplitude imbalance the presented balun achieves similar or better performance than the designs in [4] and [5].

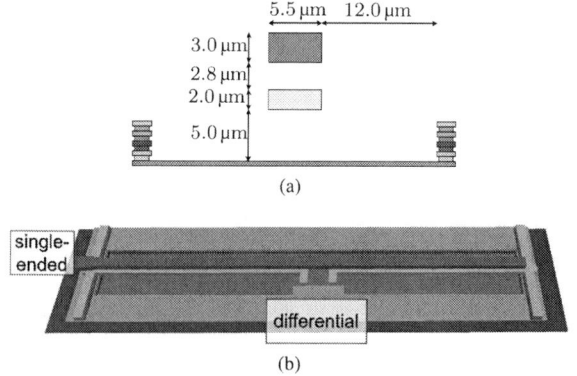

Fig. 4. (a) Cross section and (b) rendering of the straight Marchand balun using MIM capacitors as RF shorts.

D. Amplifier Design

The LNA consists of four cascaded core cells and the networks for input, interstage and output matching. The complete

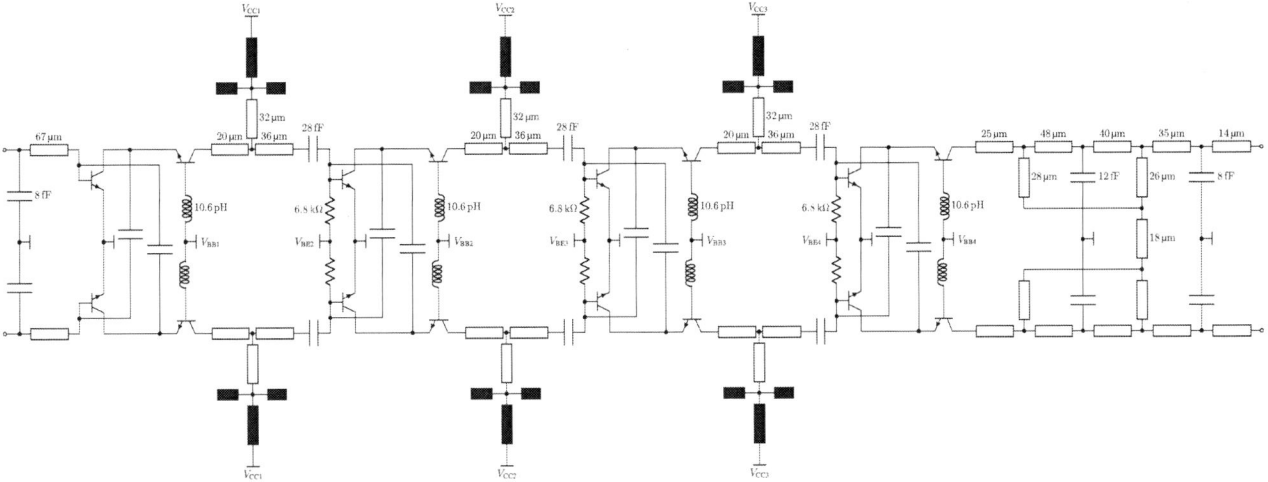

Fig. 5. Circuit diagram of the presented 4-stage LNA with T-type interstage matching, zero-Ohm line based RF shorts (black) and dual stub output matching.

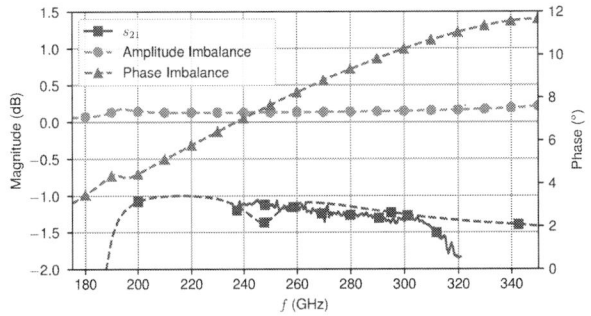

Fig. 6. Simulated (dashed) and measured (solid) insertion loss of a single balun, extracted from a back-to-back test structure, as well as simulated amplitude and phase imbalance.

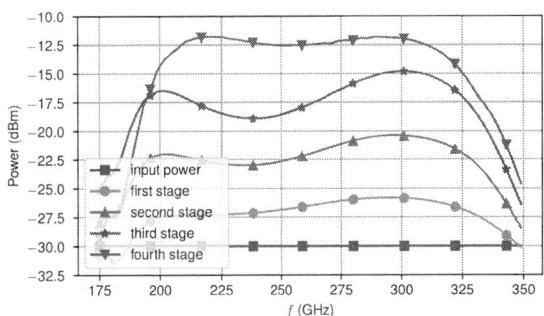

Fig. 7. Power at the output of each stage obtained from harmonic balance simulations with $-30\,\mathrm{dBm}$ input power.

circuit, except for the biasing circuits and baluns, is shown in Fig. 5. To avoid additional noise sources at the amplifier input, the Marchand balun is used to bias the first stage instead of bias resistors. While the interstage matching is optimized for the high frequency portion of the band, the output matching network needs to cover a very wide bandwidth over the whole J-Band. A matching network with dual stubs was used for this purpose since transmission line lengths are longer than those of a comparable single stub network. The power after each stage, determined by harmonic balance simulation for an input power of $-30\,\mathrm{dBm}$, is shown in Fig. 7.

III. MEASUREMENT RESULTS

A prototype of the LNA was manufactured by IHP, which is shown in Fig. 8. The presented baluns are used to connect the amplifier with $50\,\Omega$ transmission lines and GSG pads, that employ shorted stubs that resonate with the pad capacitance. Supply voltages and bias currents of all four stages can be controlled individually by the DC pads.

Measurement data was obtained by on-chip characterization using two $100\,\mu\mathrm{m}$ pitch WR-3 probes and ZVA-Z325

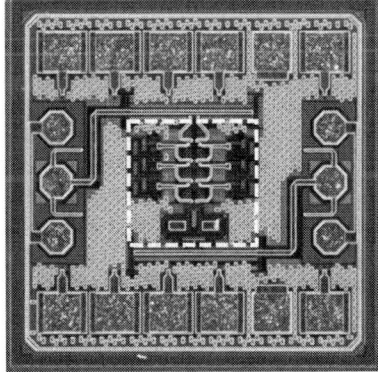

Fig. 8. Micrograph of the fabricated LNA with highlighted core area of $250 \times 230\,\mu\mathrm{m}^2$, input GSG pads (left) and output GSG pads (right).

extenders. The input power applied to the circuit during small signal measurement was set to approximately -28 dBm, while, according to simulations, the 1 dB input compression point varies from -18.5 dBm to -23 dBm over frequency. A separate chip with calibration structures was used to remove the influence of the RF pads. Additionally, in- and output transmission lines have been deembedded from the following results.

The simulated and measured frequency response, as well as the simulated NF, are shown in Fig. 9. Due to the limited frequency range of the measurement equipment, the amplifiers expected frequency range from 200 GHz to 325 GHz was only partially characterized. In the measured range, an input matching, better than -10 dB, is reported by simulation and measurement. However, at the output only moderate matching is observed, which is likely to cause the drop in gain at the center of J-Band. A good correlation of measurement and simulation results for the output matching in power off state suggests that the error could be caused by modeling inaccuracies. Despite that, the gain at the edges of the measurement range matches well with the simulation, indicating valid modeling of the core cells.

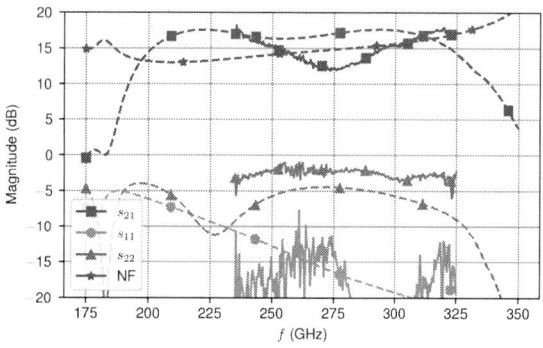

Fig. 9. Measured (solid lines) and simulated (dashed lines) small signal response and simulated NF of the LNA.

IV. CONCLUSION

This paper presents a gain-boosted LNA with a simulated bandwidth from 200 to 325 GHz, exceeding the frequency range of the WR-3 frequency extenders. It features a peak gain of 17.4 dB and a simulated NF ranging from 13.1 to 17.2 dB across the targeted J-Band. Although the outputs dual band matching is shifted in comparison to simulation results, the simulated and measured gain coincides well at the lowest and highest characterized frequency, which shows that the employed gain boosting techniques allow for improved gain at very high frequencies. With a simulated NF and gain of 17 dB the amplifier is able to enhance the noise performance of subharmonic mixers, e.g. as presented in [6], from 20 dB to a system NF of 17.2 dB.

An overview of wideband amplifiers operating close to or above 300 GHz is given in Table I. In comparison, the demonstrated amplifier achieves a slightly better NF and a much wider bandwidth than [7]. Regarding amplifiers using a superior process, the presented amplifier is not able to compete in terms of NF [8], [9].

TABLE I
WIDEBAND MILLIMETER WAVE LNAS CLOSE TO OR ABOVE 300 GHz.

	This	**[8]**	**[7]**	**[9]**
Technology	130 nm BiCMOS	130 nm BiCMOS	130 nm BiCMOS	130 nm BiCMOS
f_t/f_{max}	350/450	470/700	300/450	470/650
Min. frequency	200 GHz⁺	257 GHz	278 GHz	131 GHz
Max. frequency	325 GHz⁺	325 GHz ⁻	302 GHz	277 GHz
Peak gain	17.4 dB	10.8 dB	12.9 dB	34.6 dB
Min. NF	13.1 dB⁺	11 dB⁺	16 dB⁺	8.4 dB
NF 280 GHz	15.0 dB⁺	11.1 dB⁺*		12.5 dB*
P_{DC}	162 mW	119 mW	136 mW	152 mW
Core area (mm²)	0.058#	0.026#	0.016#	0.092#

⁻ no data provided above 325 GHz, but max. frequency is higher;
⁺ simulated; # without baluns; * extracted from diagram

ACKNOWLEDGMENT

The authors thank IHP for the production of the chips.

REFERENCES

[1] T. Maiwald, T. Li, G.-R. Hotopan, K. Kolb, K. Disch, J. Potschka, A. Haag, M. Dietz, B. Debaillie, T. Zwick, K. Aufinger, D. Ferling, R. Weigel, and A. Visweswaran, "A Review of Integrated Systems and Components for 6G Wireless Communication in the D-Band," in *Proc. IEEE*, 2023, pp. 220–256.

[2] A. C. Ulusoy, P. Song, W. T. Khan, M. Kaynak, B. Tillack, J. Papapolymerou, and J. D. Cressler, "A SiGe D-Band Low-Noise Amplifier Utilizing Gain-Boosting Technique," in *IEEE Microw. Wireless Compon. Letters*, 2015, pp. 61–63.

[3] D. Fritsche, C. Carta, and F. Ellinger, "A Broadband 200 GHz Amplifier with 17 dB Gain and 18 mW DC-Power Consumption in 0.13 μm SiGe BiCMOS," in *IEEE Microw. Wireless Compon. Lett.*, 2014, pp. 790–792.

[4] F. Ahmed, M. Furqan, and A. Stelzer, "A 200–325 GHz Wideband, Low-Loss Modified Marchand Balun in SiGe BiCMOS Technology," in *Proc. Eur. Microw. Conf. (EuMC)*, 2015, pp. 40–43.

[5] H. Sakai, K. Takano, S. Hara, A. Kasamatsu, and Y. Umeda, "A 220–330 GHz Wideband, Low-Loss and Small Marchand Balun with Ground Shields in SiGe BiCMOS Technology," in *Proc. IEEE MTT-S Int. Microw. RF Conf. (IMARC)*, 2021, pp. 1–4.

[6] A.-M. Schrotz, S. Breun, V. Issakov, M. Dietz, and R. Weigel, "A 220-325 GHz Subharmonic Receiver with 14.8 dB Peak Conversion Gain for FMCW Radar in SiGe BiCMOS Technology," in *Proc. IEEE 22nd Top. Meet. Silicon Monolithic Integr. Circuits RF Syst. (SiRF)*, 2022, pp. 74–77.

[7] S. P. Singh, T. Rahkonen, M. E. Leinonen, and A. Pärssinen, "A 290 GHz Low Noise Amplifier Operating above fmax/2 in 130 nm SiGe Technology for Sub-THz/THz Receivers," in *Proc. IEEE Radio Freq. Integr. Circuits Symp. (RFIC)*, 2021, pp. 223–226.

[8] A. Gadallah, M. H. Eissa, T. Mausolf, D. Kissinger, and A. Malignaggi, "A 300 GHz Low-Noise Amplifier in 130 nm SiGe SG13G3 Technology," in *IEEE Microw. Wireless Compon. Letters.* IEEE, 2022, pp. 331–334.

[9] M. Andree, J. Grzyb, B. Heinemann, and U. Pfeiffer, "A D-Band to J-Band Low-Noise Amplifier with High Gain-Bandwidth Product in an Advanced 130 nm SiGe BiCMOS Technology," in *Proc. IEEE Radio Freq. Integr. Circuits Symp. (RFIC)*, 2023, pp. 137–140.

Analysis of a SiGe BiCMOS Detector for a Broadband mmW-integrated EPR Spectrometer

Selina Eckel
Institute of Radio Frequency Engineering and Electronics
Karlsruhe Institute of Technology
Karlsruhe, Germany
selina.eckel@kit.edu

Ahmet Çağrı Ulusoy
Institute of Radio Frequency Engineering and Electronics
Karlsruhe Institute of Technology
Karlsruhe, Germany
cagri.ulusoy@kit.edu

Abstract—**We present the analysis of a 130-nm SiGe BiCMOS millimeter-wave (mmW) detector, which will be used in an on-chip transmission-based Electron Paramagnetic Resonance (EPR) spectrometer. The EPR effect on the transmitted RF signal (1 to 40 GHz) is modeled by amplitude modulation (AM) with a single frequency, corresponding to the modulation of the magnetic field, and a small modulation index, corresponding to the absorption of the sample. In contrast to most analyses of mmW detectors, this work focuses on AM demodulation performance of a transistor in common emitter configuration under these conditions. The proportionality between the detected EPR signal and the effective transconductance is shown. We investigate the dependency of the optimum base emitter bias on the input power and the carrier frequency.**

Index Terms—**power detector, amplitude modulation, EPR, transmission-based, broadband**

I. INTRODUCTION

Millimeter-wave (mmW) power detectors have a wide range of applications. They are used in radiometry and passive imagers, for instance making use of the Dicke radiometer topology, for various applications in space, but also security imagers [1]. Another important application is built-in self test for vast deployment of millimeter-wave transceivers, where accurate power detectors are an essential part [2]. Power detectors are also building blocks in software-defined transceivers [3]. In this paper, we present a new and interesting application area of a wideband integrated power detector for magnetic resonance spectroscopy. The paper analyses and experimentally demonstrates the nuances for the optimization of a detector for on-chip spectrometer technology.

Electron Paramagnetic Resonance (EPR) spectroscopy is used to study paramagnetic compounds, e.g. materials containing free electrons. It has a wide range of applications such as the detection of point defects in solids, dosimetry and investigation of biological samples. The EPR technique is based on the inherent magnetic moment of an electron. When an electron is placed in an external magnetic field B_0, its energy splits into two energy levels, which is called the Zeeman effect. A change of the energy level ΔE of the electron can be induced by absorbing a photon with energy $h\nu$:

$$\Delta E = g_e \mu_B B_0 = h\nu, \qquad (1)$$

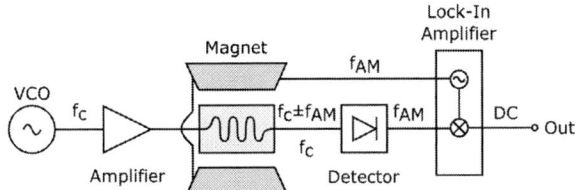

Fig. 1: Overview of the system architecture.

with the free electron g-factor g_e, the Bohr magneton μ_B and the Planck constant h. This effect is visible in the EPR spectrum as microwave absorption versus either the external magnetic field B_0 or the microwave frequency ν [4]. In conventional EPR spectrometers, the sample is placed inside a cavity or resonator, which provides the microwave energy. Due to the use of a resonator, the system only works in a limited bandwidth. To achieve a bandwidth from 1 to 40 GHz a transmission line can be used instead [5]. This enables EPR experiments at different frequencies to get better insights into unknown samples. By sweeping the frequency instead of the magnetic field it is possible to distinguish between field-dependent and field-independent processes, like the Zeeman effect and the Zero-Field splitting [6]. The system architecture for the envisioned broadband continuous wave transmission-based EPR spectrometer can be seen in Fig. 1 [7]. The sample is placed on top of a transmission line, which transports the microwave energy provided by a voltage controlled oscillator (VCO) with frequency f_c to the sample. A lock-in amplifier is used to modulate the external magnetic field with a frequency f_{AM}. As a result, two sidebands at $f_c \pm f_{AM}$ are present at the input of the detector in addition to the carrier. The amplitude of the sidebands is proportional to the absorption of the sample and the modulation amplitude of the magnetic field. After demodulation by the detector, the signal component at f_{AM} is filtered and down-converted to DC by the lock-in amplifier. In the next section, we describe the detector circuit and the simulation and measurement setup. In section III, we analyze the circuit behavior when varying the supply voltages, the input power and the carrier frequency and calculate its detection limits.

979-8-3503-4331-1/24 $31.00 © 2024 IEEE

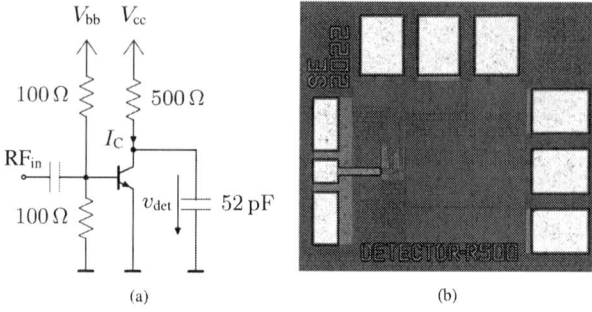

(a)   (b)

Fig. 2: (a) The schematic of the detector circuit with a two finger transistor in a common emitter configuration is displayed. (b) The photograph of the fabricated detector with a chip area of $480\,\mu\mathrm{m} \times 398\,\mu\mathrm{m}$ is seen.

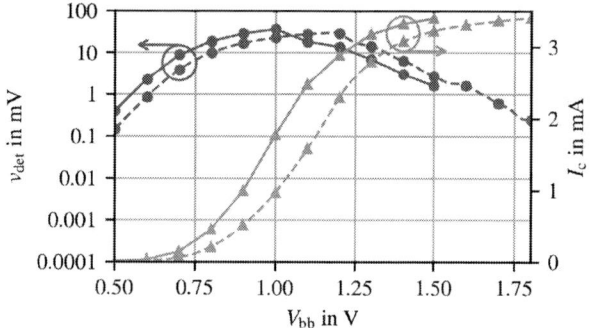

Fig. 3: The detection voltage at f_{AM} (blue curve, circles) and the collector current (orange curve, triangles) w.r.t to the base bias voltage are plotted. The curves are recorded at $P_{\mathrm{in}} = 4\,\mathrm{dBm}$, $f_{\mathrm{c}} = 1\,\mathrm{GHz}$ and $V_{\mathrm{cc}} = 2\,\mathrm{V}$. The solid curves show measurement results and the dashed curves show simulation results.

II. DETECTOR OVERVIEW

A. Circuit Implementation

The schematic of the detector can be seen in Fig. 2a which is designed in IHP $0.13\,\mu\mathrm{m}$ SiGe BiCMOS. A two finger transistor is used in common emitter configuration. To achieve a wide bandwidth, the input is matched resistively to $50\,\Omega$. At the output a large resistor of $500\,\Omega$ is used to get a high voltage amplification. The output capacitor to ground filters out the high frequency components of the output signal. The unused area between the active core and the pads at the output are used for this on-chip capacitor which results in a value of $52\,\mathrm{pF}$. In Fig. 2b the photograph of the fabricated detector is shown. The chip area including the pads is $480\,\mu\mathrm{m} \times 398\,\mu\mathrm{m}$.

B. Simulation and Measurement Setup

To perform the simulations, harmonic balance analysis in Cadence Virtuoso is used. The core of the system is electromagnetic (EM) simulated with Cadence EMX. The measurement setup consists of a signal source with a frequency up to $40\,\mathrm{GHz}$ with a maximum output power of $P_{\mathrm{in}} = 5\,\mathrm{dBm}$ over the whole band. The output voltage is measured with an oscilloscope. A Fast Fourier Transform (FFT) is performed to extract the detected voltage at f_{AM}. The bias voltages V_{bb}

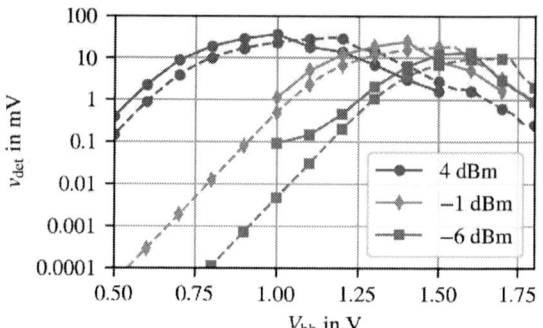

Fig. 4: The influence of the input power on the optimum base bias voltage is shown. All curves are recorded at $V_{\mathrm{cc}} = 2\,\mathrm{V}$ and $f_{\mathrm{c}} = 1\,\mathrm{GHz}$. The solid curves show measurement results and the dashed curves show simulation results.

and V_{cc} are controlled by a Source Measurement Unit. The RF power level is calibrated up to the input of the RF probe. All detection voltages shown are peak values. If not otherwise stated, a modulation depth of $m = 0.01$ and a modulation frequency of $f_{\mathrm{AM}} = 100\,\mathrm{kHz}$ is used. All measurement results are displayed in solid and the simulation results in dashed curves.

III. CIRCUIT ANALYSIS

The detection of the AM input signal is based on the non-linear characteristics of the collector current I_{C} with respect to the base emitter voltage V_{BE}. In Fig. 3, the detected output voltage at f_{AM} and the collector current I_{C} are shown w.r.t. to the supply voltage V_{bb}. It is important to note that the DC current I_{C} is measured and simulated in large signal mode, e.g. with an AM signal at the input. For this measurement a calibrated input power of $P_{\mathrm{in}} = 4\,\mathrm{dBm}$ is applied. It is observed that v_{det} reaches its maximum when the slope of I_{C} is largest. Simulation and measurement results show that v_{det} is proportional to $\mathrm{d}I_{\mathrm{C}}/\mathrm{d}V_{\mathrm{BE}}$ which is given by the effective transconductance $G_{\mathrm{m,eff}}$. The shift of the measured v_{det} and I_{C} curves towards lower V_{bb} can be explained by manufacturing tolerances of the three resistors and the additional inductance introduced by the DC probe. The maximum detection voltage depends on multiple factors, like the input power, the output resistor, the number of transistor fingers and the supply voltages. In the following, the influence of the supply voltages is further studied. The optimum V_{bb} depends strongly on the input power, as can be seen in Fig. 4. With higher input powers the optimum V_{bb} becomes smaller. This results from the transistors earlier conduction so the maximum $G_{\mathrm{m,eff}}$ is reached for lower V_{bb}. The influence of the supply voltage at the collector V_{cc}, i.e. the collector emitter voltage V_{CE}, can be understood by the transistor equation:

$$I_{\mathrm{C}} = I_{\mathrm{S}} \cdot \exp\left(\frac{V_{\mathrm{BE}}}{V_{\mathrm{T}}}\right)\left(1 + \frac{V_{\mathrm{CE}}}{V_{\mathrm{A}}}\right) \qquad (2)$$

with I_{S}: saturation current, V_{T}: temperature voltage and V_{A}: Early voltage [8]. Deriving (2) w.r.t. V_{BE} shows that an

979-8-3503-4331-1/24 $31.00 © 2024 IEEE

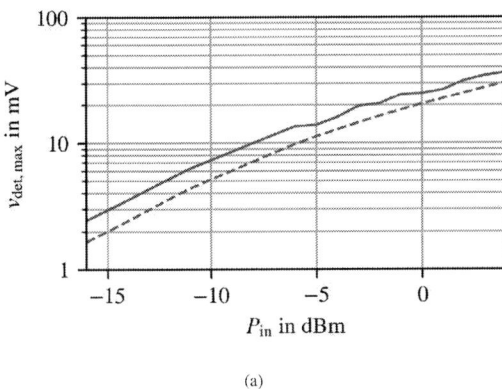
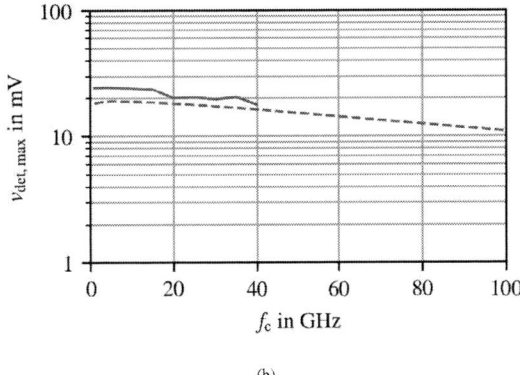

(a)

(b)

Fig. 5: Maximum detection voltage at f_{AM} versus (a) input power at $f_c = 1\,\mathrm{GHz}$ and (b) carrier frequency for $P_{in} = -1\,\mathrm{dBm}$. All curves are recorded at $V_{cc} = 2\,\mathrm{V}$, while V_{bb} was chosen for each data point for the maximum output voltage. The solid curves show measurement results and the dashed curves show simulation results.

increased V_{CE} leads to an increased $G_{m,eff}$. The maximum allowed collector current determines the maximum supply voltage V_{cc}. For the analysis of this circuit $V_{cc} = 2\,\mathrm{V}$ is chosen to operate the device safely below the breakdown.

A. Influence of Input Power

In Fig. 5a the maximum detection voltage is plotted w.r.t. the input power P_{in}. It follows the typical square law characteristic. Each data point is obtained at a different supply voltage V_{bb}, the higher the input power the lower the optimum V_{bb} (see Fig. 4). It can be observed that a high input power achieves higher maximum v_{det} because it leads to higher I_C and thus higher $G_{m,eff}$. For an EPR spectrometer, a high power is in general favorable as long as the sample is not saturated [4]. A higher power leads to a higher microwave absorption in the sample, which results in a stronger EPR signal. The saturation limit strongly depends on the sample, so the input power as well as the base bias voltage need to be adjustable.

B. Broadband Capabilities

As mentioned in the introduction, the envisioned on-chip EPR spectrometer should work in a broad range of frequencies. Fig. 5b shows the frequency behavior for the measurements up to $40\,\mathrm{GHz}$. The simulated results indicate a $6\,\mathrm{dB}$ bandwidth of around $100\,\mathrm{GHz}$. The results were obtained at a calibrated input power of $-1\,\mathrm{dBm}$ for all frequencies. At higher frequencies, the losses of the detector circuit are higher due to the parasitic components of the devices and their connections. This results in different optimum values for V_{bb} for a maximum detection voltage. Similar to Fig. 5a, an increase in frequency leads to higher losses and a higher optimum value for V_{bb}. Using the detector in the EPR spectrometer, the base bias voltage needs to be chosen according to the carrier frequency. By choosing the carrier frequency it is possible to focus on different EPR interactions. For example, the field-dependent Zeeman interaction dominates at higher carrier frequencies, while the field-independent Zero-Field splitting can dominate at lower carrier frequencies [6].

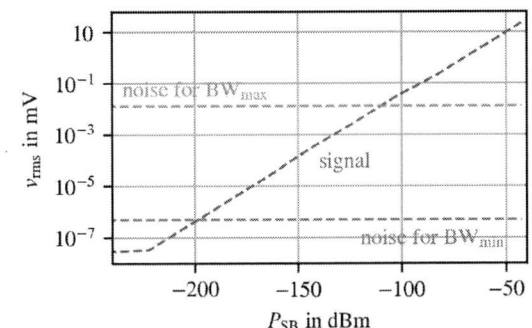

Fig. 6: The blue curve shows the rms detection voltage at f_{AM} versus the input power of a single sideband. The orange (green) curve shows the upper (lower) detection limit when using the largest (smallest) bandwidth of the lock-in amplifier. All curves are simulated at $P_{in} = 4\,\mathrm{dBm}$, $f_c = 1\,\mathrm{GHz}$, $V_{bb} = 1.18\,\mathrm{V}$ and $V_{cc} = 2\,\mathrm{V}$.

C. Detection Limit

An important parameter that determines the performance of an EPR spectrometer is its sensitivity, which in this context is given by the minimum detectable AM sideband power or modulation index m. The simulation of the output noise voltage at f_{AM} results in $v_{noise} = 28.4\,\mathrm{nV}/\sqrt{\mathrm{Hz}}$ for an input power of $P_{in} = 4\,\mathrm{dBm}$ at $f_c = 1\,\mathrm{GHz}$ and the optimum bias voltage $V_{bb} = 1.18\,\mathrm{V}$. The noise simulation is based on a harmonic balance analysis. To determine the sensitivity of the detector in combination with a lock-in amplifier, the output noise voltage is multiplied by the square root of the minimum and maximum bandwidth of a typical lock-in amplifier. The resulting output noise voltages and the simulated rms output voltage w.r.t the input power of one of the sidebands is shown in Fig. 6. The detector in combination with the lock-in amplifier has a signal to noise ratio (SNR) of 1 at $P_{SB} = -110\,\mathrm{dBm}$ ($m \approx 4 \cdot 10^{-5}$) for the largest bandwidth of $B_{max} = 206\,\mathrm{kHz}$ and $P_{SB} = -200\,\mathrm{dBm}$ ($m \approx 2 \cdot 10^{-10}$) for the smallest bandwidth $B_{min} = 276\,\mu\mathrm{Hz}$.

979-8-3503-4331-1/24 $31.00 © 2024 IEEE

IV. CONCLUSION

In this paper, we have analyzed a mmW power detector to be used in an EPR spectrometer. To model the EPR effect, an AM input signal is applied. It is shown that the detected output voltage at f_{AM} is proportional to the effective transconductance $G_{m,eff}$. The maximum v_{det} depends strongly on the base emitter voltage and needs to be adapted for specific input powers and carrier frequencies. Noise simulations are performed which show promising results for sufficient sensitivity for EPR experiments.

ACKNOWLEDGMENT

The authors would like to thank the German Research Foundation (DFG) for its funding within the Collaborative Research Center 1527 "High Performance Compact Magnetic Resonance" (HyPERiON).

REFERENCES

[1] M. A. Janssen, J. E. Oswald, S. T. Brown, S. Gulkis, S. M. Levin, S. J. Bolton, M. D. Allison, S. K. Atreya, D. Gautier, A. P. Ingersoll, J. I. Lunine, G. S. Orton, T. C. Owen, P. G. Steffes, V. Adumitroaie, A. Bellotti, L. A. Jewell, C. Li, L. Li, S. Misra, F. A. Oyafuso, D. Santos-Costa, E. Sarkissian, R. Williamson, J. K. Arballo, A. Kitiyakara, A. Ulloa-Severino, J. C. Chen, F. W. Maiwald, A. S. Sahakian, P. J. Pingree, K. A. Lee, A. S. Mazer, R. Redick, R. E. Hodges, R. C. Hughes, G. Bedrosian, D. E. Dawson, W. A. Hatch, D. S. Russell, N. F. Chamberlain, M. S. Zawadski, B. Khayatian, B. R. Franklin, H. A. Conley, J. G. Kempenaar, M. S. Loo, E. T. Sunada, V. Vorperion, and C. C. Wang, "Mwr: Microwave radiometer for the juno mission to jupiter," *Space Science Reviews*, vol. 213, pp. 139–185, 11 2017.

[2] A. Valdes-Garcia, R. Venkatasubramanian, J. Silva-Martinez, and E. Sanchez-Sinencio, "A broadband cmos amplitude detector for on-chip rf measurements," *IEEE Transactions on Instrumentation and Measurement*, vol. 57, pp. 1470–1477, 7 2008.

[3] C.-H. Wang, H.-Y. Chang, P.-S. Wu, K.-Y. Lin, T.-W. Huang, H. Wang, and C. H. Chen, "A 60ghz low-power six-port transceiver for gigabit software-defined transceiver applications." IEEE, 2 2007, pp. 192–596.

[4] V. Chechik, E. Carter, and D. Murphy, *Electron Paramagnetic Resonance.* Oxford University Press, USA, 7 2016.

[5] Y. Wiemann, J. Simmendinger, C. Clauss, L. Bogani, D. Bothner, D. Koelle, R. Kleiner, M. Dressel, and M. Scheffler, "Observing electron spin resonance between 0.1 and 67 ghz at temperatures between 50 mk and 300 k using broadband metallic coplanar waveguides," *Applied Physics Letters*, vol. 106, 5 2015.

[6] S. K. Misra, *Multifrequency Electron Paramagnetic Resonance*, 2019.

[7] B. Miksch, M. Dressel, and M. Scheffler, "Cryogenic frequency-domain electron spin resonance spectrometer based on coplanar waveguides and field modulation," *Review of Scientific Instruments*, vol. 91, 2 2020.

[8] U. Tietze, C. Schenk, and E. Gamm, *Halbleiter-Schaltungstechnik*, 16th ed. Springer Vieweg, 7 2019.

27 Gb/s PRBS Generator with In-Operation Programmable Taps for PMCW Radar

Florian Probst
Institute for Electronics Engineering
Friedrich-Alexander-Universität Erlangen-Nürnberg
Erlangen, Germany
florian.probst@fau.de

Andre Engelmann
Institute for Electronics Engineering
Friedrich-Alexander-Universität Erlangen-Nürnberg
Erlangen, Germany
andre.engelmann@fau.de

Robert Weigel
Institute for Electronics Engineering
Friedrich-Alexander-Universität Erlangen-Nürnberg
Erlangen, Germany
robert.weigel@fau.de

Abstract—This work presents an integrated pseudo-random binary sequence (PRBS) generator with programmable feedback designed in a 22 nm fully-depleted silicon on insulator (FDSOI) technology. It shall be used as a modulation signal generator for ultra-wideband phase-modulated continuous-wave (PMCW) radars. Compared to standard linear-feedback shift registers (LFSRs) exhibiting fixed feedback, our circuit can generate different PRBSs. With this, we can distinguish between channels in MIMO systems but also reduce sidelobes resulting from spill-over and Doppler shifts. The LFSR is built of true single-phase-clock (TSPC) flip-flops, whose feedback is controlled by a high-speed digital circuit operating at one-fifth of the PRBS clock frequency. Measurements of the circuit show that the generator achieves a maximum bit rate of 27 to 31 Gb/s, depending on the number of activated feedback taps. While consuming 2.44 mW, the proposed circuit is comparable to the current state of the art, offering better flexibility and radar performance.

Index Terms—Digitally modulated radars, fully-depleted silicon on insulator (FDSOI), phase-modulated continuous-wave (PMCW) radar, pseudo-random binary sequence (PRBS) generator.

I. INTRODUCTION

Ultra-wideband radar systems for human-machine gesture recognition offer several advantages over camera-based systems. For example, they can be used in the dark or behind a screen. Since the range resolution scales with radar bandwidth, systems operating in the D-band or higher have recently been in the scientific spotlight. While widely used frequency-modulated continuous-wave (FMCW) systems benefit from the low-noise oscillators of SiGe technologies, digital radar systems can profit from the high integration density of CMOS technologies. The technology must, therefore, have good millimeter-wave characteristics while providing high-performance digital cells to be suitable for digital D-band radar [1]. Both of these requirements are met by the 22 nm fully-depleted silicon on insulator (FDSOI) technology [2].

Although the FMCW waveform is widely used in the literature and industry, this work focuses on phase-modulated continuous-wave (PMCW). There, a pseudo-random binary

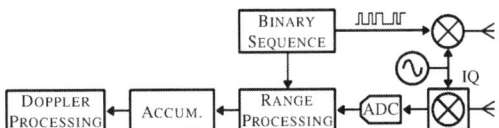

Fig. 1. PMCW radar block diagram.

sequence (PRBS) is modulated onto a fixed local oscillator (LO) frequency. At the receiver, the reflected signal is down-converted to zero-IF. Afterward, the IQ components must be digitized by an analog-to-digital converter (ADC), followed by a correlation, accumulation, and Doppler evaluation. Fig. 1 shows a PMCW radar block diagram, including the digital signal processing steps.

PMCW radar systems at 70 GHz have already been presented in the literature for automotive applications [3] and, more recently, for gesture recognition at D-band [1]. The most significant advantage over FMCW is the easy realization of multiple-input and multiple-output (MIMO) systems by distinguishing between the channels in the code domain. However, since the bit rate of the PRBS determines the radar bandwidth and, thus, the resolution, it should be chosen as high as possible, which makes its generation complex. An efficient way to generate the modulation signal regarding power and area consumption is a linear-feedback shift register (LFSR), as suggested in [1]. As the name implies, this architecture uses a shift register built from flip-flops (FFs) with a fixed XOR feedback called tap.

The top of Fig. 2 shows the typical baseband waveform for single-input and single-output (SISO) PMCW radars. The same PRBS is repeated $M \times N$ times for accumulation and range-Doppler processing, respectively. However, the results in [4] suggest that this waveform is susceptible to range sidelobes caused by spill-over and Doppler phase shifts. The authors suggest using a set of different sequences k for each accumulation block to mitigate this effect. Using the frame

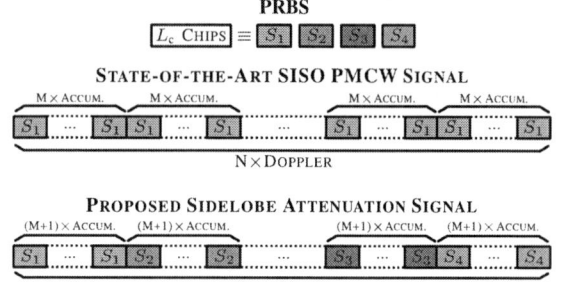

PRBS

L_c CHIPS $\equiv$ S_1 S_2 S_3 S_4

STATE-OF-THE-ART SISO PMCW SIGNAL

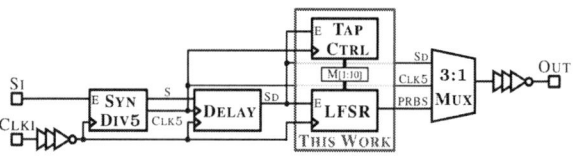

PROPOSED SIDELOBE ATTENUATION SIGNAL

Fig. 2. Transmit frame with and without code diversity as suggested in [4].

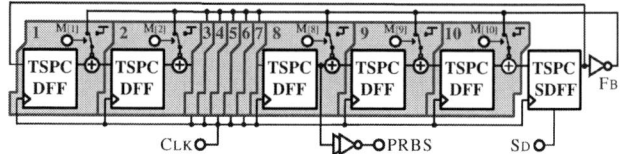

Fig. 3. Overview of the entire system, including IO buffers, synchronized divider, delay module, highlighted controllable LFSR, and MUX.

pattern in the bottom of Fig. 2 attenuates range sidelobes by $10 \log k$ dB.

The problem with the proposed radar frame is that a simple shift register with fixed feedback can no longer generate it. To this extent, this paper proposes an advanced LFSR to solve this problem, as it allows all possible feedback variants to be configured externally. Combined with a high-frequency digital circuit, the feedback can be activated or deactivated during operation so that the waveform pictured in Fig. 2 can be generated at data rates above 20 Gb/s. To the authors' knowledge, this has not yet been demonstrated in the literature.

II. CIRCUIT PRESENTATION

The proposed PRBS generator is part of the system shown in Fig. 3, which also includes a frequency divider and an adjustable delay element. Although its primary purpose is the generation of PMCW radar baseband signals, the building blocks can be evaluated individually for their functionality and power consumption. The circuit consists of two input signals: the start and clock input SI and CLKI. Both are connected to a synchronizer block, presented separately in [5], which generates a divided-by-five clock as well as a synchronized start signal. While the clock is used in the digital blocks, the start signal must pass through a delay element first. Afterward, it starts the proposed pseudo-noise generator, consisting of the programmable LFSR and the digital feedback controller. The individual output of the synchronizer, delay, and PRBS block can be connected to an output buffer via a multiplexer (MUX).

The implemented LFSR is shown in Fig. 4 and incorporates $N_{FF} = 11$ FFs with programmable feedback. Choosing a specific tap configuration can produce maximum-length sequences (MLSs) of length $L_c = 2^{N_{FF}} - 1 = 2047$, which is sufficient for our application. Mathematically, LFSRs can be

expressed by a feedback polynomial, whose coefficients are either 1 or 0. The highest existing power in the polynomial equals N_{FF}, and the coefficients describe the existence of feedback. The polynomials can be found in tables, and their number depends on N_{FF}. Our LFSR possesses 176 feedback combinations, each generating a different MLS. The last FF in the shift register is implemented as a settable DFF, whose output signal is connected to an inverter. It acts as an amplifier for the feedback signal FB to accommodate for the FF's increased fan-out and enhances its driving strength. Since FB is inverted, we need to use an XOR gate with an inverted input that can be connected to either FB or the supply voltage.

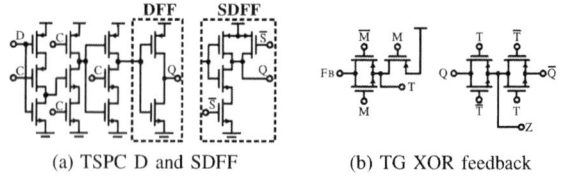

Fig. 4. LFSR's schematic with externally controllable taps.

The design only uses the dynamic true single-phase-clock (TSPC) FFs shown in Fig. 5a, whose exceptional performance in PRBS generators is demonstrated in [6]. The FF cell is merged with the controllable XOR block, which is realized through transmission gates (TGs) as shown in Fig. 5b. In several iterations of post-layout simulation, the building block's clock-to-output-delay is minimized. This procedure is intended to maximize the achieved data rate of an LFSR consisting of them [6]. The FF with controllable feedback eventually used exhibits a core area of $3.89 \times 1.47 \, \mu m^2$ and the overall LFSR, including two inverter buffers, $24.46 \times 3.21 \, \mu m^2$.

Finally, we will briefly discuss the control circuit for setting the LFSR feedback. As it is supposed to generate the radar frames shown in Fig. 2, changing the feedback during operation is required. Thus, the control circuit must be included in the timing considerations of the system. To this extent, we use the circuit pictured in Fig 6, which is similar to the outer-code generator used in [1]. It consists of a digitally realized programmable counter that controls the shift-enable signal for ten parallel ring buffers storing the desired feedback configuration. In our implementation, each ring buffer has eight memory elements, so that eight different sequences can be generated in succession during a radar frame. However, this number can easily be increased in the hardware description language (HDL) code. Ten additional TSPC FFs are placed at the ring buffer outputs for synchronization. The digital circuit

(a) TSPC D and SDFF   (b) TG XOR feedback

Fig. 5. Hardware realization of the high-speed building blocks.

979-8-3503-4331-1/24 $31.00 © 2024 IEEE

uses an internally created synchronized clock at one-fifth of the PRBS generator's frequency. The LFSR and the tap controller are started synchronously by the delay circuit's output signal S_D. The digital block, which also includes part of the delay circuit, comprises $62 \times 40.3\,\mu m^2$.

Fig. 6. Feedback control circuit realization.

Fig. 7 shows a micrograph of the $600 \times 530\,\mu m^2$ breakout circuit fabricated in 22 nm FDSOI. The design features only super-low-threshold-voltage (SLVT) transistors, exhibiting the technology's best timing performance. Applying forward body biasing can further increase their performance at the expense of power consumption. The auxiliary supply net V_{DDM} can be connected to the divider, delay, or PRBS circuit for an independent power measurement. The body bias voltage can also be externally tuned for complete circuit characterization. However, it must be noted that the digital circuit cannot be forward body biased.

III. MEASUREMENT RESULTS

During the measurement on a wafer-prober, an *E9275D* PSG vector signal generator delivers the input clock, and a *UXR0804A* high-speed oscilloscope captures the system's output signal and stores a snapshot. Afterward, it is post-processed utilizing a correlation to evaluate whether the LFSR operates correctly.

First, the PRBS generator's maximum achieved bit rate R_{Bmax} is determined as a function of the applied supply voltage. This procedure is repeated for different body biasing and tap configurations. LFSRs with eleven FFs have correct feedback configurations for one, three, five, and seven activated taps. The results in Fig. 8 show that increasing the number of taps and, thus, the load of the feedback inverter lowers R_{Bmax}. Only the results with the lowest and highest number of taps are displayed for clarity. The others are in between. Although the data rate drops significantly, the measured power consumption remains relatively constant due to the additional switching current in the TG. Since the generator will be used in any feedback situation, we define the maximum achievable

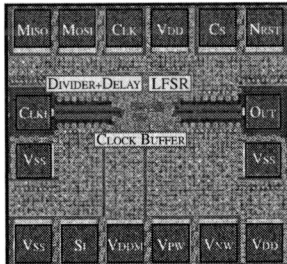

Fig. 7. Micrograph of the measured chip with a total area of $600 \times 530\,\mu m^2$ containing the PRBS generator with programmable taps.

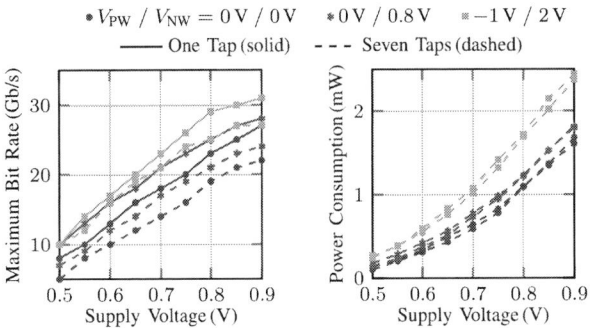

Fig. 8. Measured maximum bit rate and corresponding power consumption as a function of the supply voltage and number of taps.

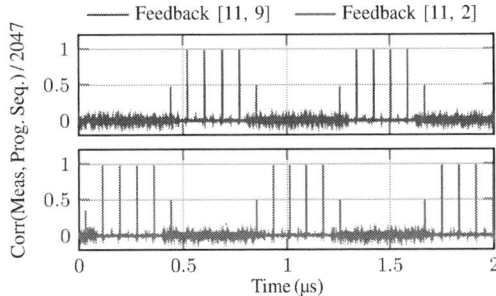

Fig. 9. Normalized correlation result of a single measurement at a bit rate of 25 Gb/s with the two configurations with minimum feedback.

data rate as 27 Gb/s. The core power consumption of 2.44 mW with seven activated taps decreases by 5.4, 11.5, and 15.1% when using fewer taps.

Finally, the functionality of the digital control circuit in combination with the LFSR must be demonstrated by measurement. Therefore, the two configurations with only a single activated tap are programmed into the ring buffer. At the same time, the counter is set to 2047, which must result in five repetitions of each sequence. Fig. 9 shows the correlation result of a single snapshot with the two different sequences at the data rate of 25 Gb/s. Each peak represents one sequence repetition, which clearly shows that both sequences are generated alternately. At the same time, it becomes evident that between each accumulation block, one sequence repetition must be dropped, or periodicity errors will occur.

Tab. I compares the proposed PRBS generator with state-of-the-art CMOS realizations, which primarily include earlier work by the authors due to the absence of other competitive publications. It also introduces the common figure of merit (FoM) for PRBS generators, defined as the power consumption divided by the word length times the maximum data rate. In order to keep the comparison fair, only the programmable LFSR is considered for the specified active area. Although the proposed baseband generator cannot compete with the circuits published in [6] in size and FoM, it still outperforms them regarding the number of different sequences possible. The fact

979-8-3503-4331-1/24 $31.00 © 2024 IEEE   77

TABLE I
STATE-OF-THE-ART CMOS PRBS GENERATORS

		[7]	[6]		This Work
Node	—	40 nm CMOS	22 nm FDSOI		22 nm FDSOI
f_T/f_{max}	GHz	260/−	350/370		350/370
R_{Bmax}	Gb/s	15	33	43	27
N_{FF}	—	31	11	11	11
V_{DD}	V	1.1	0.88	0.88	0.9
$P_{DC,LFSR}$	mW	8.8	1.72	2.63	2.44
Area	μm^2	3700	42	34	78.5
FoM	fJ/b	42	4.74	5.56	8.21
Arch.	—	TSPC	TSPC	E-TSPC	TSPC
Seq. Number	—	4	1	1	176

$\text{FoM} = P_{DC,LFSR}/(N_{FF} * R_{Bmax})$

that sequence selection can also be adjusted at runtime gives the circuit great potential for use in PMCW radar systems. In the authors' eyes, this justifies the higher power and space requirements.

IV. SUMMARY

This work presented a PRBS generator with programmable taps for PMCW radar transmitters. It is based on TSPC FFs optimized on the minimum clock-to-output delay. The feedback and, thus, the desired sequence can be edited during operation to generate advanced baseband signals that increase the radar performance. The circuit achieves a maximum bit rate of 27 Gb/s at a power consumption of 2.44 mW that can compete with the state of the art.

ACKNOWLEDGMENT

The authors thank GLOBALFOUNDRIES for silicon fabrication. This work is funded by the Federal Ministry of Education and Research (BMBF), Germany, through the research project "ForMikro-REGGAE" under grant 16E106S1.

REFERENCES

[1] F. Probst, A. Engelmann, M. Koch, and R. Weigel, "A Dual-Channel 15 Gb/s PRBS Generator for a D-Band PMCW Radar Transmitter in 22 nm FDSOI," in *2023 IEEE Wireless and Microwave Technology Conference (WAMICON)*, 2023, pp. 129–132.

[2] S. Ong *et al.*, "A 22nm FDSOI Technology Optimized for RF/mmWave Applications," in *2018 IEEE Radio Frequency Integrated Circuits Symposium (RFIC)*, 2018, pp. 72–75.

[3] D. Guermandi *et al.*, "A 79-GHz 2 × 2 MIMO PMCW Radar SoC in 28-nm CMOS," *IEEE Journal of Solid-State Circuits*, vol. 52, no. 10, pp. 2613–2626, 2017.

[4] M. Bauduin and A. Bourdoux, "Code Diversity for Range Sidelobe Attenuation in PMCW and OFDM Radars," in *2021 IEEE Radar Conference (RadarConf21)*, 2021, pp. 1–5.

[5] F. Probst, A. Engelmann, and R. Weigel, "A Synchronized 35 GHz Divide-by-5 TSPC Flip-Flop Clock Divider in 22 nm FDSOI," in *2023 Asia-Pacific Microwave Conference (APMC)*, 2023, pp. 1–3.

[6] F. Probst, J. Weninger, A. Engelmann, and R. Weigel, "Design of an E-TSPC Flip-Flop for a 43 Gb/s PRBS Generator in 22 nm FDSOI," in *2023 18th European Microwave Integrated Circuits Conference (EuMIC)*, 2023, pp. 353–356.

[7] J. Hu, Z. Zhang, and Q. Pan, "A 15-Gb/s 0.0037-mm² 0.019-pJ/Bit Full-Rate Programmable Multi-Pattern Pseudo-Random Binary Sequence Generator," *IEEE Transactions on Circuits and Systems II: Express Briefs*, vol. 67, no. 9, pp. 1499–1503, 2020.

A 23-30 GHz Low-Phase-Noise 5-Bit Voltage-Controlled Oscillator in 90-nm CMOS Process

Po-Yuan Chen
Department of Electrical Engineering
National Central University
Jhongli, Taiwan
111521112@cc.ncu.edu.tw

Jun-Liang Chen
Department of Electrical Engineering
National Central University
Jhongli, Taiwan
s110521113@g.ncu.edu.tw

Hong-Yeh Chang
Department of Electrical Engineering
National Central University
Jhongli, Taiwan
hychang@ee.ncu.edu.tw

Abstract—In this paper, a *Ka*-band 5-bit voltage-controlled oscillator (VCO) is presented using a 90-nm CMOS process. The proposed VCO is composed of 5-bit switched-capacitor array (SCA) and 1-bit switchable varactor array to achieve a wide tuning range of 27.5% and an excellent tuning figure-of-merit (FoM$_T$) of -193.1 dBc/Hz. The proposed 5-bit VCO also demonstrates a lowest phase noise of -103.5 dBc/Hz at 1-MHz offset frequency, corresponding to a FoM$_{PN}$ of -184.4 dBc/Hz. The measured oscillation frequency is from 22.6 to 29.8 GHz with a differential output power of higher than -3 dBm. The chip size is 0.94×0.81 mm^2. The core dc power consumption is merely 3 mW with a supply voltage of 0.6 V. The circuit performances can be compared to the reported silicon-based advanced VCOs.

Keywords—*CMOS, low phase noise, microwave, millimeter-wave, RFIC, voltage-controlled oscillator (VCO), wide tuning range.*

I. INTRODUCTION

To date, a low-phase-noise voltage-controlled oscillator (VCO) is an indispensable building block in microwave and millimeter-wave phase-locked loop (PLL) or frequency synthesizer. For further enhancing circuit performance of the frequency synthesizer, there are several specifications are used to evaluate the VCO, including phase noise, jitter, tuning range, tuning sensitivity, output power, and dc power consumption [1]. Besides, with the trend of higher data-rate, higher carrier frequency and higher order of modulation, the requirement for low phase noise and low jitter becomes even more crucial in high-frequency designs.

The phase noise of the VCO generally degrades with increasing tuning sensitivity, but the tuning range increases under certain control voltage. To compromise the design tradeoff between the phase noise and tuning range, a switching-mode VCO demonstrates the possibility of achieving low phase noise and wide tuning range simultaneously, but it consumes more dc power [2]. A wideband VCO in [3] with small VCO gain (K$_{VCO}$) variation, employed a switchable capacitor bank to extend the frequency locking range. In [4], a VCO with an extra cross-coupling path realized by a novel transformer-based resonator is presented, and it helps the switching transistors in the cross-coupled pair are closer to the ideal switches, which improves the phase noise and start-up time naturally. A VCO

with 4-bit switched-capacitor array (SCA) in [5] achieves wide tuning range percentage while the dc power consumption is high.

In [6]-[7], some switched-inductors techniques are presented with wide tuning range while the phase noise is high. Another switched-band topologies like switched-capacitor (SC) are presented in [8]-[12]. In [8], a gm-boosted Colpitts VCO with SC technique is presented to widen the tuning range while the 65-nm CMOS process is more expensive. In [9]-[11], the binary-weighted SC techniques achieve low-phase-noise characteristics in high frequency. To further widen tuning range with good linearity, a segment-weighted SC technique can be utilized [12]. In this paper, a wide-tuning-range low-phase-noise 5-bit VCO is proposed using segment-weighted SC technique in a 90-nm CMOS process, and the design methodology is completely presented with simulated and experimental results. Instead of using a few advanced technologies, this work not only achieves the best figure of merit with tuning range (FoM$_T$), and it also features low phase noise, good flatness of output power, and low manufacturing cost.

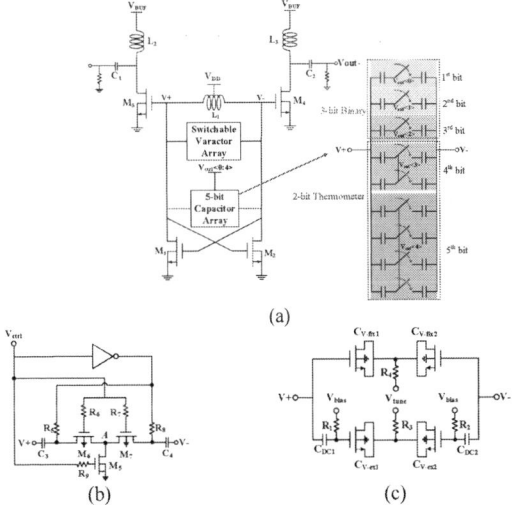

Fig. 1. Schematics of (a) the proposed 5-bit VCO, (b) one unit SCA, and (c) switchable varactor array.

II. CIRCUIT DESIGN

The schematic of the proposed 5-bit VCO is shown in Fig. 1(a). There are a few parts of this work, including the conventional cross-coupled pair M1-2, a 5-bit switched-capacitor array, a switchable varactor array, and the common source (CS) buffer. The design considerations of the VCO with multi-bit SCA are detailed as follows. The reason why the VCO core using conventional cross-coupled pair for the proposed 5-bit VCO is because of the design simplicity and broad bandwidth. Besides, instead of the complicated ways to create oscillation, the conventional cross-coupled pair ensures the robustness of the startup oscillation condition over the tuning range. The main feature of the proposed 5-bit VCO is the oscillation core with 5-bit SCA for wide tuning range with good linearity. By using SCA in the inductance-capacitance (LC) tank, the K_{VCO} of the VCO can be further reduced with good flatness, and also wide tuning range can be achieved. The schematic of one unit SCA is shown in Fig. 1(b), and the proposed 5-bit SCA is composed of 3 bits of the binary-weighted unit and 2 bits of the thermometer-weighted unit as shown in the right hand side of Fig. 1(a). The pros and cons between these two methods is that the binary-weighted unit occupies small chip area but the deviations of the capacitance difference (ΔC) among every switching frequency bands are high. However, the ΔC deviations in the thermometer-weighted unit is low, but the occupied chip area is large. To compromise the design tradeoff for the coarse frequency tuning, this work is realized using both two methods, which is called the segment-weighted SC technique. The reason why we utilized two series switching NMOSs M6-7 to realize the function of switches is because that if there is only one switching transistor, the voltage swing at both ends of transistor would be large, leads to a concern of reliability. The circuit node "A" in Fig. 1(b) is a virtual ground due to the differential network. The phase imbalance of the differential outputs would be caused with asymmetric circuit layout. To balance the phase difference between the differential outputs, an additional NMOS M5 is included in the node "A" to make sure the virtual-ground characteristic. Besides, the gate length of the NMOS is longer than the other NMOSs, so as to further reduce the noise contribution to the output of the VCO.

An 1-bit switchable varactor array is utilized for the fine frequency tuning, and the schematic is shown in Fig. 1(c). Normally, the varactors C_{V-fix1} and C_{V-fix2} are turned on, while the varactors C_{V-ex1} and C_{V-ex2} are off. However, if the tuning range among the control bits is not overlapped, it can be maintained by turning on the varactors C_{V-ex1} and C_{V-ex2}. The control mechanism ensures the frequency continuity over the tuning range with various control bits. From this perspective, high K_{VCO} maintains the frequency continuity while low K_{VCO} is good for the practical PLLs design in the future. Finally, this work is not only widen the tuning range by utilizing the 5-bit SCA, it also enhances the quality factor of the resonance cavity by thickening the thickness of the inductor in LC-tank. That way, the circuit can reduce the phase noise and maintain the wide tuning range characteristic as well. The circuit is fabricated using a commercial standard-bulk general-purpose 90-nm CMOS process provided by the Taiwan Semiconductor Manufacturing Company (TSMC). The chip photograph of the proposed 5-bit VCO is shown in Fig. 2 with a chip size of 0.94

$\times$ 0.81 mm^2. All the passive components are simulated using a full-wave electromagnetic simulator [13].

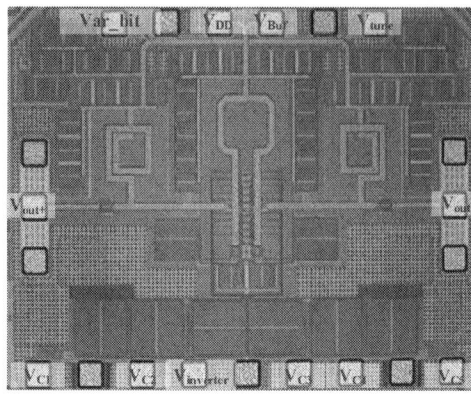

Fig. 2. Chip photograph of the proposed 5-bit VCO with a chip size of 0.94 $\times$ 0.81 mm^2.

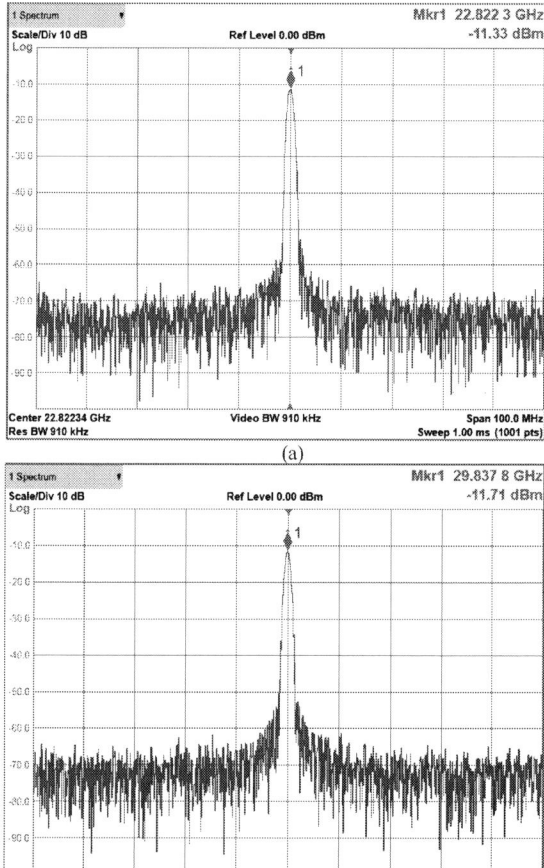

Fig. 3. Measured output spectra of the proposed 5-bit VCO at the (a) lowest (state: 11111, Vtune = 0 V), (b) highest (state: 00000, Vtune = 1.2 V) frequency.

III. EXPERIMENTAL RESULTS

The proposed 5-bit VCO is measured via on-wafer probing. Fig. 3(a) and (b) show the measured output spectra of the

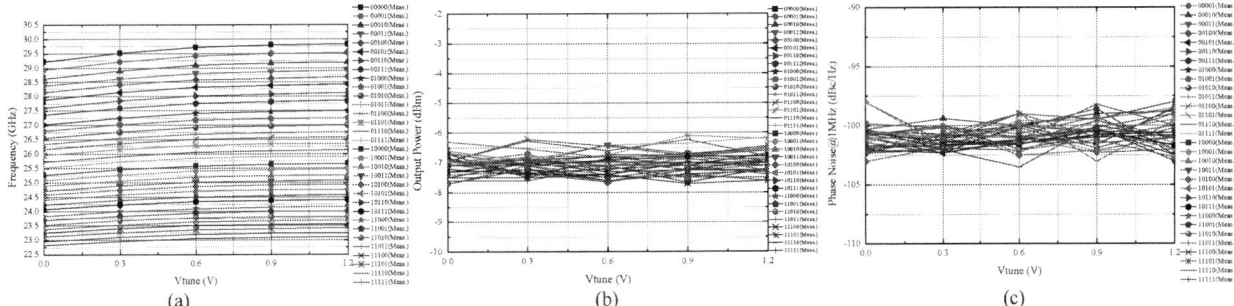

Fig. 4. Measured (a) oscillation frequency, (b) output power, and (c) phase noises versus control voltage and bit for the proposed 5-bit VCO.

proposed 5-bit VCO at the lowest (state: 11111, Vtune = 0 V) and highest (state: 00000, Vtune = 1.2 V) oscillation frequencies, respectively. In order to realize wide tuning range characteristic, the measured tuning range of this work has 32 states using 5 control bits. The measured oscillation frequency is plotted Fig. 4(a), where the measured overall tuning range is from 22.6 to 29.8 GHz, corresponding to a fractional bandwidth of 27.5%. The measured single-end output power versus control voltage and bit is shown in Fig. 4(b), and the output power is from -6 to -7.7 dBm. The measured phase noises versus control voltage and bit are plotted in Fig. 4(c), and all the measured phase noises are lower than -96 dBc/Hz.

$$FoM_{PN} = L\{\Delta f\} - 10\log[(\frac{f_0}{\Delta f})^2 \frac{1mW}{P_{DC}}], \quad (1)$$

$$FoM_T = FoM_{PN} - 20\log(\frac{TR\%}{10}), \text{ and} \quad (2)$$

$$FoM_{TA} = FoM_T + 10\log(\frac{Area}{1mm^2}), \quad (3)$$

where $L\{\Delta f\}$ is the phase noise at offset frequency (Δf) in Hz, f_0 is the oscillation frequency in Hz. P_{DC} is the dc power consumption in mW, TR% is the percentage of frequency tuning range. Area stands for the chip size. Among the reported VCOs, this work achieves the best FoM_T, and the overall performance can be compared with prior art.

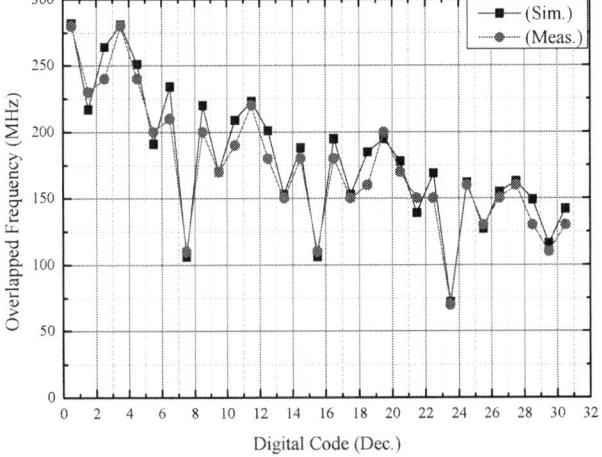

Fig. 5. Simulated and measured overlapped frequency versus control bit for the proposed 5-bit VCO.

Fig. 5 shows the simulated and measured overlapped frequency versus control bit. The amount of the measured overlapped frequency is less than the simulation, leads to a wider tuning range in the measured result. The simulated and measured phase noises versus offset of the proposed 5-bit VCO at the lowest and highest oscillation frequencies are shown in Fig. 6(a) and (b), respectively, and the measured phase noise at 1-MHz offset frequency is from -98 to -103.5 dBc/Hz. The performance comparison of the prior art and this work is summarized in Table I. Some performance merits, including FoM_PN, FoM_T, and FoM_TA, are also included in Table I for fair comparison, and they are expressed as

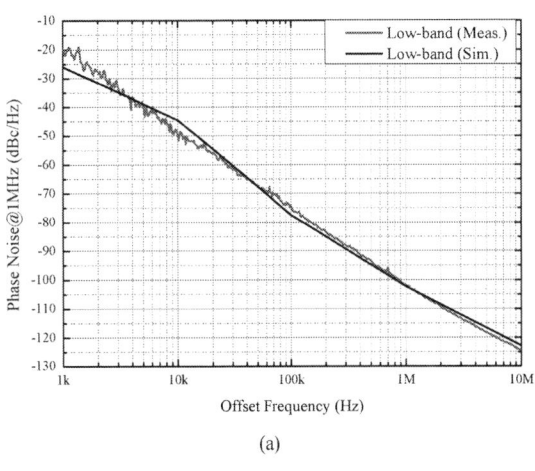

(a)

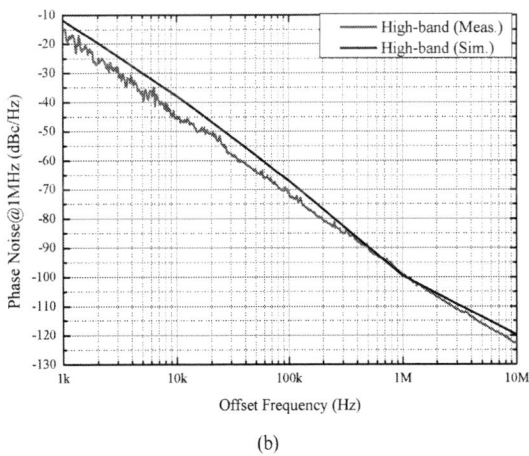

(b)

Fig. 6. Simulated and measured phase noises versus offset of the proposed 5-bit VCO at the (a) lowest and (b) highest oscillation frequency.

979-8-3503-4331-1/24 $31.00 © 2024 IEEE

TABLE I.
PERFORMANCE COMPARISON OF THE PREVIOUSLY REPORTED VCOS AND THIS WORK

Ref.	Tech.	Switched Band Topology	Freq. (GHz)	TR (%)	Min. Overlap. (%)	Band Number/Bit	Supply Vol. (V)	P_{DC} (mW)	PN @1 MHz (dBc/Hz)	Core Area (mm²)	FoM$_{PN}$ (dBc/Hz)	FoM$_T$ (dBc/Hz)	FoM$_{TA}$ (dBc/Hz)
[6]	65 nm CMOS	Switched Inductors	21.5~33.4	43.3	12*	4/3	1.2	7.2	-90	0.084	-171	-183.7	-194.5
[7]	65 nm CMOS	Switched Inductors	21~31.6	40.3	34	2/1	1	4.3	-89.5	0.05	-171.6	-183.7	-196.7
[8]	65 nm CMOS	#SC	23.7~29	20.1	13.7*	3/2	1	2.3	-98	0.22	-185.8	-191.8	-198.4
[9]	65 nm CMOS	#SC with $BW	21.9~24.2	9.5	12*	4/2	1	8	-100.8	0.118	-179.3	-179.1	-188.4
[10]	90 nm CMOS	#SC with $BW	24.2~29.1	20	45*	4/2	0.8	2.8	-101	N/A	-184	-187	N/A
[11]	130 nm BiCMOS	#SC with $BW	31.8~39.6	21.9	15*	64/6	2.3	30	-103.9	0.06	-180.2	-187	-199.2
[12]	45 nm SOI CMOS	#SC with &SW	22.4~31.2	33	40	32/5	1	6	-101.4	0.018	-182.2	-193	-210.4
This work	90 nm CMOS	#SC with &SW	22.6~29.8	27.5	28	32/5	0.6	3	-98~ -103.5	0.06	-184~ -184.4	-192.8~ -193.1	-205~ -205.3

*: estimated from the graph, #: switched-capacitor, &: segment-weighted, $: binary-weighted

IV. CONCLUSION

This paper presents a *Ka*-Band low-phase-noise 5-bit VCO using 90-nm CMOS process. To widen the frequency tuning range and lower tuning sensitivity, a 5-bit SCA and a 1-bit switchable varactor array are adopted in the LC tank of the oscillation core. The VCO features compact chip area, low dc power consumption, low phase noise, good tuning linearity over the control voltage and bit. Moreover, this work is suitable for some advanced PLLs in millimeter-wave regime owing to its competitive performance.

ACKNOWLEDGMENT

The chip was fabricated by TSMC in Hsinchu City, Taiwan. This work was supported by the National Science and Technology Council (NSTC), Taiwan, under Grant MOST 110-2221-E-008-029-MY3, and the Taiwan Semiconductor Research Institute (TSRI), Hsinchu City, Taiwan. The dc and RF probes are supported by GGB Inc., FL, USA.

REFERENCES

[1] H.-Y. Chang, C.-C. Chan, S.-M. Li, H.-N. Yeh, I. Y.-E. Shen, and G.-L. Huang, "Design and analysis of CMOS low phase noise low quadrature error *V*-band subharmonically injection-locked quadrature FLL," *IEEE Trans. Microw. Theory Techn.*, vol. 66, no. 6, pp. 2851–2866, June 2018.

[2] G. Li, L. Liu, Y. Tang, and E. Afshari, "A low-phase-noise wide-tuning-range oscillator based on resonant mode switching," *IEEE J. Solid-State Circuits*, vol. 47, no. 6, pp. 1295-1308, June 2012.

[3] Y. -J. Moon, Y. -S. Roh, C. -Y. Jeong, and C. Yoo, "A 4.39–5.26 GHz LC-tank CMOS voltage-controlled oscillator with small VCO-gain variation," *IEEE Microw. Wireless Compon. Lett.*, vol. 19, no. 8, pp. 524-526, Aug. 2009.

[4] C. Wan, T. Xu, X. Yi, and Q. Xue, "A VCO with extra cross-coupling path," *IEEE Microw. Wireless Compon. Lett.*, vol. 31, no. 10, pp. 1130-1133, Oct. 2021.

[5] W. -C. Lai, "Design of 1V CMOS 5.8 GHz VCO with switched capacitor array tuning for intelligent sensor fusion," *in 2020 Int. Conf. on Advanced Robotics and Intelligent Systems (ARIS)*, 2020, pp. 1-4.

[6] J. Zhang, N. Sharma, and K. K. O, "21.5-to-33.4 GHz voltage-controlled oscillator using NMOS switched inductors in CMOS," *IEEE Microw. Wireless Compon. Lett.*, vol. 24, no. 7, pp. 478-480, July 2014.

[7] P. Agarwal, *et al.*, "Switched substrate-shield-based low-loss CMOS inductors for wide tuning range VCOs," *IEEE Trans. Microw. Theory Techn.*, vol. 65, no. 8, pp. 2964-2976, Aug. 2017.

[8] M. Haghi Kashani, R. Molavi, and S. Mirabbasi, "A 2.3-mW 26.3-GHz gm-boosted differential colpitts VCO with 20% tuning range in 65-nm CMOS," *IEEE Trans. Microw. Theory Techn.*, vol. 67, no. 4, pp. 1556-1565, April 2019.

[9] W. Tan, T. Wu, Z. Xing, Y. Peng, H. Liu, and K. Kang, "A 21.95-24.25 GHz class-c VCO for 24 GHz FMCW radar applications," *in Proc. IEEE MTT-S Int. Wireless Symp.*, Guangzhou, China, 2019, pp. 1-3.

[10] D. Yang, *et al.*, "Design of a 24 GHz low phase-noise, wide tuning-range VCO with optimized switches in capacitor array and bias filtering technique," *IEEE Solid-State and Integrated Circuit Technology (ICSICT)*, Shanghai, China, 2010, pp. 696-698.

[11] B. Sadhu, T. Anand and S. K. Reynolds, "A fully decoupled LC tank VCO topology for amplitude boosted low phase noise operation," *IEEE J. Solid-State Circuits*, vol. 53, no. 9, pp. 2488-2499, Sept. 2018.

[12] S. Alzahrani, *et al.*, "Analysis and design of the tank feedline in millimeter-wave VCOs," *IEEE Trans. Microw. Theory Techn.*, vol. 70, no. 5, pp. 2668-2679, May 2022.

[13] *Sonnet User's Guide*, 14th ed., Sonnet Software, Syracuse, NY, USA, 2013.

Low Phase Noise 104 GHz Oscillator Using Self-Aligned On-Chip Voltage-Tunable Spherical Dielectric Resonator in 130-nm SiGe BiCMOS

Yu Zhu
Chair for Circuit Design and Network Theory
Technische Universität Dresden
Dresden, Germany
yu.zhu2@tu-dresden.de

Georg Sterzl
Institute of Radio Frequency Technology
University of Stuttgart
Stuttgart, Germany
georg.sterzl@ihf.uni-stuttgart.de

Jan Hesselbarth
Institute of Radio Frequency Technology
University of Stuttgart
Stuttgart, Germany
jan.hesselbarth@ihf.uni-stuttgart.de

Tilo Meister
Chair for Circuit Design and Network Theory
Technische Universität Dresden
Dresden, Germany
tilo.meister@tu-dresden.de

Frank Ellinger
Chair for Circuit Design and Network Theory
Technische Universität Dresden
Dresden, Germany
frank.ellinger@tu-dresden.de

Abstract—This paper studies a low phase noise voltage-controlled oscillator that is based on a self-aligned on-chip voltage-tunable spherical dielectric resonator. The proposed resonator has been designed for millimeter-wave applications, provides a high quality factor and is voltage controlled. To prove the concept, the circuit is implemented in a 130-nm SiGe BiCMOS technology. It consists of a two stage amplifier and a microstrip feedback path which couples to the resonator. Measurement results demonstrate a phase noise of -95.9 dBc/Hz at 10 MHz offset from the oscillation frequency at 104.03 GHz and a frequency tuning range of 88 MHz. A maximum output power of -9.9 dBm from 32.5 mW dc power is achieved. Simulations based on measurements of the on-chip spherical dielectric resonator indicate that circuit optimizations will lead to an excellent phase noise of -114.8 dBc/Hz at 10 MHz offset. To the best of the authors' knowledge, this circuit is the first reported silicon-based MMIC voltage-controlled oscillator using an on-chip dielectric resonator at millimeter-wave band.

Index Terms—SiGe, dielectric resonator, Q-factor, phase noise (PN), mm-wave oscillators

I. INTRODUCTION

Millimeter-wave (mm-wave) and sub-terahertz (sub-THz) silicon-based integrated systems are becoming more attractive for applications like high data rate wireless communication, close-range radar and ultra-wideband imaging. As a result of the low quality factor (Q-factor) of integrated LC tanks, limited tuning range of the varactors, and increasing operating frequencies the required mm-wave oscillator sources with low phase noise (PN) and excellent frequency stability are increasingly difficult to obtain.

Due to significant improvement of PN performance, which is largely determined by using high quality factor (high-Q) resonators, well-known high-Q dielectric resonators have been

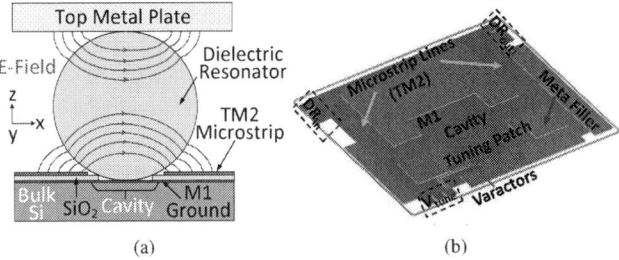

Fig. 1. VTSDR structure with the coupling microstrip lines: (a) cross-section of the on-chip VTSDR with a top metal plate, (b) 3D-layout view.

used at high frequencies to overcome the above shortcomings of mm-wave VCOs [1]. A dielectric resonator usually has properties such as high-Q and excellent resonance frequency stability against temperature change, thus a dielectric resonator oscillator (DRO) offers good PN and high frequency stability performance compared to other lumped element LC tank based VCOs in mm-wave frequencies. In this work, a spherical dielectric resonator (SDR) is investigated and used as an on-chip high-Q resonator [2], [3] for mm-wave VCOs. The SDRs operating in higher-order mode are used to solve problems relate to substrate resistivity, mechanical alignment precision and chip-area consumption, and have been demonstrated for mm-wave applications in recent years.

II. DESIGN AND IMPLEMENTATION OF OSCILLATOR

A. Voltage-Tunable Spherical Dielectric Resonator (VTSDR)

The VTSDR comprises a spherical resonator and a metal patch with varactors underneath it for center frequency tuning. Low-loss alumina ceramic spheres with a relative permittivity

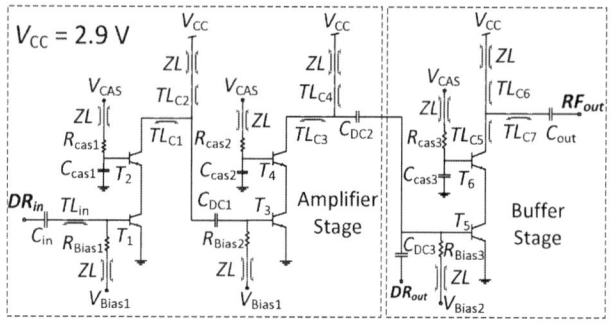

Fig. 2. Schematic of the designed amplifier and buffer.

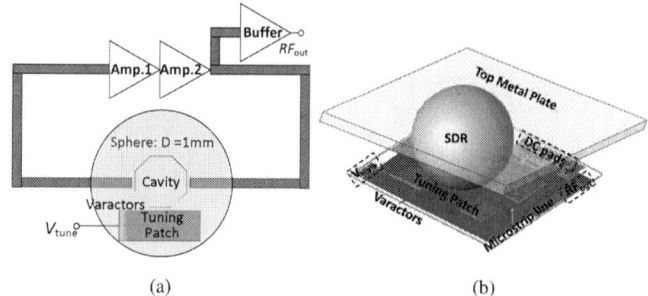

Fig. 3. The proposed VCDRO chip including the top metal plate: (a) block diagram, (b) 3D-layout view.

of around 10.15 and a diameter of 1 mm are used as dielectric resonators. A measured loaded Q-factor of 143 at 103 GHz has been achieved for a same SDR structure in [3]. The sphere is accurately self-aligned on the chip by placing it into an octagonally-shaped shallow cavity inside the back-end-of-line (BEOL) dielectric layers of silicon-based semiconductor circuit, as can be seen in Fig. 1(a). Therefore, the self-alignment of the resonator is achieved by the sphere falling into the cavity and then settling there. To prevent radiation leakage from the resonator while maintaining a high Q-factor, the sphere is covered with a top metal plate that is parallel to the substrate. Fig. 1(a) illustrates a high-Q resonance mode of the VTSDR structure orthogonal to the parallel-plate mode.

Furthermore, the sphere is loosely coupled to two open-ended microstrip lines from both sides to offer a high Q-factor, thus the loss of such resonators can be higher than 18 dB above 100 GHz. The ground planes (M1) of the microstrip lines (TM2) prevent leakage into the high-loss silicon substrate. In order to obtain a frequency-tunable dielectric resonator, a rectangular tuning patch is placed in the middle of two microstrip lines but below them, and carefully coupled to the sphere, as shown in Fig. 1(b). The varactors under the tuning patch are used for the voltage-controlled frequency tuning. All simulations here were performed with the 3D EM-solver of CST Microwave Studio.

B. Design of Amplifier and Buffer

In order to fulfill the oscillator conditions and to achieve sufficient gain above 20 dB, a two stage cascode amplifier is proposed. On the basis of negative resistance theory, the two stage amplifier is configured in parallel feedback generating the negative resistance to over compensate the loss from the resonator. To retain the advantages of the high-Q resonator and its crucial impact on the PN of the oscillator, the active part of the circuit has to isolate it from the 50 Ω output load that could otherwise lower its Q-factor and thus decrease the PN. Therefore, a high-impedance output buffer is implemented to match the output of the oscillator for the high-Q resonator.

The circuit schematic of the designed amplifier and buffer is shown in Fig. 2. According to the post-layout simulations, this two stage amplifier achieves a peak gain of 27.4 dB at 103.4 GHz with a 3-dB bandwidth from 92.5 GHz to

111.5 GHz. The output buffer provides a simulated maximum output power of 5.5 dBm.

C. Design of Voltage-Controlled Dielectric Resonator Oscillator (VCDRO)

A corresponding circuit block diagram representing the proposed VCDRO is given in Fig. 3(a). It consists of a tunable resonator network in the parallel feedback loop, a two stage amplifier, and an output buffer. To realize the oscillation and fulfill the oscillation condition, the loop gain at the resonant frequency must be greater than one and the phase should equal zero. Two microstrip feed lines coupled to the on-chip SDR component serve not only as a physical connection, but also to adjust the phase between the resonator and the amplifier. The feedback gain is controlled by the coupling from the amplifier output to the amplifier input. Basically, the parallel feedback based dielectric resonator oscillator can be considered as a positive feedback amplifier with a VTSDR, which is used as a frequency-selective feedback element. An output buffer employed to isolate the resonator from the 50 Ω load is connected at the output of the oscillator to maintain the high-Q characteristics of the resonator. The frequency tuning is achieved by using the varactors connected to the tuning patch. Fig. 3(b) presents the complete 3D-layout view of the proposed VCDRO chip including the top metal plate. All co-simulations, which combine the circuit simulations and the EM model of VTSDR, were performed in Keysight ADS.

III. EXPERIMENTAL RESULTS

To prove and investigate the concept, the circuit was fabricated in a 130-nm SiGe BiCMOS technology, which has a maximum f_T/f_{max} of 300 GHz/500 GHz. Fig. 4(a) shows the chip micrograph and the focused ion beam (FIB) micrograph of the on-chip cavity. The whole circuit occupies an area of 3.74 mm^2 containing all dc/rf pads. The total dc power is 32.5 mW, under a supply voltage of 2.9 V. The on-wafer measurement setup of the contacted chip is illustrated in Fig. 4(b). The chip is embedded into a crate of an aluminium plate and its surface is leveled with metal plate surface, to prevent electromagnetic leakage into the high-loss silicon substrate [3]. Due to mechanical limitations by the on-wafer probes, a small top metal plate with a width of 6 mm is directly placed above the sphere. Ideally this plate should be

979-8-3503-4331-1/24 $31.00 © 2024 IEEE

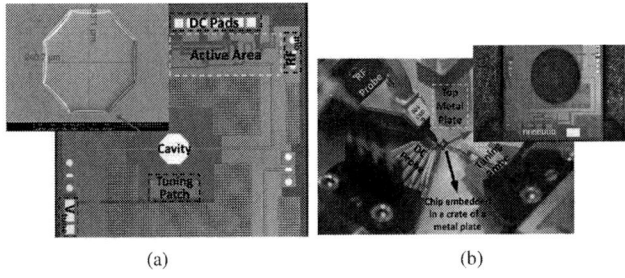
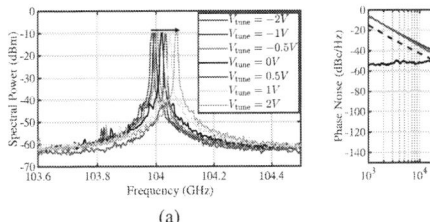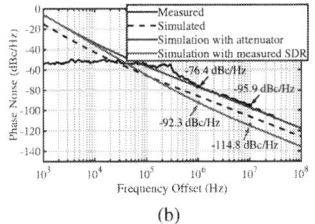

| (a) | (b) |

Fig. 4. (a) Micrograph of the fabricated VCDRO chip and FIB image of the on-chip cavity, (b) on-wafer measurement setup of the VCDRO chip.

Fig. 6. (a) Measured output spectral power of the VCDRO under different V_{tune}, (b) simulated and measured PN of the VCDRO.

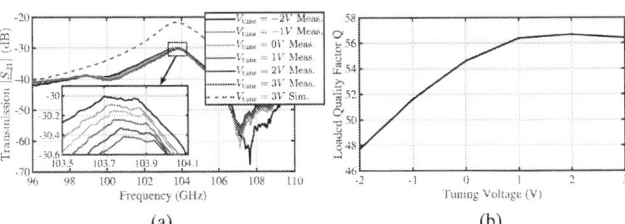

| (a) | (b) |

Fig. 5. (a) Simulated and measured $|\underline{S}_{21}|$ of the VTSDR under different V_{tune}, (b) measured loaded Q-factor of the VTSDR as a function of V_{tune}.

much larger. Moreover, it should be taken into account that the relative permittivity and the diameter of fabricated spheres vary slightly from sphere to sphere.

Fig. 5(a) presents the simulated and measured transmission coefficients at resonance frequencies of the VTSDR under different tuning voltage V_{tune} from -2 V to 3 V. The VTSDR demonstrates measured resonance frequencies between 103.7 GHz to 103.8 GHz with a forward transmission $|\underline{S}_{21}|$ of around -30.3 dB. The simulated resonance frequency under a tuning voltage of 3 V appears at 103.8 GHz with a forward transmission $|\underline{S}_{21}|$ of -21.6 dB. In order to realize a very high Q-factor, the dielectric resonator is usually loosely coupled to the open-ended microstrip lines for the on-chip resonator applications, thus the transmission loss is pretty high in some cases. A good compromise between loss and Q-factor has been investigated during the design. According to [3], all measured resonator transmission losses are higher than 30 dB at around 102 GHz. The reasons for the higher measured losses than the simulations are suspected to be caused by the radiation leakage out of two small parallel metal plates and excess loss in the thick bulk silicon through the chip edges. The measured loaded Q-factor (Q_L) of the VTSDR as a function of tuning voltage V_{tune} is shown in Fig. 5(b). Compared with the simulated highest loaded Q-factor of 52.9 with V_{tune} = 3 V, the measured highest loaded Q-factor of 56.7 with V_{tune} = 2 V is provided. The corresponding Q_L is defined as [4]:

$$Q_L = \frac{f_c}{BW_{3dB}} \quad (1)$$

where f_c is the resonance frequency, and BW_{3dB} is the 3dB-bandwidth.

The measured output spectral power of the VCDRO with different tuning voltages V_{tune} is given in Fig. 6(a). Measurement results show a tuning range of 88 MHz from

103.986 GHz to 104.074 GHz by changing V_{tune} from -2 V to 2 V, and a maximum output power of -9.9 dBm. Moreover, original simulations show an expected PN of -85.9 dBc/Hz at 1 MHz offset and -106 dBc/Hz at 10 MHz offset from the oscillation frequency of 103.804 GHz in Fig. 6(b). However, measurements present a higher PN of -76.4 dBc/Hz at 1 MHz offset and -95.9 dBc/Hz at 10 MHz offset from the oscillation frequency of 104.03 GHz. We attribute this to a higher loss in the fabricated VTSDR, when compared to the losses predicted for the VTSDR by simulation. It leads to non-optimal loop gain and phase in the VCDRO circuit, and directly causes the resonator to operate in a region with a reduced Q-factor. To prove this assumption in simulation, we inserted a 10 dB attenuator in the microstrip feedback path of Fig. 3(a), which effectively emulates an increased loss by the VTSDR. The PN simulated with this setup is shown in Fig. 6(b) and agrees well with the measurement result. This increased losses and the resulting unwanted shift to a non-optimal Q-factor can be compensated in a future revision of the chip, by increasing the loop gain with a third amplifier stage. Simulations taking these higher losses into account, using a measured high-Q on-chip spherical dielectric resonator model with the loss higher than 30 dB from [3], and having a higher gain due to an added third amplifier stage, predict an excellent PN of -114.8 dBc/Hz at 10 MHz offset from the oscillation frequency of 103.242 GHz. These results are shown by the red curve in Fig. 6(b).

IV. CONCLUSION

In this feasibility study, a concept for low phase noise silicon-based integrated voltage-controlled oscillator using high-Q self-aligned on-chip voltage-tunable spherical dielectric resonator was investigated, fabricated and measured. The designed resonator consists of a high-Q on-chip dielectric sphere and a tuning patch with varactors underneath it. An amplifier, which is combined with the proposed resonator as a positive feedback, is designed to provide the oscillation. An output buffer is used to isolate it from the 50 Ω output load. In Table I, the proposed VCDRO is compared against the state of the art for LC tank based MMIC VCOs considering frequencies above 90 GHz. According to simulations based on measurements of all its blocks, after a simple redesign with increased loop gain, this oscillator can achieve a very promising PN of -114.8 dBc/Hz at 10 MHz offset.

979-8-3503-4331-1/24 $31.00 © 2024 IEEE

TABLE I
COMPARISON WITH STATE-OF-THE-ART OF SILICON-BASED VCOs
OPERATING ABOVE 90 GHz

Ref.	This work	[5]	[6]	[7]	[8]
Tech.	130nm SiGe	65nm CMOS	65nm CMOS	130nm CMOS	130nm SiGe
Topology	Spherical Dielectric Resonator	Coupled Oscillators	Fund. Colpitts	Fund. Cross-Coupled	Push-Push
$f_{osc,center}$ (GHz)	104.03	105	98	91	103.5
Δf_{osc} (GHz)	0.088	9.975	8.64	0.455	25
$P_{out,max}$ (dBm)	-9.9	4.5	-1	4.5	4.2
PN @ 1 MHz (dBc/Hz)	-76.4/ -92.3^a	-92.83	-90.85	-87	-108.7
PN @ 10 MHz (dBc/Hz)	-95.9/ -114.8^a	-100^b	-110.95	-107.7	-111^b
P^c_{DC} (mW)	32.5	54	21.6	46	50

^{a}: simulated with measured SDR, ^{b}: graphically estimated, ^{c}: with buffer.

ACKNOWLEDGMENT

This research was supported by the German Research Foundation (DFG) under grant EL 506/36-1 (project "SPHERE").

REFERENCES

[1] K. Hosoya, K. Ohata, M. Funabashi, T. Inoue, and M. Kuzuhara, "V-band HJFET MMIC DROs with low phase noise, high power, and excellent temperature stability," *IEEE Transactions on Microwave Theory and Techniques*, vol. 51, no. 11, pp. 2250–2258, 2003.

[2] D. L. Cuenca, G. Alavi, and J. Hesselbarth, "On-chip mm-wave spherical dielectric resonator bandpass filter," in *2017 IEEE MTT-S International Microwave Symposium (IMS)*, 2017, pp. 1460–1463.

[3] G. Sterzl, Y. Zhu, J. Hesselbarth, C. Carta, M. Lisker, and F. Ellinger, "Self-Aligned on-Chip Spherical Dielectric Resonators and Antennas for SiGe MMIC," in *2022 Asia-Pacific Microwave Conference (APMC)*, 2022, pp. 686–688.

[4] D. M. Pozar, *Microwave engineering*. John wiley & sons, 2011.

[5] M. Adnan and E. Afshari, "A 105-GHz VCO With 9.5% Tuning Range and 2.8-mW Peak Output Power in a 65-nm Bulk CMOS Process," *IEEE Transactions on Microwave Theory and Techniques*, vol. 62, no. 4, pp. 753–762, 2014.

[6] S. Jameson and E. Socher, "A 93.9-102.5 GHz Colpitts VCO utilizing magnetic coupling band switching in 65nm CMOS," in *2015 IEEE International Conference on Microwaves, Communications, Antennas and Electronic Systems (COMCAS)*, 2015, pp. 1–5.

[7] A. Tarkeshdouz, A. Mostajeran, S. Mirabbasi, and E. Afshari, "A 91-GHz Fundamental VCO With 6.1% DC-to-RF Efficiency and 4.5 dBm Output Power in 0.13-μm CMOS," *IEEE Solid-State Circuits Letters*, vol. 1, no. 4, pp. 102–105, 2018.

[8] K. Wu and M. Hella, "A 103-GHz Voltage Controlled Oscillator with 28% Tuning Range and 4.2 dBm Peak Output Power Using SiGe BiCMOS Technology," in *2018 IEEE/MTT-S International Microwave Symposium - IMS*, 2018, pp. 606–609.

A 34 GHz CMOS VCO with Transformer Tail-Node Filter and TSPC Frequency Divider in 22 nm FDSOI

Andre Engelmann
Institute for Electronics Engineering
Friedrich-Alexander-Universität Erlangen-Nürnberg
Erlangen, Germany
andre.engelmann@fau.de

Florian Probst
Institute for Electronics Engineering
Friedrich-Alexander-Universität Erlangen-Nürnberg
Erlangen, Germany
florian.probst@fau.de

Philip Hetterle
Institute for Electronics Engineering
Friedrich-Alexander-Universität Erlangen-Nürnberg
Erlangen, Germany
philip.hetterle@fau.de

Robert Weigel
Institute for Electronics Engineering
Friedrich-Alexander-Universität Erlangen-Nürnberg
Erlangen, Germany
robert.weigel@fau.de

Abstract—This work proposes a compact, low-power CMOS voltage-controlled oscillator (VCO) with distributed transformer tail-node filter for phase noise reduction. The circuit is integrated into a 22 nm fully-depleted silicon-on-insulator (FDSOI) technology, offering high-performance n- and p-MOS transistors. Using a combination of SOI varactors and switched capacitors, a wide tuning range of 22.8%, covering 30.4 to 38.24 GHz, is reached while maintaining a phase noise below -97 dBc/Hz at a 1 MHz carrier offset. An area- and power-efficient true single-phase-clock (TSPC) frequency divider is implemented to ease the use of the VCO within a low-frequency PLL. The overall circuit consumes 9.4 mW supplied from a 0.8 V source and $170 \times 370\,\mu m^2$ core area, including an output buffer and the divider. The circuit performance results in a corresponding FOM and FOM_T of -178.3 and -185.3 dBc/Hz.

Index Terms—Cross-coupled VCO, CMOS, FDSOI, frequency divider, low phase-noise, oscillator, signal generation, tail filter.

I. INTRODUCTION

The vast available bandwidth in the mm-wave frequency domain is a driving factor for the implementation of high-data-rate communication and high-resolution radar systems. Due to the increasing performance of available silicon technologies, the systems trend to D-band ($110 - 170$ GHz) and even higher operation frequencies [1]. A co-integration of the mm-wave front-end circuitry together with the digital signal processing blocks in the same chip leads to a significant cost reduction of the overall system. For this purpose, the advanced 22 nm fully-depleted silicon-on-insulator (FDSOI) technology is a viable option due to the high maximum oscillation frequency f_{max} of n- and p-MOS transistors of 370 and 300 GHz [2]. A fundamental component of integrated mm-wave systems is the frequency synthesizer which in many applications comprises a voltage-controlled oscillator (VCO) and frequency multiplication stages. The VCO must fulfill challenging requirements, such as low power consumption, a compact chip footprint, and a wide tuning range (TR) while simultaneously maintaining low phase noise (PN). Low PN is required to enable efficient

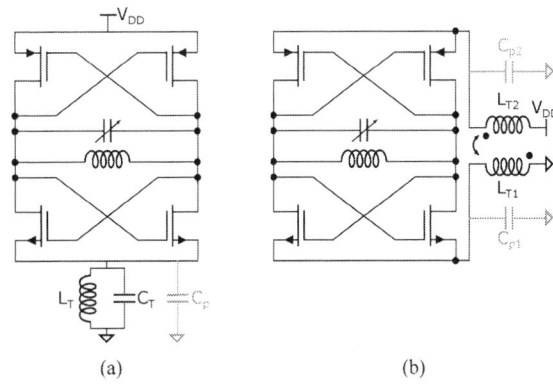

Fig. 1. CMOS VCO with (a) conventional tail-node filter and (b) proposed transformer tail-node filter.

modulation schemes with large symbol rates in communication systems and achieve good target discrimination in radar systems. The high losses of the resonator tank at mm-wave frequencies and the increased $1/f$ noise in small-scale CMOS nodes make it difficult to integrate low PN VCOs [3]. Many techniques, like the usage of higher operation classes [4] and waveform-shaping with harmonic extraction [3], are reported to encounter the previously mentioned challenges. For class-B oscillators, second harmonic LC filters at the tail node, shown in Fig. 1a, are widely used to prevent the up-conversion of noise to the output signal, as described in [5].

This work presents a voltage-biased class-B CMOS VCO intended to operate within a phase-modulated continuous-wave (PMCW) radar system for joint-communication and sensing (JCAS) applications. A distributed transformer-based tail-node filter (Fig. 1b) is employed between the source nodes of the cross-coupled n- and p-MOS transistors to increase the PN performance. An area- and power-efficient true single-phase-clock (TSPC) frequency divider is used to incorporate

979-8-3503-4331-1/24 $31.00 © 2024 IEEE

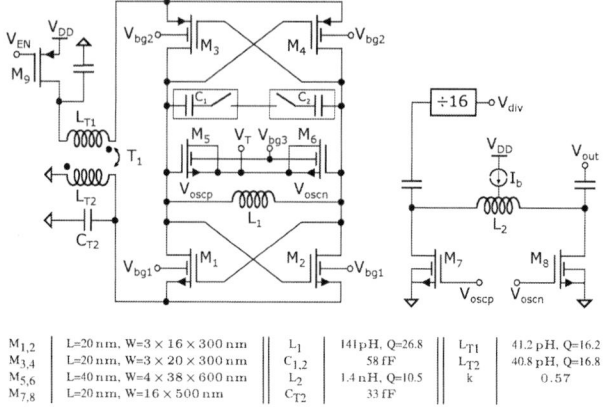

$M_{1,2}$	L=20 nm, W=3×16×300 nm	L_1	141 pH, Q=26.8	L_{T1}	41.2 pH, Q=16.2
$M_{3,4}$	L=20 nm, W=3×20×300 nm	$C_{1,2}$	58 fF	L_{T2}	40.8 pH, Q=16.8
$M_{5,6}$	L=40 nm, W=4×38×600 nm	L_2	1.4 nH, Q=10.5	k	0.57
$M_{7,8}$	L=20 nm, W=16×500 nm	C_{T2}	33 fF		

Fig. 2. VCO schematic with transformer tail-node filter and output buffer.

the VCO inside a low-frequency phase-locked loop (PLL). The detailed circuit design concept is described in the subsequent section. Afterward, the obtained measurement results are presented, followed by a comparison to other recent state-of-the-art implementations.

II. Circuit Design

Fig. 2 shows the schematic of the complete VCO together with the corresponding component values. The used FD-SOI technology offers p-MOS transistors, which performance is comparable to their n-MOS counterpart [2]. Therefore, the complementary MOS transistors are used in the cross-coupled pairs compensating for the losses of the resonator tank. The CMOS implementation reaches twice the voltage swing, achievable with an n-MOS-only implementation using the same bias current, resulting in higher attainable efficiency [6]. Furthermore, the maximum voltage at the cross-coupled transistors is limited to the supply voltage, ensuring an operation within the safe operating area. All transistors are super-low-threshold-voltage (SLVT) devices to achieve best performance. The VCO operates in the voltage-biased region and omits an additional current source. Thereby, the degradation of the phase noise caused by the up-conversion of the current source's thermal noise to the carrier frequency is avoided [3]. The bias current of the cross-coupled pairs can still be precisely adjusted by tuning the SOI transistors' back-gate voltages ($V_{bg1,2}$).

The resonator tank is formed by the single-turn inductor L_1 and the SOI varactors $M_{5,6}$ for fine and switchable alternate-polarity metal-oxide-metal (AP-MOM) capacitors banks ($C_{1,2}$) for coarse frequency control. L_1 is implemented on the top copper layer of the metal stack and features 141 pH differential inductance with a quality factor Q of 26.8 at 35 GHz determined by EM simulation. With a diameter of 78 μm and a turn width of 15 μm, L_1 mainly contributes to the overall size of the VCO. The SOI varactor transistors are chosen with a gate length of 40 nm as a compromise between a wide TR and high quality factor. Due to the substrate isolation of these transistors, the tuning voltage can also be set to

a negative voltage, thus enhancing the achievable capacitive TR compared to conventional MOSCAP realizations. The back-gate voltage (V_{bg3}) of $M_{5,6}$ can be used additionally to the regular gate tuning voltage (V_T) for fine capacitance control [7].

For PN improvement, the proposed transformer tail-node filter generates a high impedance path for the 2nd harmonic current [5]. Therefore, the two stacked transformer coils ($L_{T1,2}$) built on the two available thick top copper layers are distributed between the source nodes of the n- and p-MOS cross-coupled pairs and the ground and supply connection, respectively, forming a filter together with the parasitic source capacitance and the additional AP-MOM capacitor C_{T2}. A combination of EM simulations and parasitic extraction (PEX) is used to model the parasitics of interconnections between passive components and transistor devices. Fig. 3 visualizes the simulated PN using the proposed transformer and conventional LC filters at the source of n- and p-MOS pairs compared to the filter-less case. The stacked transformer variant achieves the lowest PN of the tested implementations and and exhibits a lower chip area than single inductors.

A common source output buffer is added to achieve better load isolation for the resonator tank and provide increased output power. The output matching is formed by the symmetric three-winding inductor L_2 having an inner diameter of 90 μm and AP-MOM capacitors. To adjust the output power, the bias current of the buffer transistors $M_{7,8}$ can be controlled by a current source connected via the center-tap connection of L_2.

The 22 nm technology enables the integration of high-performance, low-power TSPC flip-flops (FF) operating at clock rates up to the mm-wave frequency range [8]. Four of those FF are connected to build a frequency divider by the factor of 16. The divider omits area-consuming peaking inductors and occupies a compact chip area of $24 \times 5 \, \mu m^2$, while consuming only 1.8 mW of DC power, including inverter input and output buffers.

III. Measurement and Results

The VCO is fabricated in a 22 nm FDSOI CMOS technology. The complete chip, shown in Fig. 4, occupies an area of $0.34 \, mm^2$, while the core cell of the circuit, including the output buffer and divider, requires only $170 \times 370 \, \mu m^2$. The DC supply and control pads are wire-bonded to a test PCB. The

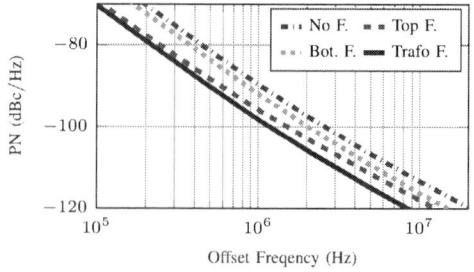

Fig. 3. Simulated phase noise for different tail filter typologies.

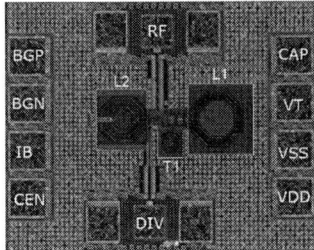

Fig. 4. Micrograph of the VCO chip with a total area of $520 \times 660\,\mu m^2$.

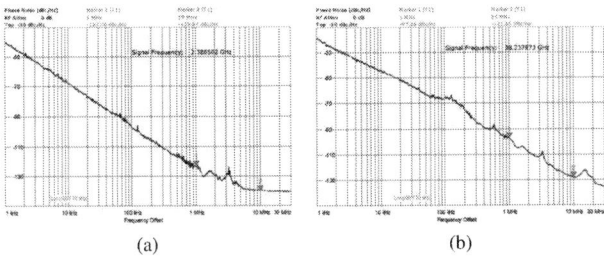

(a)   (b)

Fig. 5. Measured phase noise over offset frequency at (a) divider output and (b) RF output at 0.8 V tuning voltage and capacitor bank off.

RF and divider outputs are connected at a probe station using 40 GHz Infinity (GSG, 2.92 mm) probes. PN measurement is performed with an R&S FSUP signal source analyzer, which provides the supply, tuning, and bias voltages. A K_a-band $(26.5 - 40\,\text{GHz})$ harmonic mixer is used to extend the frequency range of the FSUP. Fig 5 shows the PN over offset frequency measured at the divider and the RF output for the highest f_0 setting. The VCO achieves a PN of -97.28 dBc/Hz and -127.85 dBc/Hz at an offset of 1 and 10 MHz from the carrier frequency.

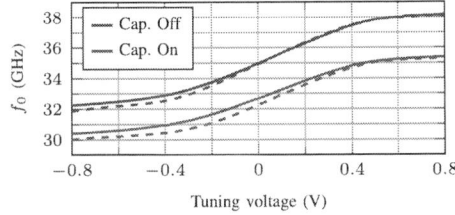

Fig. 6. Measured (solid) and simulated (dashed) frequency tuning range for upper (Cap. Off) and lower (Cap. On) tuning mode.

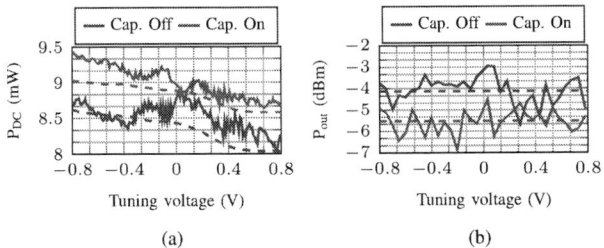

(a)   (b)

Fig. 7. (a) Measured (solid) and simulated (dashed) DC power consumption. (b) Measured output power (mean dashed).

TABLE I
COMPARISON OF STATE-OF-THE-ART INTEGRATED CMOS VCO.

Ref.	[3]	[4]	[6]	[9]	[10]	This work
Tech.	28 nm CMOS	28 nm FDSOI	22 nm FDSOI	45 nm CMOS	65 nm CMOS	**22 nm FDSOI**
f_0 (GHz)	29.25	38.95	39.6	36.3	25.4	**34.32**
TR (%)	13.3	17.1	12.1	24.8	11.8	**22.8**
P_{DC} (mW)	13/23	7.3	17.4/45.3	20/24	4.8	**9.4**
V_{DD} (V)	1	1	1.6	1.7	0.6	**0.8**
PN* ($\frac{dBc}{Hz}$)	-104	-88.3	-93.4	-93.5	-110	**-97.3**
Area (mm²)	0.15	0.016	0.015	0.027	0.012	**0.063**
FOM ($\frac{dBc}{Hz}$)	-182.2	-171.5	-172.9	-171.7	-191.3	**-178.3**
FOM$_T$ ($\frac{dBc}{Hz}$)	-184.7	-176.2	-174.6	-179.6	-192.7	**-185.5**

$$\text{FOM} = \text{PN}(\Delta f) - 20\log_{10}\left(\frac{f_0}{\Delta f}\right) + 10\log_{10}\left(\frac{P_{DC}}{1\,\text{mW}}\right)$$
$$\text{FOM}_T = \text{FOM} - 20\log_{10}\left(\frac{\text{TR}}{10}\right), \text{ *at 1 MHz offset}$$

Fig. 6 displays the measured oscillation frequency f_0 over the tuning voltage, which coincides remarkably with the simulation results. f_0 covers a frequency range from 30.4 to 35.45 GHz with an activated capacitor bank and 32.25 to 38.24 GHz while the same is switched off. This results in a total TR of 22.8% centered around 34.3 GHz. The consumed DC power of the complete VCO is measured with an N6705B power analyzer and is shown in Fig. 7a. The maximal power consumption stays below 9.4 mW over the complete TR at $V_{DD} = 0.8\,\text{V}$. A PXA N9030A spectrum analyzer records the power delivered from the buffer (Fig. 7b). Table I gives an overview of other recently integrated VCO implementations in a similar frequency range. The circuit achieves excellent performance regarding DC-power consumption and wide TR while maintaining a low PN, reflected by the high figure of merits (FOM) and FOM$_T$ of -178.3 and -185.5 dBc/Hz.

IV. CONCLUSION

This work proposed a 34 GHz CMOS VCO using a transformer-based distributed tail-node filter for PN improvement, integrated into 22 nm FDSOI technology. The circuit requires a core chip area of only 0.063 mm², including an output buffer and a power- and area-saving TSPC-FF frequency divider. A wide TR of 22.8% covering 30.4 to 38.24 GHz was achieved using SOI varactors together with a switched capacitor bank. A PN lower than -97.3 dBc/Hz at 1 MHz carrier offset could be reached due to the 2nd harmonic tail-node filter while only consuming a total DC power of 9.4 mW. Due to its compact and power-saving realization, the design is well-suited for integration into mm-wave JCAS radar systems.

ACKNOWLEDGMENT

The authors thank GlobalFoundries for silicon fabrication. This work is funded by BMBF, Germany, through the research project "ForMikro-REGGAE" under grant 16E106S1.

REFERENCES

[1] F. Probst, A. Engelmann, M. Koch, and R. Weigel, "A Dual-Channel 15 Gb/s PRBS Generator for a D-Band PMCW Radar Transmitter in 22 nm FDSOI," in *IEEE Wireless and Microwave Technology Conference (WAMICON)*, 2023, pp. 129–132.

979-8-3503-4331-1/24 $31.00 © 2024 IEEE

[2] S. Ong *et al.*, "A 22nm FDSOI Technology Optimized for RF/mmWave Applications," in *IEEE Radio Frequency Integrated Circuits Symposium (RFIC)*, 2018, pp. 72–75.

[3] Y. Hu, T. Siriburanon, and R. B. Staszewski, "A Low-Flicker-Noise 30-GHz Class-F23 Oscillator in 28-nm CMOS Using Implicit Resonance and Explicit Common-Mode Return Path," *IEEE Journal of Solid-State Circuits*, vol. 53, no. 7, pp. 1977–1987, 2018.

[4] A. A. Ihda, Y. Deval, H. Lapuyade, F. Rivet, M. Gastaldi, and S. Rochette, "A 40 GHz Varactor-less Class-C VCO with 17.1% Tuning Range and Long-Term Reliability in 28nm FD-SOI for Satellite Communications," in *29th IEEE International Conference on Electronics, Circuits and Systems (ICECS)*, 2022, pp. 1–4.

[5] E. Hegazi, H. Sjoland, and A. Abidi, "A Filtering Technique to Lower LC Oscillator Phase Noise," *IEEE Journal of Solid-State Circuits*, vol. 36, no. 12, pp. 1921–1930, 2001.

[6] L. Szilagyi, S. Li, X. Xu, P. V. Testa, A. Seidel, C. Carta, and F. Ellinger, "37.2-to-42.0 GHz VCO with -93.4 dBc/Hz Phase Noise for FMCW Radar in 22 nm FDSOI," in *16th European Microwave Integrated Circuits Conference (EuMIC)*, 2022, pp. 221–224.

[7] C. Zhang and M. Otto, "A Wide Range 60 GHz VCO Using Back-Gate Controlled Varactor in 22 nm FDSOI Technology," in *IEEE SOI-3D-Subthreshold Microelectronics Technology Unified Conference (S3S)*, 2017, pp. 1–3.

[8] F. Probst, A. Engelmann, M. Dietz, V. Issakov, and R. Weigel, "An Area Efficient Low-Power mmWave PRBS Generator in FDSOI," in *IEEE/MTT-S International Microwave Symposium (IMS)*, 2022, pp. 283–286.

[9] J. Lee, D.-W. Kang, Y. Baek, and B. Koo, "A 31.8–40.8GHz Continuously Wide-Tuning VCO Based on Class-B Oscillator using Single Varactor and Inductor," in *IEEE Radio Frequency Integrated Circuits Symposium (RFIC)*, 2018, pp. 204–207.

[10] S. Guo, P. Gui, T. Liu, T. Zhang, T. Xi, G. Wu, Y. Fan, and M. Morgan, "A Low-Voltage Low-Phase-Noise 25-GHz Two-Tank Transformer-Feedback VCO," *IEEE Transactions on Circuits and Systems I: Regular Papers*, vol. 65, no. 10, pp. 3162–3173, 2018.

D-Band VCO with Uniformly Low Phase Noise versus Frequency and Temperature

Isabel Kraus
Ruhr University Bochum
Bochum, Germany
Infineon Technologies AG
Neubiberg, Germany
Isabel.Kraus@infineon.com

Herbert Knapp
Infineon Technologies AG
Neubiberg, Germany
Herbert.Knapp@infineon.com

Nils Pohl
Ruhr University Bochum
Bochum, Germany
Fraunhofer Institute for High Frequency Physics and Radar Techniques FHR
Wachtberg, Germany
nils.pohl@ruhr-uni-bochum.de

Abstract—**This paper presents a D-band VCO with special focus on low phase noise and high-temperature stability which is enabled by adjusting the oscillation frequency over tuning voltage to its respective impedance minimum according to the negative resistance analysis. Its structure is based on a differential push-push Colpitts oscillator with second harmonic transformer output. The consumed current amounts to 40 mA with a 3.3 V source and additional 32 mA for the integrated frequency divider. A uniform differential output power of about -1.5 dBm without buffer and low phase noise of -98.7 dBc/Hz at 1 MHz offset is achieved with only little variation over tuning voltage. Phase noise measurements were conducted with particular attention to load pulling effects. In on-wafer measurements, the VCO remains fully functional for temperatures of more than 200°C which demonstrates the efficacy of the proposed design approach.**

Keywords—D-Band, low phase noise VCO, load pulling

I. INTRODUCTION

For high-resolution and reliable FMCW radar systems, the design of appropriate signal sources is of great interest. In terms of FMCW radar performance, the phase noise of the transmitted signal is a crucial characteristic [1]. For example, up-to-date automotive radar sensors operating at 77 GHz typically exhibit a single-side-band phase noise of less than -95 dBc/Hz at 1 MHz offset [2], which is therefore a benchmark for state-of-the-art performance.

For competitive VCO design, the specific frequency properties of the used technology must be considered and exploited in order to reach a high quality factor for the resonant elements, which is increasingly crucial for higher frequencies. The quality factors Q of on-chip inductors or capacitors show a contrary dependency vs. frequency, leading to a process-specific frequency where the overall Q reaches a maximum. These frequencies are favorable for low-noise signal generation and are therefore typically used as fundamental frequency in commercial low-noise signal sources. These low fundamental frequencies are then scaled up to reach the target frequency which also facilitates to obtain higher relative tuning ranges.

However, starting with such low frequencies inevitably comes along with several negative impacts. Besides the space requirement and current consumption of one or more cascaded multiplier stages, the spectral purity of the target frequency is unfavorably affected. These disadvantages considerably increase with the total multiplication factor making it unattractive exploiting higher frequency bands based on this approach. Nevertheless, due to higher resolution capability, there is a rising interest in radar systems operating at higher frequencies [3].

This paper aims for exploring the opportunities of an established BiCMOS technology to provide a D-band VCO around 140 GHz without any explicit multiplication stage, providing state-of-the-art features like very low phase noise, offering a sufficiently wide bandwidth and complying with or even surpassing typical requirements in terms of temperature stability. Furthermore, the load-pulling effect is considered in detail, how it affects or possibly impedes the phase noise measurement, and how to deal with it.

II. CIRCUIT DESIGN

The oscillator is based on a differential Colpitts topology with a transformer to obtain a push-push output where the second harmonic is extracted as shown in Fig. 1. This comes along with a significantly reduced achievable maximum output power compared to fundamental-output VCOs but then again shows the advantages of higher quality factors, good decoupling, as well as lower demands on the chip-included frequency divider.

For reliability and design considerations, the VCO was represented according to the negative resistance model to describe its oscillation capability. The effective impedance at the base of the VCO transistors was simulated over frequency and temperature. Additionally, the corresponding resonance frequency was simulated by evaluating the $\pm 180°$ passage of the phase which has to be adjusted preferably close to the input resistance minimum for reliable operation over temperature. Since this investigation of stable oscillation was conducted under small signal conditions, the nonlinear parasitic transistor capacitance is not modeled properly, whereby a slightly higher simulated oscillation frequency is expected than under real operating conditions. The simulation shown in Fig. 2 exhibits a reliably fulfilled oscillation condition for temperatures up to 200°C with the resonance frequency consistently close to the minimum input resistance. This high temperature, which considerably exceeds typical industrial standards, is chosen to validate the proposed design method of aligning impedance minimum and phase transition close to the physical limits of the used process technology. Furthermore, this alignment has the additional benefit of very little phase noise variation over tuning range, as well as general operational robustness.

For output frequency and phase noise analysis the nonlinear parasitic transistor capacitance was considered. Depending on the varactor's tuning voltage it is simulated to generate a fundamental frequency around 70 GHz in order to reach the D-band center frequency of 140 GHz at the push-push output with ≥ 10 GHz tuning range. Further objectives in simulation were low phase noise competitive to current systems operating around 78 GHz, as well as constant output power vs. frequency.

The fundamental frequency signal is tapped to connect a frequency divider chain where the line ratio is determined by simulation so that the parasitic impact on the VCO core is preferably low while ensuring a sufficient input level to the first buffer stage of the divide-by-16 chain for at least 125°C and all tuning voltages. The collector common node, corresponding to virtual ground with respect to the fundamental frequency, is directly connected to a transformer which provides decoupling between VCO core and the output signal. This allows to forego a cascode stage and therefore enables a higher voltage swing at the transistor, which contributes to lower phase noise. The secondary output of the transformer provides the doubled fundamental frequency where it can be extracted as a differential signal. Thus, the total ratio of VCO output and divider output frequency amounts to 32. The layout structure was designed with a series transmission line to the varactor which avoids any signal line crossing on different metal planes. The transformation of the varactor impedance caused by the transmission line leads to enhanced phase noise behavior in simulation.

III. MEASUREMENT RESULTS

The proposed circuit was realized in a 130 nm SiGe BiCMOS process with an f_T/f_{max} of 250 GHz/370 GHz [4]. A chip micrograph of the VCO is shown in Fig. 3 with the size of $810 \times 785\ \mu m^2$ including pads. With a 3.3 V supply the consumed current amounts to 72 mA, which can be split into a share of 32 mA for the divider chain and the remaining 40 mA consumed by the VCO core itself including its bias current.
The phase noise was measured at the divider output using an E5052B Signal Source Analyzer (SSA). The frequency is also taken from the divider output from which the output frequency of oscillation versus tuning voltage is determined. In Fig. 4a, the frequency factor of 32 is already considered. For a tuning voltage of 0 to 7 V a frequency range from 133.3 GHz to 146.7 GHz (corresponding to 13.4 GHz absolute or 9.6% relative span) is obtained which could be further extended down to 126.4 GHz for slightly negative tuning voltage. Fig. 4a shows the very close agreement as obtained by the periodical steady-state analysis for both room temperature as well as a chuck temperature of 125°C or 200°C, respectively. Further measurements showed the VCO to remain functional up to chuck temperatures of even more than 200°C and also down

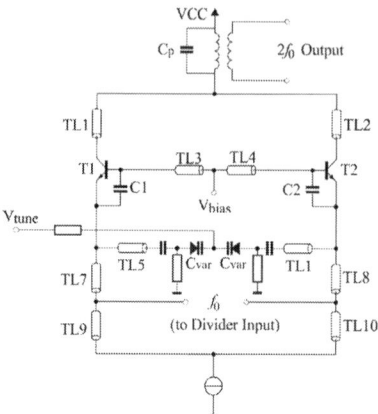

Fig. 1. Simplified schematic of the push-push Colpitts VCO.

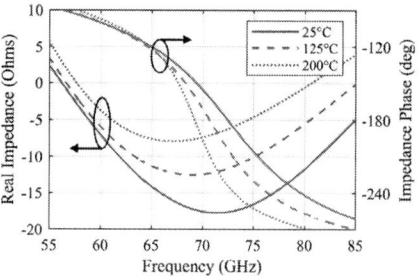

Fig. 2. Simulation of effective impedance at the VCO base transistor for a tuning voltage of 2 V over three temperatures: real part (left axis) and phase (right axis).

Fig. 3. Photograph of the VCO chip.

to -40°C which demonstrates the effectiveness of the chosen design approach according to Fig. 2. For output power measurement, an Erickson PM5B power meter was used with a D-band waveguide probe connected to one of the output signal pads while the other one was terminated with 50 Ohms to ground on chip. Since a single-ended RF probe was used, 3 dB are added. The additional losses due to probe and waveguide transitions where determined to 3.4 dB and de-embedded. This leads to a maximum expected differential VCO output power of about -1.3 dBm with a total variation within about 1 dB for all tuning voltages from 0 to 7 V.

IV. PHASE NOISE AND LOAD PULLING CONSIDERATIONS

In order to assess the signal decoupling of the VCO core to the output, the variance of frequency and phase noise depending

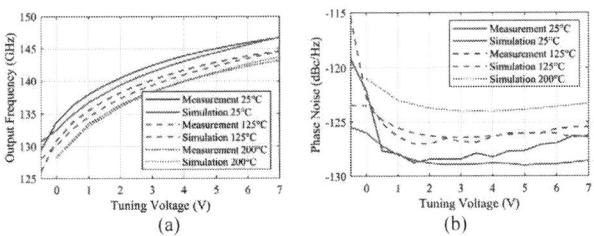

Fig. 4. VCO output frequency (a) and phase noise (1 MHz offset) at divider output (b) over tuning voltage, comparison of measurement and simulation.

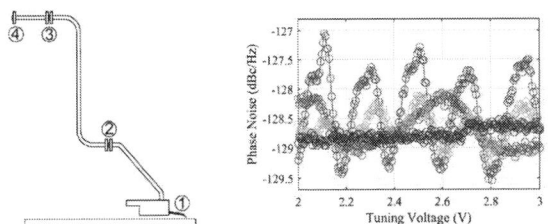

Fig. 5. Phase noise at divider output for 5 mV steps in tuning voltage for four different load scenarios: open pad (1, purple), with contacted RF probe (2, red), with additional S-bend waveguide (3, blue), and with additional attenuator (4, yellow).

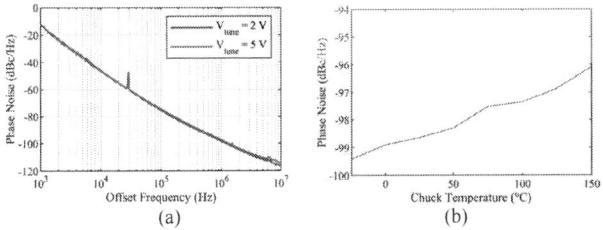

Fig. 6. Measured phase noise over frequency at 25°C for two different tuning voltages (a) and measured phase noise at 1 MHz offset over temperature for a tuning voltage of 2 V (b), both scaled to VCO output frequency by adding 30 dB offset.

on the presence of an external load and its variation was examined. This load pulling can significantly affect the phase noise performance of an VCO, as it alters the operating conditions and can cause its resonant frequency and output power to deviate from their nominal values. Load pulling can also cause unintentional discrepancies in general phase noise measurements as demonstrated in [5]. In order to evaluate load pulling effects during phase noise measurements, it is expedient to perform measurements under different load conditions. A fine step size of the tuning voltage is necessary to identify faulty measuring points due to load-induced resonance effects which would lead to overly optimistic phase noise values.

The phase noise and output frequency measurements were conducted with a step of only 5 mV for the tuning voltage of which the section between 2 V and 3 V is shown in Fig. 5 at the divider output for four different load scenarios at 1 MHz offset. In case of the blue line, a moderate fluctuation can be observed with a periodicity of approximately 430 MHz. This is attributed to the RF probe tip contacting the pad which is followed by a waveguide transition and an open-ended S-shaped waveguide.

TABLE I: COMPARISON OF PREVIOUSLY PUBLISHED D-BAND VCOs.

	[6]	[7]	[8]	[9]	This work
Technology	350 nm SiGe	130 nm SiGe	22 nm CMOS SOI	40 nm CMOS	130 nm SiGe
Center Frequ.	144 GHz	153 GHz	146.6 GHz	132.9 GHz	140 GHz
Tuning Range	39 GHz	12 GHz	17.1 GHz	20 GHz	13 GHz
Topology	VCO+doubler	fundamental	push-push	VCO+doubler	push-push
V_{DD}	5 V	5 V	0.9 V	1 V	3.3 V
P_{DC}[1]	240 mW+75 mW	132 mW	12.2 mW	51 mW	132 mW
Output Power	3 dBm	9 dBm	-16 dBm	-2 dBm	-1.3 dBm
Phase Noise (1 MHz offset)	-93 dBc/Hz	-96 dBc/Hz	-90.1 dBc/Hz (10 MHz offset)	-96.5 dBc/Hz	-98.7 dBc/Hz
FOM [2]	-171.2 dBc/Hz	-178.5 dBc/Hz	-162.6 dBc/Hz	-181.9 dBc/Hz	-180.4 dBc/Hz

[1] For signal generation only
[2] FOM = $\mathscr{L}(\Delta f) - 20\lg(f_0/\Delta f) + 10\lg(P_{DC}/\text{mW})$

The estimated total electrical length matches well with the observed periodicity. By removing the S-shaped waveguide both the amplitude of the fluctuation and its periodicity is decreased as shown in red. For the yellow line, the S-bend is

added again followed by an additional attenuator which caused a further slight decrease in amplitude of the variation.

Since all these measurement curves are very even and sinusoid in shape it can be assumed that variations on the external load exhibited to the VCO output pad cause only moderate fluctuations which indicates good decoupling of the VCO core. Therefore, the steady measurement results without contacted probe pictured in purple at around -128.7 dBc/Hz can be assumed as reliable results for the shown tuning voltage.

Fig. 4b shows the measured phase noise at the divider output at a carrier offset frequency of 1 MHz as a function of the tuning voltage. There is a minimum value of -129 dBc/Hz around the center of the tuning range. Considering the 30 dB offset, the expected phase noise at the VCO output is -96 dBc/Hz or better at 1 MHz offset within a frequency range of 135.6 to 146.6 GHz or tuning voltage range of 0.5 to 7 V, respectively. Fig. 6a depicts two exemplary curves of phase noise over offset frequency which are measured with the SSA at the divider output. In this case, the 30 dB offset with respect to the VCO output is already considered. The heating of the chuck up to 125°C led to moderate and uniform increase of the noise as shown in Fig. 4b. For 200°C, only simulation data is available since the on-chip divider failed for temperatures above 150°C. As shown in Fig. 6b, the phase noise only varies by about 3.5 dB within a temperature range from -25°C to +150°C.

V. CONCLUSIONS

In comparison with state-of-the-art implementations shown in Table I, the proposed push-push VCO is characterized by exceptionally flat low phase noise behavior within a competitive tuning range even for high temperatures which is rarely shown in literature for SiGe BiCMOS circuits. The results confirm that this design approach enables excellent performance close to the limits imposed by the available RF performance of the technology. The VCO entirely meets or even exceeds typical phase noise standards for 77 GHz FMCW radar, despite of nearly twice the frequency, with a minimum value of less than -98 dBc/Hz (1 MHz offset) at room temperature which is among the best published in literature.

ACKNOWLEDGMENT

This work is part of the Eureka XECS project InnoStar which is funded by the German Federal Ministry of Education and Research under grant 16ME0464.

REFERENCES

[1] K. Siddiq, R. J. Watson, S. R. Pennock, P. Avery, R. Poulton and B. Dakin-Norris, "Phase noise analysis in FMCW radar systems," 2015 European Microwave Conference (EuMC), Paris, France, 2015, pp. 1523-1526.

[2] J. Vovnoboy, R. Levinger, N. Mazor and D. Elad, "A fully integrated 75–83 GHz FMCW synthesizer for automotive radar applications with −97 dBc/Hz phase noise at 1 MHz offset and 100 GHz/mSec maximal chirp rate," 2017 IEEE Radio Frequency Integrated Circuits Symposium (RFIC), Honolulu, HI, USA, 2017, pp. 96-99.

[3] T. Weber, "ITS, automotive sensors and future railway radiocommunications," Eur. Commun. Office, Int. Telecommun. Union, Saint Petersburg, Russia, Jun. 2018.

[4] J. Böck et al., "SiGe HBT and BiCMOS process integration optimization within the DOTSEVEN project," 2015 IEEE Bipolar/BiCMOS Circuits and Technology Meeting - BCTM, 2015, pp. 121-124.

[5] M. Schott, F. Lenk, and P. Heymann, "On the load-pull effect in MMIC oscillator measurements," in Proc. 33rd European Microwave Conference, Oct. 2003, pp. 367–370.

[6] C. Bredendiek, N. Pohl, K. Aufinger and A. Bilgic, "An ultra-wideband D-Band signal source chip using a fundamental VCO with frequency doubler in a SiGe bipolar technology," 2012 IEEE Radio Frequency Integrated Circuits Symposium, Montreal, QC, Canada, 2012, pp. 83-86.

[7] F. Ahmed, M. Furqan, B. Heinemann and A. Stelzer, "A SiGe-Based D-Band Fundamental-Wave VCO with 9 dBm Output Power and -185 dBc/Hz FoMT," 2015 IEEE Compound Semiconductor Integrated Circuit Symposium (CSICS), New Orleans, LA, USA, 2015, pp. 1-4.

[8] Y. Shafiullah et al., "A Low-Power Push-Push D-Band VCO with 11.6% FTR utilizing Back-gate Control in 22nm FDSOI," 2023 IEEE Radio and Wireless Symposium (RWS), Las Vegas, NV, USA, 2023, pp. 20-23.

[9] Y. -T. Chang and H. -C. Lu, "A D-band wide tuning range VCO using switching transformer," 2017 IEEE MTT-S International Microwave Symposium (IMS), Honololu, HI, USA, 2017, pp. 1353-1356.

979-8-3503-4331-1/24 $31.00 © 2024 IEEE

Voltage-Controlled-Oscillator Using an 8-shaped Transformer-coupled Transmission Line

Sheng-Lyang Jang
Department of Electronic Engineering
National Taiwan University of Science and
Technology
Taipei, Taiwan, R.O.C
sljjj@mail.ntust.edu.tw

Yi-Ping Hsieh
Department of Electronic Engineering
National Taiwan University of Science and
Technology
Taipei, Taiwan, R.O.C
m10702342@mail.ntust.edu.tw

Wen-Cheng Lai
Department of Electrical Engineering
Ming Chi University of Technology,
Taiwan, R.O.C.
Wenlai@mail.mcut.edu.tw

Abstract—**This letter presents a 3:1 8-shaped transformer used to form an 8-shaped transformer-coupled transmission line consisting of four back-to-back connected transformers. The stacked transformers form a closed loop. By placing cross-coupled inverter pairs in the loop, a VCO is built. The die area of the VCO in the TSMC 0.18 μm CMOS process is 1.2×1.2 mm². The measured phase noise of the eight/four-phase VCO at 5.43 GHz is -114.82 dBc/Hz and the VCO figure of merit (FOM) is -180.128 dBc/Hz. The wave on the loop is a traveling wave. The 8-shaped transformer reduces magnetic field coupling and limits the Electro-Magnetic Compatibility (EMC) issue.**

Keywords—voltage-controlled oscillator (VCO), 8-shaped transformer-coupled transmission line, magnetic field coupling.

I. INTRODUCTION

Nowadays, the demand for wireless communication continues to expand, and more and more on-chip spiral inductors on fully integrated radio frequency integrated circuits (RFIC). Coupled inductors can create multiple resonant frequencies in a compact high-order resonator and crosstalk coupling effects. Due to the latter, co-planar spiral inductors are a potential source of Electromagnetic (EM) interference. The range of applications for RF inductors includes impedance matching, resonant circuits, filters, bias circuits, etc. *LC*-tank voltage-controlled oscillator (VCO) is subject to injection pulling by a power amplifier and produces interference leakage to the low noise amplifier.

This letter designs a VCO using the transformer-coupled transmission line (TL) as shown in Fig. 1. The wave can propagate on the TL from the left input to the right output. When we tie together the inputs and outputs of the TL, the wave on the closed loop transformer-coupled TL becomes a rotary traveling wave (RTW). The VCO uses four cross-coupled inverters embedded in the closed-loop TL in uniform spacing to provide the energy to sustain the oscillation. Similar to other RTW TL, multiple-phase signals are easily available from different positions on the transmission line [1], [2]. In the proposed VCO, we implement the transformer as a 3:1 8-

Fig. 1. Schematic of transformer-coupled T-line.

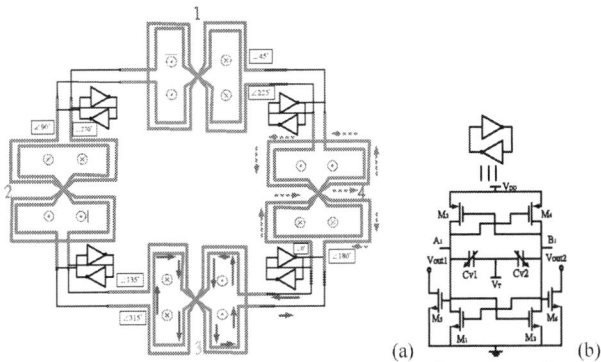

Fig. 2. Schematic of the proposed CMOS QVCO with four 1:1 transformers. The insert shows the cross-coupled inverter. The real transformer uses a 3:1 transformer. The arrow shows the current direction.

shaped transformer for suppressing the magnetic field coupling so the whole closed-loop TL is EM noise suppressive. The construction of an 8-shaped primary inductor has a twisted loop to direct the current flow in such a way that the net magnetic fields generated in the two lobes offset each other.

II. CIRCUIT DESIGN

Fig. 2 shows the circuit schematic of the VCO using the transformer-coupled T-line with four transformers in a twisted connection. The four concentric 1:1 transformers are identical. Accumulation-mode varactors C_{v1}, C_{v2}, and ($M_1 \sim M_4$) form the cross-coupled inverter pair. V_T is the control bias. V_{DD} is the supply, and V_O is the drain bias voltage of buffer FET. NMOSFETs (M_5, M_6) form the output buffers. The currents injected by the cross-coupled pairs travel in both clockwise and counter-clockwise directions to form a rotary traveling wave. Fig. 3(a) shows the layout of the implemented 8-shaped transformer [3], the secondary is a one-turn 8-shaped inductor

This work is supported by the project MOST 111-2221-E-011-161-, Taiwan, R.O.C.

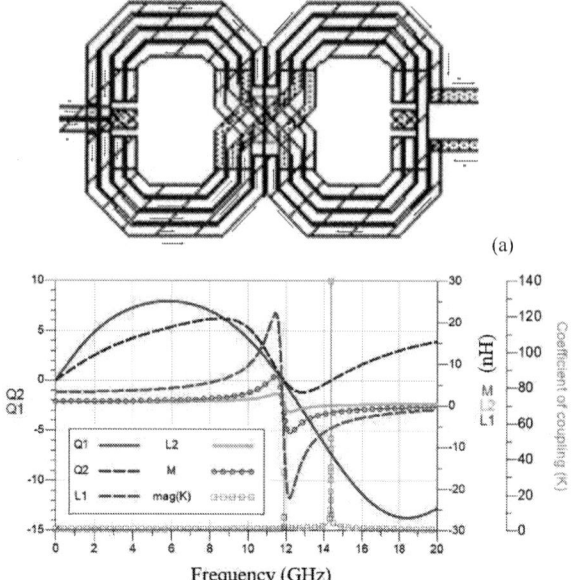

(a)

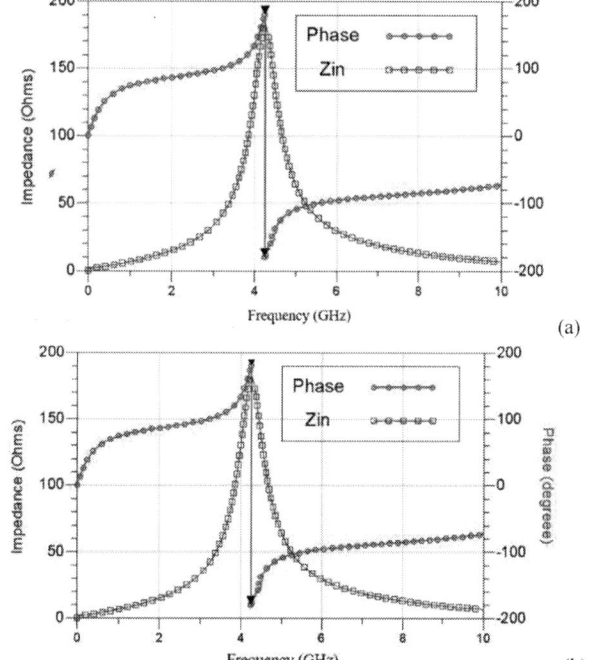

Fig. 3. (a) Layout of 8-shaped transformer. (b) Simulated transform property.

(a)

(b)

Fig. 4. Simulated impedance of one core (a) and QVCO (b).

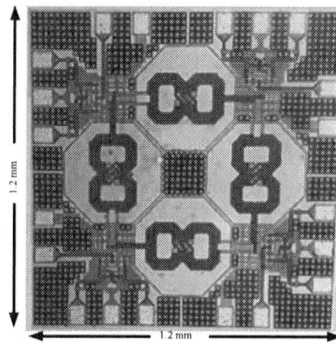

Fig. 5. Chip photograph of the proposed QVCO.

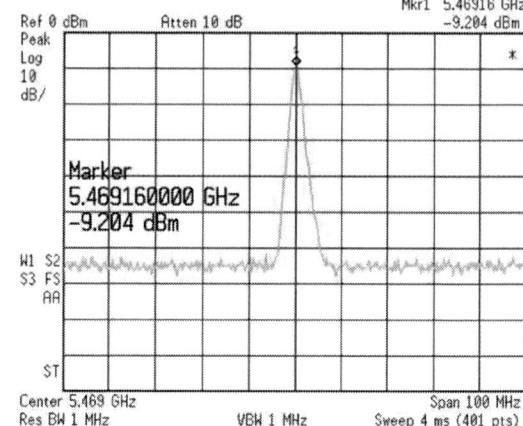

Fig. 6. Measured spectrum of the VCO. V_{DD} = 1.24 V, V_O = 1.2 V, V_T = 0.0 V.

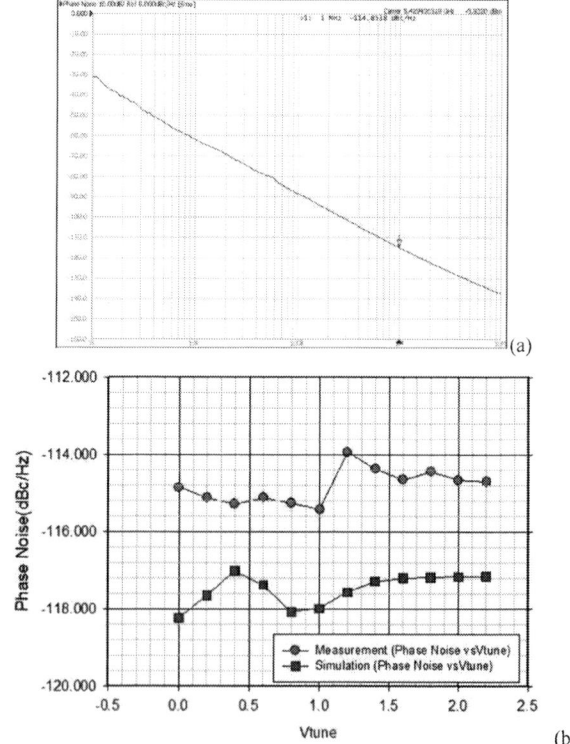

(a)

(b)

Fig. 7. (a) Measured phase noise at V_T = 0.0 V. (b) Measured/simulated phase noise at 1MHz offset versus V_T. V_{DD} = 1.18 V, V_O = 1.2 V.

occupying the most outer turn. The primary consists of a 3-turn octagonal inductor in series with a 3-turn octagonal inductor. Fig. 3(b) shows the simulated transformer property. At 5.4 GHz, the simulated transformer parameters are L_1=3.903nH, L_1=0.902nH, Q_1=7.884, Q_2=5.064, K=0.608. Peak Q_2 /Q_1 is 7.9/8.55.

Fig. 4(a) shows the simulated impedance of sub-VCO, the resonant frequency is 4.24GHz. Fig. 4(b) shows the simulated impedance of QVCO which consists of four sub-VCOs. The resonant frequency 4.03GHz. The transformer coupling

979-8-3503-4331-1/24 $31.00 © 2024 IEEE   96

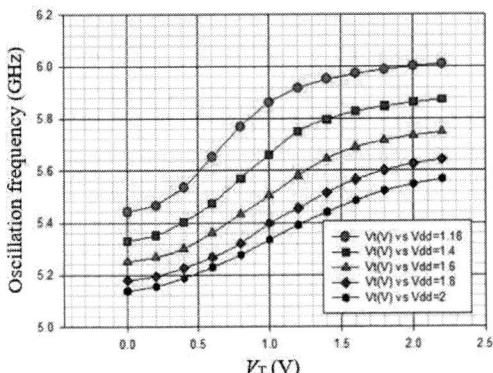

Fig. 8. Measured/simulated phase noise at 1MHz offset versus V_T. V_{DD} = 1.18~2 V, V_O = 1.2 V.

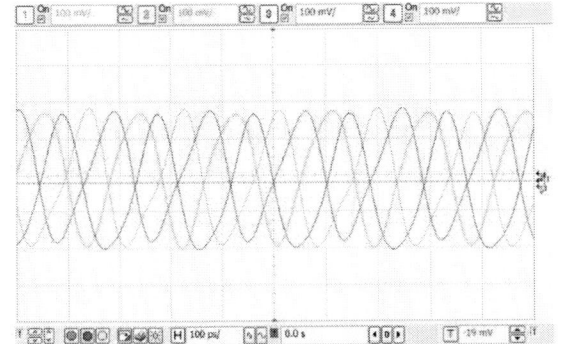

Fig. 9. Measured four outputs of the VCO.

Table I Performance Comparison of standalone CMOS RTWOs

Ref.	Tech (μm)	f_{osc} (GHz)	Vdd(V) Pdis,mW	Area (mm^2)	PN @ 1MHz dBc/Hz	FOM dBc/ Hz
[9]	0.18	5.28	1.8/129	1.5×1.5	-102.0	-155.35
[8]	0.25	18	4/54	-	-98.0	-165.8
[7]	0.13	12.2	1.2/30	0.3×0.35	-105.2	-171.95
[6]	0.18	23.6	1.8/70.2	0.8	-105.0	-164
[5]	0.18	2.79	1.1/4.78	1.2×1.2	-121.4	-181.8
[4]	0.18	3.26	1.0/5.33	1.07×1.07	-122.14	-185.11
This	0.18	5.43	1.18/8.73	1.2×1.2	-114.82	-180.13

increases the parasitic and reduces the resonant frequency.

III. MEASUREMENT RESULTS

Fig. 5 shows the chip photograph of the QVCO. The die area of the VCO in the TSMC 0.18 μm CMOS process is 1.2×1.2 mm^2. Fig. 6 shows the measured spectrum of the VCO. The oscillation frequency is 5.47 GHz and the output power is -9.204 dBm. Fig. 7(a) shows measured phase noise at V_T = 0 V. At the carrier 5.43 GHz, the phase noise at 1 MHz offset is -114.85 dBc/Hz. The phase noise is owing to $1/f^3$ and $1/f^2$-shaped noises with corner frequency around 300KHz. Fig. 7(b) shows measured phase noise at 1MHz offset versus V_T. At V_{DD} = 1.18 V the power consumption is 8.732 mW. The figure of merit (FOM) is a benchmark parameter of VCOs and takes into account the oscillator phase noise, power consumption, and operating frequency and it is equal to:

$$FOM = \varsigma(\omega_0, \Delta\omega) + 10 \cdot \log(P_{DC}) - 20 \cdot \log\left(\frac{\omega_o}{\Delta\omega}\right) \qquad (1)$$

The VCO FOM is -180.128dBc/Hz without counting for buffer power. Fig. 8 shows the measured tuning range of the VCO versus V_T. At V_{DD} = 1.18 V, the oscillation frequency is tunable from 5.435 to 6.01 GHz. As V_{DD} increases, the oscillation frequency decreases, this is caused by the voltage swing effect on the tank capacitance. As V_T increases, the oscillation frequency increases, this is caused by the varactor capacitance reduction with V_T. Fig. 9 shows four measured outputs of the VCO, quadrature signals are extracted.

IV. CONCLUSION

This letter presents a new CMOS eight/quadrature-phase VCO in the TSMC 0.18 μm CMOS using a closed loop 8-shaped transformer-coupled TL. The coupled 8-shaped transformer TL shows the suppression of magnetic field coupling noise. This letter also shows stacked transformer can be used to design an oscillator. The measured VCO FOM is -180.128 dBc/Hz. The proposed TL is a possible transmission media as verified by the functional QVCO.

ACKNOWLEDGMENT

The authors would like to thank the Staff of the TSRI for the technical support.

REFERENCES

[1] K. -W. Cheng and Y. -R. Tseng, "5 GHz CMOS quadrature VCO using trifilar-transformer-coupling technology," *IEEE Microw. Compon. Lett.*, vol. 26, no. 9, pp. 717-719, Sept. 2016, doi: 10.1109/LMWC.2016.2598225.

[2] K. Takinami, R. Walsworth, S. Osman, and S. Beccue, "Phase noise analysis in rotary traveling wave oscillators using simple physical model," *IEEE Trans. Microw. Theory Tech.*, vol. 58, no. 6, pp. 1465–1474, Jun., 2010.

[3] H. -C. Lee, S. -L. Jang and R. -X. Yang, "Low power CMOS VCO using an 8-shaped transformer," *2023 IEEE 23rd Topical Meeting on Silicon Monolithic Integrated Circuits in RF Systems*, Las Vegas, NV, USA, 2023, pp. 19-21, doi: 10.1109/SiRF56960.2023.10046255.

[4] W. -C. Lai, S. -L. Jang and J. -W. Syu, "Quadrature VCO via transformer-coupled transmission line," *2019 12th International Workshop on the Electromagnetic Compatibility of Integrated Circuits (EMC Compo)*, 2019, pp. 2-5, doi: 10.1109/EMCCompo.2019.8919938.

[5] W. -C. Lai, S. -L. Jang and J. -J. Wang, "Dual-band quadrature VCO using switched-transformer coupling for wireless robot applications," *2019 4th Asia-Pacific Conference on Intelligent Robot Systems (ACIRS)*, 2019, pp. 109-112, doi: 10.1109/ACIRS.2019.8935963.

[6] Y. Y. Peng, X. P. Yu, J. M. Gu, W. M. Lim and W. Q. Sui, "An area-efficient CRLH (Composite Right/Left-Handed)-TL approach to the design of rotary traveling-wave oscillator," *IEEE Microw. Wireless Compon. Lett.*, vol. 23, no. 10, pp. 560-562, Oct. 2013, doi: 10.1109/LMWC.2013.2252424.

[7] F. Ben Abdeljelil, W. Tatinian, L. Carpineto, and G. Jacquemod, "Design of a CMOS 12 GHz rotary travelling wave oscillator with switched capacitor tuning," *2009 IEEE Radio Frequency Integrated Circuits Symposium*, 2009, pp. 579-582, doi: 10.1109/RFIC.2009.5135608.

[8] K. T. Ansari, T. N. Ross and C. Plett, "Ku-band high output power multiphase rotary travelling-wave VCO in SiGe BiCMOS," *Microwave Integrated Circuits Conference (EuMIC)*, 2013 European, Nuremberg, 2013, pp. 97-100

[9] X. Hu, Y. Dai, H. Zhang, J. Zhou, and K. Chen, " Rotary traveling-wave oscillator design using 0.18 μm CMOS," *J. Semicond.* 2010, 31(6).

Single-Voltage-Supply pHEMT/mHEMT 2.4 and 5.8 GHz LNAs Using Power Constrained Design

Chinchun Meng
Department of Electronics and Electrical Engineering
National Yang Ming Chiao Tung University
Hsinchu, Taiwan
ccmeng@nycu.edu.tw

Chung-Yo Lin
Department of Electronics and Electrical Engineering
National Yang Ming Chiao Tung University
Hsinchu, Taiwan

Guo-Wei Huang
Taiwan Semiconductor Research Institute
National Applied Research Laboratories
Hsinchu, Taiwan

Abstract—2.4-GHz and 5.8-GHz single-voltage-supply LNAs using depletion-mode pHEMT and mHEMT technologies are demonstrated in this paper. Both pHEMT and mHEMT inductively source-degenerated LNAs with a common-drain output stage are designed around 2.4-GHz and 5.8-GHz for a single-band application. The current consumption of 11 mA at 3 V supply voltage is intentionally designed for all of the LNAs to make a performance comparison. At 2.4-GHz, the 2.4-GHz pHEMT LNA has 19 dB gain and NF= 1.64 dB while the 2.4-GHz mHEMT LNA has 19 dB gain and NF=1.18 dB. At 5.8-GHz, the 5.8-GHz pHEMT LNA has 16 dB gain and NF= 1.93 dB while the 5.8-GHz mHEMT LNA has 19 dB gain and NF=1.87 dB. The experimental results imply that LNAs with mHEMT technology have better noise and gain performance than those with pHEMT technology.

Keywords—LNA, metamorphic high electron mobility transistor (mHEMT), pseudomorphic high electron mobility transistor (pHEMT), single-voltage-supply.

I. INTRODUCTION

The ability of introducing higher indium mole fraction in the channel of a HEMT structure has made the metamorphic high electron mobility transistor (mHEMT) technology of 40% indium mole fraction a better alternative choice than the pseudomorphic high electron mobility transistor (pHEMT) of 15 - 20% indium mole fraction [1][2]. The 0.15-μm mHEMT and pHEMT technologies employed in this paper have f_T of 110 GHz and 85-GHz and f_{max} of 200 GHz and 183 GHz, respectively, with the detailed characteristics of transconductance (g_m) and drain current (I_{ds}) described in reference [2]. However, the low channel breakdown voltage caused by the higher indium mole fraction in the channel prevents the mHEMT technology from the power amplifier applications. For the stand-alone LNA applications, the high f_T of an mHEMT technology can offer an LNA with lower noise figure and higher gain. The paper demonstrates the merits of the mHEMT technology over the pHEMT technology for LNA applications at 2.4-GHz and 5.8-GHz.

II. NOSIE OPTIMIZATION AND NOISE PARAMETERS OF A SINGLE-BAND INDUCTIVELY SOURCE-DEGENERATED LNA

The noise figure of an inductively source-degenerated LNA is determined by both extrinsic noise caused by the low-Q on-chip inductors and the intrinsic noise of the FET device. The semi-insulating GaAs substrate offers high-Q on chip inductors and thus the LNA noise figure of HEMT technologies depends on the intrinsic device property. Thus, a lossless input match network is assumed to discuss the circuit design of an inductively source-degenerated LNA in this paper.

An inductively source-degenerated LNA achieves the simultaneous noise and input match (SNIM) condition at the operating frequency by inserting a degeneration inductor between the source and ground of the common source amplifier [3][4][5]. A single-band LNA under the SNIM match condition at the operating frequency has the noise figure the same as the device minimum noise figure. The FET device width is selected for the noise match while the device noise figure minimum is determined by the cut-off frequency and thus the current density [4][5]. Thus, the operating current of a single-band SNIM LNA is determined with no flexibility.

The power (current) constrained condition can be added in two ways. The first way is to reduce the operating current while maintaining the same current density and source degeneration inductor as the case of the SNIM condition. This design methodology is called power constrained noise optimization (PCNO) because an input match condition is maintained while the noise match condition is not achieved [3][4]. Still, a PCNO LNA achieves a local minimum noise figure at the operating frequency by resonating the device gate-to-source capacitor with the gate and source inductors. However, the local minimum noise figure of a PCNO LNA is higher than the device minimum noise figure at the given current density.

The second way is to add a parallel capacitor between the gate and source while keeping the same current density and source degeneration inductor as the case of the SNIM condition. The simultaneous noise and input match can be achieved at a reduced device width when compared to the device width of a SNIM condition. Thus, a power-constrained simultaneous noise and input match (PCSNIM) condition is achieved [4][5]. However, the total capacitance C_t (the summation of gate-source capacitance C_{gs} and the added parallel capacitance) becomes larger and then a large gate inductor is needed for the resonant operating frequency [5]. Thus, the device width cannot be reduced indefinitely because the extrinsic noise associated with the large gate inductor sets a lower bound for the device width.

For a single-band FET LNA, the FET device size based on PCSNIM is determined by the following equation [5].

$$g_m Z_o K_g = \frac{\omega_T}{\omega_o} \sqrt{K_g K_r} \qquad (1)$$

Here, g_m is the device conductance, ω_T is the device cut-off frequency, Z_o is the port termination impedance, K_g and K_r are the dimensionless proportional constants of noise conductance and uncorrelated noise resistance of the FET device, respectively [5][6][7]. The product of K_g and K_r is independent of C_{gs}/C_t ratio [7].

The optimum noise resistance $R_{opt}(\omega)$ and reactance $X_{opt}(\omega)$ of a single-band FET LNA with the PCSNIM design is written as

$$\frac{R_{opt}(\omega)}{Z_o} = \frac{\omega_T}{\omega}\frac{\sqrt{K_r K_g}}{g_m Z_0 K_g} = \frac{\omega_o}{\omega} \qquad (2)$$

$$X_{opt}(\omega) = -\omega L_1 + \frac{K_c}{\omega C_1}. \qquad (3)$$

Here, K_c is the dimensionless proportional constant of the correlated noise reactance of the FET device and is close to one for any C_{gs}/C_t ratio. L_1 is the summation of the gate inductor and source inductor [5][6][7].

The noise conductance $G_n(\omega)$ of a single-band FET LNA with PCSNIM design is obtained as follows:

$$G_n(\omega)Z_0 = \left(\frac{\omega}{\omega_T}\right)^2 g_m Z_o K_g = \frac{\omega^2}{\omega_T \omega_o}\sqrt{K_g K_r} \quad (4)$$

The minimum noise figure $F_{min}(\omega)$ of a single-band FET LNA with PCSNIM design is expressed as follows:

$$F_{\min}(\omega) = 1 + 2\frac{\omega}{\omega_T}\sqrt{K_g K_r} = 1 + \sqrt{\frac{3}{5}}\frac{\omega}{\omega_T} = 1 + 0.77\frac{\omega}{\omega_T} \quad (5)$$

Thus, all the noise parameters of a single-band FET LNA with PCSNIM are independent of the C_{gs}/C_t ratio [8].

The local minimum noise figure of a PCNO LNA is expressed as follows [3][9][10].

$$F_{\min,P_D}(\omega) = 1 + \sqrt{\frac{16}{15}}\delta\gamma\frac{\omega}{\omega_T}\left(1 - 2|c|\sqrt{\frac{\delta\alpha^2}{5\gamma}} + \frac{\delta\alpha^2}{5\gamma}\right) \quad (6)$$

$$= 1 + \frac{8\sqrt{3}}{15}\frac{\omega}{\omega_T} = 1 + 0.924\frac{\omega}{\omega_T}$$

III. 2.4-GHz LNA DESIGN AND MEASURED RESULTS

Fig. 1 and 2 illustrate the schematics and die micrographs of 2.4-GHz pHEMT and mHEMT LNAs, respectively, using the gate-source-connected biasing transistors for the single-supply-voltage operation. As expected, a narrow band input match of the inductively source-degenerated CS stage at 2.4-GHz and a wideband output match of the CD stage are shown in Fig. 3.

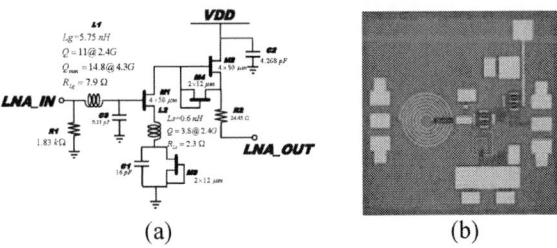

Fig. 1. (a) Schematic and (b) die micrograph of the 2.4-GHz single-voltage-supply pHEMT LNA. The die size is 1mm x 1mm

The power gain at 2.4-GHz is 19 dB for the pHEMT LNA and 19 dB for the mHEMT LNA as shown in Fig. 4. Power performance is shown in Fig. 5. The 2.4-GHz pHEMT LNA has IP1dB of -24 dBm and IIP3 of -12.5 dBm while the 2.4-GHz mHEMT LNA has IP1dB of -20 dBm and IIP3 of -8 dBm. The SNIM design condition is employed for the 2.4-GHz pHEMT LNA and 11 mA biasing current at 3V supply voltage is thus obtained. To make a fair comparison between all of the LNAs in this paper, a power (current) constrained condition of 11 mA at 3 V supply voltage is exclusively employed.

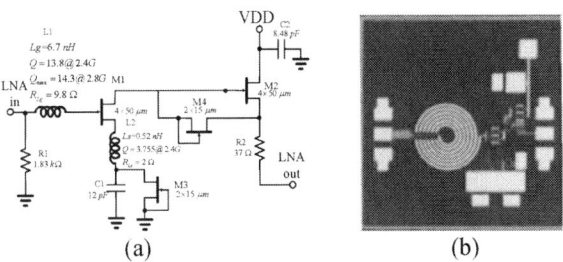

Fig. 2. (a) Schematic and (b) die micrograph of the 2.4-GHz single-voltage-supply mHEMT LNA. The die size is 1mm x 1mm.

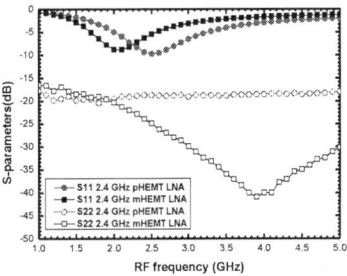

Fig. 3. Input and output return loss of the 2.4-GHz single-voltage-supply pHEMT and mHEMT LNAs.

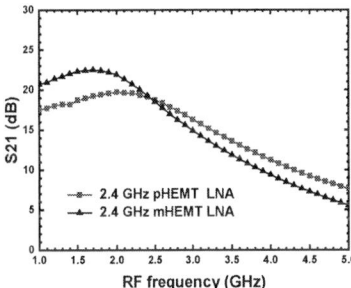

Fig. 4. Power gain of the 2.4-GHz single-voltage-supply pHEMT and mHEMT LNAs.

The device width of 4X50 μm without any parallel capacitor and thus the current density of 55 μA/μm is adopted in the pHEMT and mHEMT LNAs in Fig. 1 and 2. Thus, a PCNO condition is employed for the 2.4-GHz mHEMT LNA. At 2.4-GHz, the 2.4-GHz pHEMT LNA has noise figure of 1.64 dB with NFmin=1.23 dB while the 2.4-GHz mHEMT LNA has noise figure of 1.18 dB with NFmin=0.88 dB. As seen in Fig. 3, the input match frequency is higher for the pHEMT LNA and thus the PCSNIM condition for the 2.4-GHz pHEMT LNA shifts to a higher frequency. The PCSNIM condition of the 2.4-

979-8-3503-4331-1/24 $31.00 © 2024 IEEE

GHz pHEMT LNA is manifested by the fact that there exists a noise match at 3.5 GHz with the noise figure of 1.29 dB and minimum noise figure of 1.27 dB in Fig. 6. On the contrary, the PCNO condition of the 2.4-GHz mHEMT LNA is manifested by the fact that there is no noise match over all the frequencies and the best noise figure is achieved at 3-GHz with the noise figure of 1.08 dB and the corresponding minimum noise figure of 0.9 dB in Fig. 6. Still, there is a gap between the noise figure and the noise figure minimum at 3-GHz.

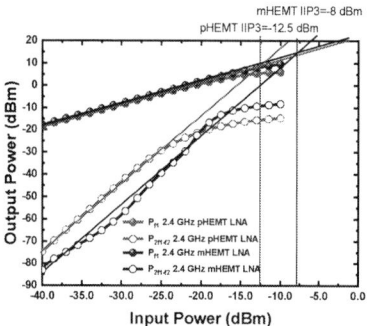

Fig. 5. Power performance of the 2.4-GHz single-voltage-supply pHEMT andIV. mHEMT LNAs.

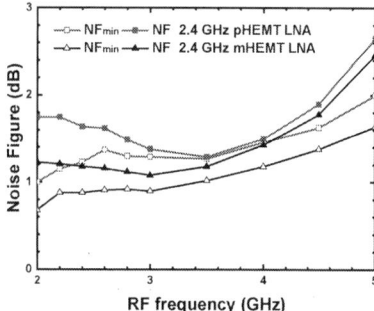

Fig. 6. Noise performance of the 2.4-GHz single-voltage-supply pHEMT and mHEMT LNAs.

IV. 5.8-GHz LNA DESIGN AND MEASURED RESULTS

Fig. 7 and 8 illustrate the schematics and die micrographs of 5.8-GHz pHEMT and mHEMT LNAs. As expected, a narrow band input match of the inductively source-degenerated CS stage around 5.8-GHz and a wideband output match of the CD stage are shown in Fig. 9. The power gain at 5.8-GHz is 16 dB for the pHEMT LNA and 19 dB for the mHEMT LNA as shown in Fig. 10. Power performance is shown in Fig. 11. The 5.8-GHz pHEMT LNA has IP1dB of -17.5 dBm and IIP3 of -6 dBm while the 5.8-GHz mHEMT LNA has IP1dB of -17 dBm and IIP3 of -7.5 dBm.

The PCSNIM design condition with a parallel capacitor between gate and source is employed for the 5.8-GHz pHEMT LNA with 11 mA biasing current at 3V supply voltage. The device width of 2X50 μm and thus the current density of 110 μA/μm is adopted in the pHEMT and mHEMT LNAs in Fig. 7 and 8. The higher current density to provide a higher cut-off frequency is need for the 5.8 GHz design. There is no parallel capacitor in Fig. 8 and thus a PCNO condition is employed for the 5.8-GHz mHEMT LNA. At 5.8-GHz, the 5.8-GHz pHEMT LNA has noise figure of 1.93 dB with NFmin=1.67 dB while the

5.8-GHz mHEMT LNA has noise figure of 1.87 dB with NFmin=1.48 dB. PCSNIM condition of the 5.8-GHz pHEMT LNA is manifested by the fact that there exists a noise match at 7 GHz with the noise figure of 1.86 dB and minimum noise figure of 1.84 dB in Fig. 12. On the contrary, the PCNO condition of the 5.8-GHz mHEMT LNA is manifested by the fact that there is no noise match over all the frequencies and the best noise figure is achieved at 5-GHz with the noise figure of 1.74 dB and the corresponding minimum noise figure of 1.26 dB in Fig. 6. Still, there is a gap between the noise figure and the noise figure minimum at 5-GHz.

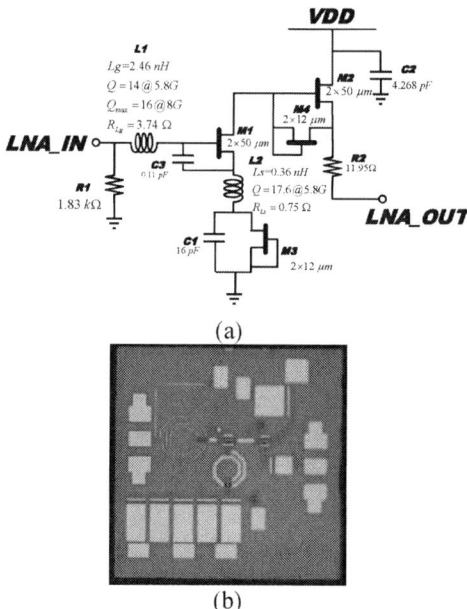

Fig. 7. (a) Schematic and (b) die micrograph of the 5.8-GHz single-voltage-supply pHEMT LNA. The die size is 1mm x 1mm.

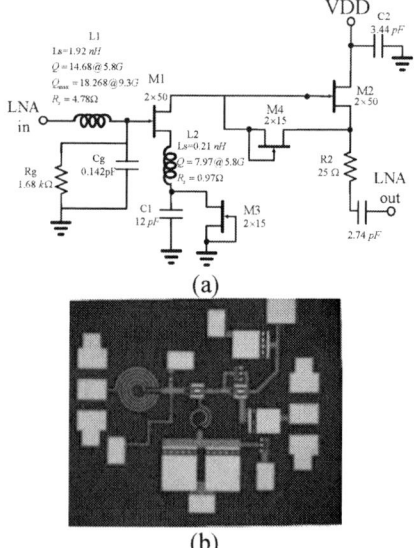

Fig. 8. (a) Schematic and (b) die micrograph of the 5.8-GHz single-voltage-supply mHEMT LNA. The die size is 1mm x 1mm.

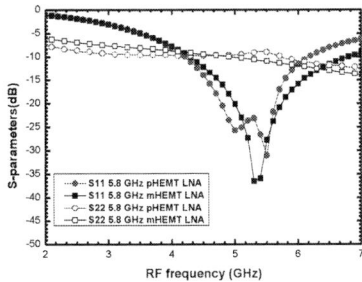

Fig. 9. Input and output return loss of the 5.8-GHz single-voltage-supply pHEMT and mHEMT LNAs.

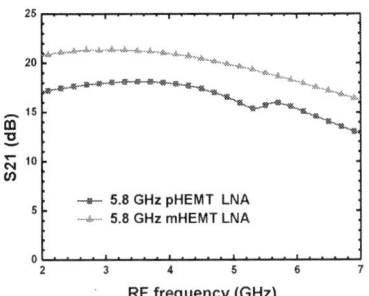

Fig. 10. Power gain of the 5.8-GHz single-voltage-supply pHEMT and mHEMT LNAs.

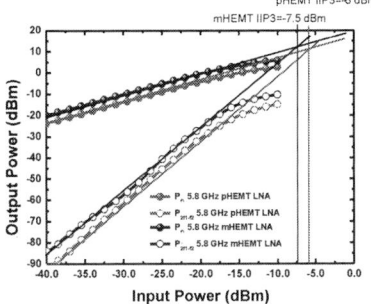

Fig. 11. Power performance of the 5.8-GHz single-voltage-supply pHEMT and mHEMT LNAs.

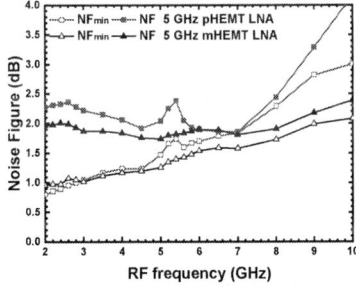

Fig. 12. Noise performance of the 5.8-GHz single-voltage-supply pHEMT and mHEMT LNAs.

V. CONCLUSION

SNIM and PCSNIM design methodologies are employed to design the 2.4-GHz and 5.8-GHz pHEMT LNAs, respectively, while PCNO design methodology is employed to design 2.4-GHz and 5.8-GHz mHEMT LNAs. All of the LNAs are designed at a supply current of 11 mA and a 3V supply voltage for comparison. The PCNO 2.4-GHz mHEMT LNA has the minimum noise figure of 0.9 dB with the noise figure of 1.08 dB at 3-GHz while the PCNO 5.8-GHz mHEMT LNA has the minimum noise figure of 1.26 dB with the noise figure of 1.74 dB at 5-GHz. Thus, it is very promising to design a PCSNIM mHEMT LNA with lower noise figure and higher gain than the PCSNIM pHEMT LNA.

ACKNOWLEDGMENT

This work is supported by National Science and Technology Council of Taiwan, Republic of China under contract numbers MOST 111-2221-E-A49-062-MY3. The authors would like to thank Taiwan Semiconductor Research Institute (TSRI) for technical support.

REFERENCES

[1] C. Kärnfelt, R. Kozhuharov, H. Zirath and I. Angelov, "High-purity 60-GHz-band single-chip ×8 multipliers in pHEMT and mHEMT technology," *IEEE Trans. Microwave Theory Tech.*, vol.54, no. 6, pp. 2887 - 2898, Jun. 2006.

[2] Jen-Yi Su, Chinchun Meng, and Po-Yi Wu, "Q-Band pHEMT and mHEMT Subharmonic Gilbert Upconversion Mixers," IEEE Microwave and Wireless Components Letters, vol. 19, no. 6, pp. 392-394 June 2009.

[3] D. K. Shaeffer, and T. H. Lee, "A 1.5-V, 1.5-GHz, CMOS low noise amplifier," *IEEE J. Solid-State Circuits*, vol.32, no. 5, pp. 745–759, May 1997.

[4] T.-K. Nguyen, C.-H. Kim, G.-J. Ihm, M.-S. Yang, and S.-G. Lee, "CMOS low-noise amplifier design optimization techniques," *IEEE Trans. Microw. Theory Techn.*, vol. 52, no. 5, pp. 1433–1442, May 2004.

[5] Yu-Chih Hsiao, Chinchun Meng, and Chun Yang, "Design Optimization of Single-/Dual-Band FET LNAs Using Noise Transformation Matrix," IEEE Trans. Microw. Theory Tech., Vol.64, No. 2, pp. 519-532, Feb. 2016.

[6] R. A. Pucel, D. J. Masse, and C. F. Krumm, "Noise performance of gallium arsenide field-effect transistors," *IEEE J. Solid-State Circuits*, vol. 11, no. SSC-2, pp. 243–255, Apr. 1976.

[7] R. A. Pucel, H. A. Haus, and H. Statz, "Signal and noise properties of Gallium arsenide field effect transistors," in *Advances in Electronics and Electron Physics*, L. Morton, Ed. New York: Academic, 1975, vol. 38, pp. 195–265.

[8] Wei Ling Chang, Chinchun Meng, Jung-Hung Ni, Kai-Chun Chang, Chih-Kai Chang, Po-Yi Lee and Yen-Lin Huang, "Analytical Noise Optimization of Single-/Dual- Band MOS LNAs With Substrate and Metal Loss Effects of Inductors," IEEE Trans. Circuits Syst. I, Reg. Papers, vol. 66, no. 7, pp. 2454–2467, July 2019.

[9] D. K. Shaeffer and T. H. Lee, "Corrections to "A 1.5-V, 1.5-GHz CMOS low-noise amplifier"," *IEEE J. Solid-State Circuits*, vol. 40, no. 6, pp. 1397–1398, Jun. 2005.

[10] D. K. Shaeffer and T. H. Lee, "Comment on Corrections to 1.5-V, 1.5-GHz CMOS Low noise amplifier," *IEEE J. Solid-State Circuits*, vol. 41, no. 10, pp. 2359, Oct. 2006.

A Comprehensive Approach to Extracting Coupling Matrix From Filtenna Measurements

Sara Javadi
Institute of Microwave and Photonic Engineering
Graz University of Technology
Graz, Austria
sara.javadi@tugraz.at

Behrooz Rezaee
Institute of Microwave and Photonic Engineering
Graz University of Technology
Graz, Austria
b.rezaee@tugraz.at

Manfred Stadler
Qualcomm Austria RFFE GmbH in Deutschlandsberg
Graz, Austria
mstadler@qti.qualcomm.com

Michael Leitner
Qualcomm Austria RFFE GmbH in Deutschlandsberg
Graz, Austria
mleitner@qti.qualcomm.com

Wolfgang Bösch
Institute of Microwave and Photonic Engineering
Graz University of Technology
Graz, Austria
wbosch@tugraz.at

Abstract— This paper presents a novel and accurate methodology for extracting the coupling matrix of lossy microwave Filtennas from experimentally measured scattering parameters. Filtennas, renowned for their hybrid integration of filter and antenna functionalities, necessitates precisely determining the coupling matrix to achieve optimal design and performance. The proposed method leverages the Simulated Annealing algorithm to effectively fit the extracted Filtenna parameters to the measured ones. Two diverse Filtenna characterized by distinct degrees and symmetric radiation nulls are employed to demonstrate the efficiency of the approach. This research significantly advances Filtenna design and optimization, facilitating improved application performance.

Keywords—coupling matrix extraction, lossy Filtenna, optimization

I. INTRODUCTION

Recently, there has been increasing attention on integrating filters with radiation performance. Filtennas combine the functionalities of antennas and filters, offering improved performance, reduced size, and increased efficiency. Integrating filtering and radiation capabilities in these modules opens up new possibilities for advanced wireless communication systems and other applications. Researchers and industry professionals are actively exploring innovative approaches to realize the full potential of these integrated multifunction modules [1]-[6].

Extracting the coupling matrix of Filtennas provides several significant benefits, such as obtaining detailed characterization of the coupling behavior between different resonators of the Filtenna and providing valuable insights into the device's interactions which enables designers to optimize the Filtenna's performance by adjusting the coupling matrix to achieve desired filter response, antenna radiation patterns, and impedance matching. Accurately extracting the coupling matrix enhances design flexibility, enables performance prediction, and ultimately leads to developing efficient and reliable Filtenna designs for various wireless communication applications.

Coupling matrix extraction for microwave filters and Filtenna is an essential area of research for microwave

engineers. Various methods have been proposed to accurately determine the coupling matrices of filters with two ports network [7]-[11]. Traditional approaches involve nonlinear optimization, fitting measured scattering parameters with calculated parameters from an unknown equivalent circuit using sequential parameter estimation and systematic tuning. However, despite their limitations and time-consuming computational demands, these methods are not applicable for coupling matrix extraction in Filtennas, a one-port device that introduces new challenges. Filtennas combine the functionalities of both filters and antennas, introducing additional complexities in characterizing their coupling matrices.

This paper presents a novel methodology to address the inherent challenge of coupling matrix extraction in Filtennas. The proposed approach involves fitting the measured scattering and radiating parameters of a Filtenna device with an ideal Filtenna model, enabling the extraction of the corresponding coupling matrix. By integrating considerations of both the filtering and radiating attributes inherent to Filtennas, this technique offers a comprehensive framework for accurately determining the coupling matrix. The method provides valuable insights into the intricate coupling behavior exhibited by Filtennas and facilitates the efficient optimization of these devices to enhance their overall performance.

II. EVALUATION OF COUPLING MATRIX

The proposed methodology begins with establishing an initial value for the coupling matrix. This initial value is derived from the ideal general coupling matrix method based on the synthesized filter's topology. The approach draws upon previous works cited in the literature to determine the appropriate initial value for the coupling matrix. By utilizing this initial value, the subsequent steps of the methodology can be initiated to refine and optimize the coupling matrix extraction process for Filtennas.

Once the ideal coupling matrix is obtained, the values for Scattering and radiating parameters are extracted from this coupling matrix. Considering the topology and the ideal coupling matrix, we selected the variables within the coupling matrix to achieve a suitable fit between the obtained parameter values and the corresponding measured parameters [11].

In this method, the optimization algorithm adjusts the variables in the coupling matrix to minimize the difference between the measured response and the response predicted by the coupling matrix. This aim was accomplished by defining a cost function that quantifies the difference between the two responses within the frequency band of interest.

$$C_1 = |S_{11}^2(f) - S_{11,m}^2(f)|$$
$$C_2 = |RG^2(f) - RG_m^2(f)| \qquad (1)$$
$$C = C_1 + C_2$$

Where S_{11} and RG are the reflection coefficient and the realized gain associated with the extracted coupling matrix, and $S_{11,m}$ and RG_m refer to the measured scattering parameter and the realized gain, respectively.

We used Simulated Annealing (SA) to find the global minimum of the cost function. The diagram of the process is shown in Fig. 1.

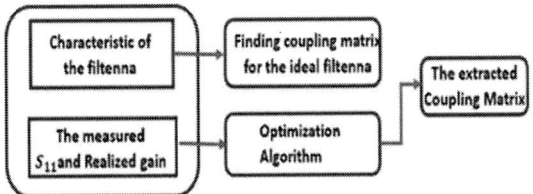

Fig. 1. Flow diagram of the coupling matrix extraction.

The main purpose of this method is to optimize the elements in the ideal coupling matrix, focusing on the elements that have the most significant impact on the desired filter Filtenna properties. The constraint of keeping the rest of the elements fixed is the most important, as it guarantees to preserve the intended Filtenna response. By fine-tuning the relevant elements, this procedure ensures compliance with the specified design requirements and determines the practicality and feasibility of the resulting coupling matrix.

III. INVESTIGATION OF THE PROPOSED METHOD

This section extensively analyses a procedure designed to showcase its effectiveness in working with a wide range of network topologies. We present two examples demonstrating the procedure's versatility, each utilizing a distinct network topology. Through thoroughly examining the results obtained from these examples, our goal is to highlight the adaptability and reliable operation of the method on different network topologies.

A. Example A

To showcase the procedure's effectiveness across different network topologies, we commenced our study by examining a Filtenna with distributed configuration depicted in Fig. 2. This specific topology corresponds to a 4th-order Filtenna with a remarkable return loss of 15 dB, realized gain of 10 dB, and operating frequency of 9.7-10.3 GHz [5].

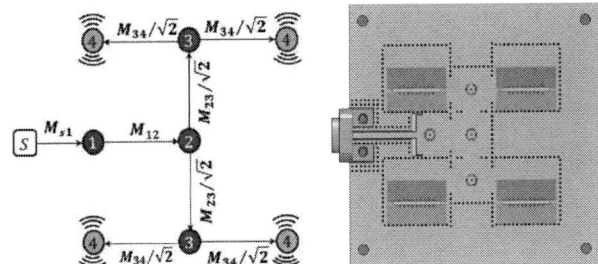

Fig. 2. The topology and layout of the proposed Filtenna in [5]

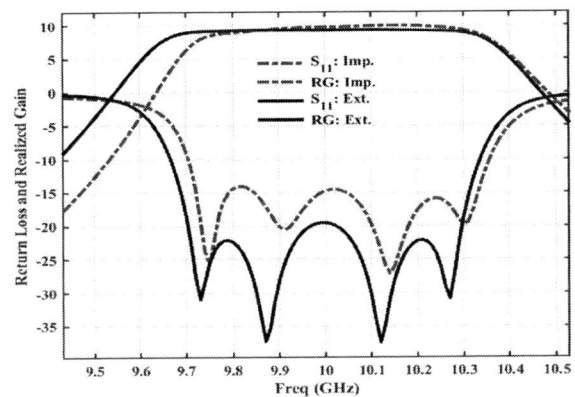

Fig. 3. Comparison of the imported (dotted line) and extracted (solid line) return loss (S_{11}) and realized gain (RG) of example B in the first step of the procedure.

Fig. 3 shows the first step in the optimization process. The procedure begins with extracting the return loss (S_11) and realized gain from the ideal Filtenna. The optimization procedure uses the ideal Filtenna parameters as a reference. It iteratively refines the coupling matrix to achieve a Filtenna response that matches the desired specifications as closely as possible.

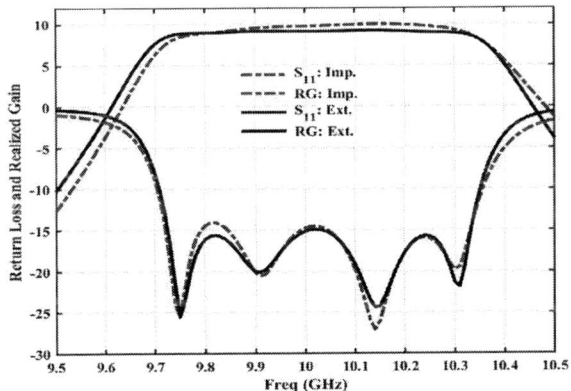

Fig. 4. Comparison of the Filtenna response after optimization.

The plots in Fig. 4 compare the measured S_11 and realized gain responses with the correspondingly optimized return loss and realized gain plots obtained by optimization. The extracted coupling matrix is as below:

$$M_E=\begin{bmatrix} 0 & 0.9892 & 0 & 0 & 0 & 0 \\ 0.9892 & -0.1072 & 0.8629 & 0 & 0 & 0 \\ 0 & 0.8629 & -0.1533 & 0.6558 & 0 & 0 \\ 0 & 0 & 0.6558 & -0.0430 & 0.7651 & 0 \\ 0 & 0 & 0 & 0.7651 & -0.1123 & 0.852 \\ 0 & 0 & 0 & 0 & 0.852 & 0 \end{bmatrix}$$

B. Example B

As the second example, we will examine a specific topology that introduces a filtering power divider loaded by two slot antennas presented in [6]. The design employs microstrip open-loop resonators to achieve a 5th-degree Chebyshev filtering response with two symmetric radiation nulls. The operating frequency of this Filtenna is 3.3-3.8 GHz, and in-band return loss and realized gain are 15 dB and 4.5 dBi, respectively [6].

Fig. 6 represents the initial step using the coupling matrix of an ideal Filtenna similar to our proposed power divider. Fig. 7 compares the imported return loss and realized gain responses with the optimized plots obtained through the optimization.

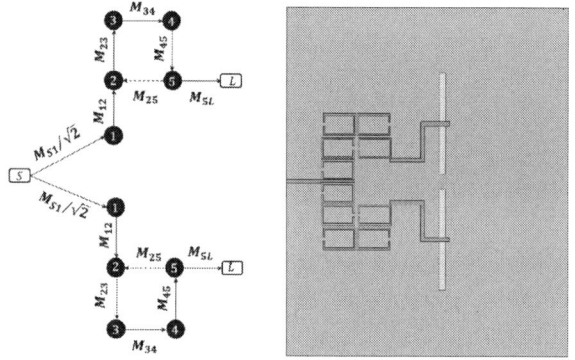

Fig. 5. The Topology and layout of the proposed Filtenna in[6]

As shown in Fig. 7, we have a great fitting in the pass band of the Filtenna; however, for out of band, the main lobe of the Filtenna is in other directions, which causes these discrepancies to appear.

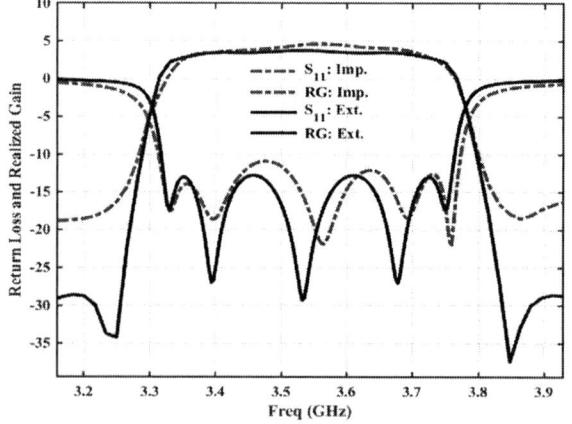

Fig. 6. Comparison of the imported (dotted line) and extracted (solid line) return loss (S_{11}) and realized gain (RG) of example B in the first step of the procedure.

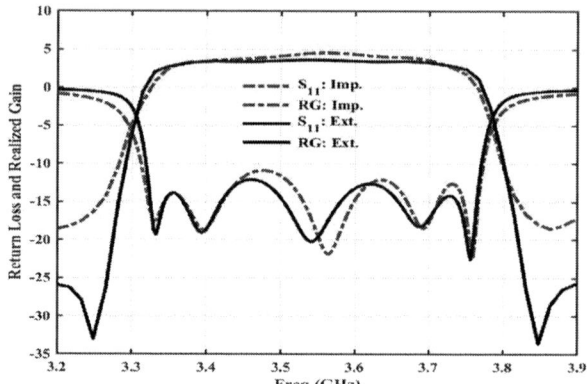

Fig. 7. filtering power divider response after optimization

The Extracted coupling matrix is as below:

$$M_E=\begin{bmatrix} 0 & 0.8515 & 0 & 0 & 0 & 0 & 0 \\ 0.8515 & -0.0293 & 0.7422 & 0 & 0 & 0 & 0 \\ 0 & 0.7422 & -0.0136 & 0.5673 & 0 & -0.2186 & 0 \\ 0 & 0 & 0.5673 & -0.0256 & 0.6791 & 0 & 0 \\ 0 & 0 & 0 & 0.6791 & -0.0132 & 0.6691 & 0 \\ 0 & 0 & -0.2186 & 0 & 0.6691 & -0.03 & 0.7920 \\ 0 & 0 & 0 & 0 & 0 & 0.7920 & 0 \end{bmatrix}$$

IV. CONCLUSION

In this study, a curve-fitting-based method was used to evaluate the coupling matrix of Filtenna based on measured return loss and realized gain. This technique was used in a frequency framework that considers losses and assumes that all resonators have the same unloaded Q. The proposed method provided a simple and accurate means of evaluating the Filtenna coupling matrix, which was not considered by previous coupling matrix methods. Finally, this method was verified by applying it to two Filtenna with different topologies.

ACKNOWLEDGMENT

The financial support by the Austrian Federal Ministry for Digital and Economic Affairs, the National Foundation for Research, Technology, and Development, and the Christian Doppler Research Association is gratefully acknowledged.

REFERENCES

[1] H. S. Farahani, B. Rezaee and W. Bösch, "Ka-band Coupled-resonator Filtering Magneto-Electric Dipole Antenna," 2020 50th European Microwave Conference (EuMC), Utrecht, Netherlands, 2021, pp. 722-725, doi: 10.23919/EuMC48046.2021.9337943.

[2] J. -F. Qian, F. -C. Chen, Y. -H. Ding, H. -T. Hu and Q. -X. Chu, "A Wide Stopband Filtering Patch Antenna and its Application in MIMO System," in IEEE Transactions on Antennas and Propagation, vol. 67, no. 1, pp. 654-658, Jan. 2019, doi: 10.1109/TAP.2018.2874764.

[3] H. S. Farahani, B. Rezaee and W. Bösch, "High Gain Filtering Lens Antenna," 2022 International Symposium on Antennas and Propagation (ISAP), Sydney, Australia, 2022, pp. 11-12, doi: 10.1109/ISAP53582.2022.9998640

[4] K. -Z. Hu, M. -C. Tang, D. Li, Y. Wang and M. Li, "Design of Compact, Single-Layered Substrate Integrated Waveguide Filtenna With Parasitic

Patch," in IEEE Transactions on Antennas and Propagation, vol. 68, no. 2, pp. 1134-1139, Feb. 2020, doi: 10.1109/TAP.2019.2938574.

[5] B. Rezaee, H. S. Farahani and W. Bösch, "Compact Cavity-backed Magneto-Electric Dipole Array Filtenna Using Hybrid Coupled-resonators," 2021 IEEE International Symposium on Antennas and Propagation and USNC-URSI Radio Science Meeting (APS/URSI), Singapore, Singapore, 2021, pp. 249-250, doi: 10.1109/APS/URSI47566.2021.9704471

[6] B. Rezaee, H. S. Farahani and W. Bösch, "An Integrated Frequency Selective Power Divider with Double-Slot Antenna Radiation and Wide Stopband for Sub-6 GHz 5G," 2021 IEEE Texas Symposium on Wireless and Microwave Circuits and Systems (WMCS), Waco, TX, USA, 2021, pp. 1-5, doi: 10.1109/WMCS52222.2021.9493298.

[7] MacChiarella, G. (2010). Extraction of unloaded Q and coupling matrix from measurements on filters with large losses. IEEE Microwave and Wireless Components Letters, 20(6), 307–309. https://doi.org/10.1109/LMWC.2010.2047455

[8] B. Liu, H. Yang, and M. J. Lancaster, "Global Optimization of Microwave Filters Based on a Surrogate Model-Assisted Evolutionary Algorithm," in IEEE Transactions on Microwave Theory and Techniques, vol. 65, no. 6, pp. 1976-1985, June 2017, doi: 10.1109/TMTT.2017.2661739.

[9] K. Yang, Y. Li, and J. Wang, "Design of a Millimeter-wave Bandpass Filter with Computer-aided Tuning Method," 2021 13th Global Symposium on Millimeter-Waves & Terahertz (GSMM), Nanjing, China, 2021, pp. 1-3, doi: 10.1109/GSMM53250.2021.9511984.

[10] M. Meng and K. -L. Wu, "An Analytical Approach to Computer-Aided Diagnosis and Tuning of Lossy Microwave Coupled Resonator Filters," in IEEE Transactions on Microwave Theory and Techniques, vol. 57, no. 12, pp. 3188-3195, Dec. 2009, doi: 10.1109/TMTT.2009.2033868

[11] S. Javadi and B. Rezaee, " Coupling Matrix Extraction From Lossy Filter Measurements," In Proceedings of the Asia-Pacific Microwave Conference (APMC), 2023 (submited)

Design of a Six-stage *W*-band Low-Noise Amplifier Using a 90-nm CMOS Technology

Rou-Yin Huang
Department of Electrical Engineering
National Central University
Jhongli, Taiwan
a109521101@g.ncu.edu.tw

Yu-Chia Su
Department of Electrical Engineering
National Central University
Jhongli, Taiwan
syc110521173@g.ncu.edu.tw

Hong-Yeh Chang
Department of Electrical Engineering
National Central University
Jhongli, Taiwan
hychang@ee.ncu.edu.tw

Abstract—This paper presents a W-band six-stage cascade, low-power, low-noise amplifier (LNA) implemented in a 90-nm CMOS process. The proposed LNA achieves gain improvement and broadband matching by cascading six common-source stages. To enhance both noise figure (NF) and the stability, a source degeneration inductor is employed at the first stage. The proposed W-band LNA demonstrates a maximum small-signal gain of 13 dB at 86.7 GHz, with a 3-dB bandwidth of 9.4 GHz (83.4-92.8 GHz), while consuming only a DC power consumption of 11.6 mW with a supply voltage of 1.2-V. The proposed LNA achieves a minimum noise figure of 8 dB at 89 GHz, remaining below 9 dB across a bandwidth of 86-90 GHz. Furthermore, the LNA exhibits an input 1-dB compression point (P_{1dB}) of -18.8 dBm and an input third-order intercept point (IIP3) of -11.2 dBm at 89 GHz. The chip size, including the RF and DC pads, is 0.73×0.81mm².

Keywords— *CMOS, low noise amplifier (LNA), millimeter-wave, RFIC, W-band*

I. INTRODUCTION

The *W*-band, spanning 75 to 110 GHz, has drawn considerable attention due to its high bandwidth, high-data rate. There are some emerging applications in *W*-band such as communication, military radar, wireless transmission, and radio astronomy systems [1]. As the primary component of the receivers in these systems, the low-noise amplifier (LNA) is required to possess sufficiently high gain and low noise figure (NF) to meet the sensitivity requirements. In general, LNAs achieve the lowest noise performance when utilizing III-V semiconductors at millimeter-wave (mm-wave) frequencies. However, they also lack the ability for high levels of integration, have higher power consumption, and are more costly compared to the CMOS process. As a result, several mm-wave LNAs have been proposed using CMOS process [2]-[6]. A 28-nm LNA with a bandwidth of 64 GHz, achieved by employing a simple *T*-type matching network is mentioned in reference[2]. Despite its wideband characteristics, it still consumes a considerable amount of dc power. In [3], a 90-nm CMOS LNA with two current-reused stages followed by a cascode stage and a current-steering cascode stage with g_m-boosting are utilized to reduce power consumption and obtain a gain higher than 20 dB. Although several 90-nm CMOS LNAs have been published in [3]-[4], covering the *W*-band, the circuit performance is still limited by the process.

In order to further challenge the limitations of the 90-nm CMOS process, this paper adopts *T*-type matching network at input and output stages to reduce component losses caused by

matching, and utilizes a six-stage cascade-cascade common-source (CS) topology to enhance gain, bandwidth, and lower dc power consumption. Additionally, the proposed LNA employs the 90-nm CMOS process instead of more expensive advanced CMOS processes, thereby reducing manufacturing costs.

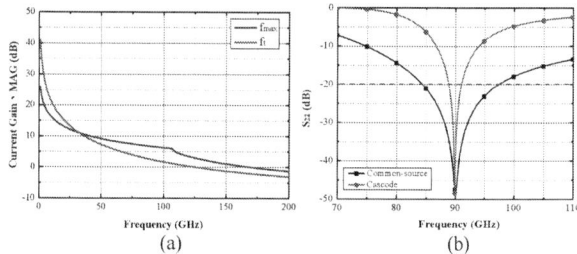

Fig. 1. (a) Simulated current gain and maximum available gain versus frequency for total gate peripheries of 8 (2×4) μm, and (b) output matching of the cascode and CS topology.

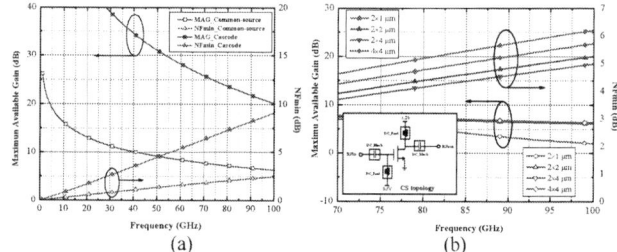

Fig. 2. Simulated MAG and NF_{min} of (a) CS versus cascode, and (b) CS with different total gate peripheries.

II. *W*-BAND LNA DESIGN

The proposed *W*-band six-stage LNA was designed and fabricated in TSMC GUTM 90-nm CMOS process. There are nine metals and one poly layer for interconnection. Fig. 1(a) shows a unity current gain frequency (f_T) of 126 GHz and a maximum a maximum oscillation frequency (f_{max}) of 171 GHz. Fig. 2(a) shows the simulated maximum available gain (MAG) and minimum noise figure (NF_{min}) for CS and cascode topologies. Within the *W*-band, although the CS topology exhibits lower gain compared to the cascode topology, it still demonstrates a lower NF_{min}. As a result, the first stage of the proposed *W*-band LNA adopts the CS topology owing to the tradeoff considerations among gain, noise figure, and power consumption. Under the aforementioned considerations, we selected a device size with a total gate periphery of 8 (2×4) μm, as shown in Fig. 2(b). Furthermore, we incorporate the source

979-8-3503-4331-1/24 $31.00 © 2024 IEEE

SiRF 2024

degeneration inductor in the CS topology, which brings the gain and noise circles closer, leading to a further reduction in the noise figure and enhancing the overall circuit stability, as shown in Fig. 4(b).

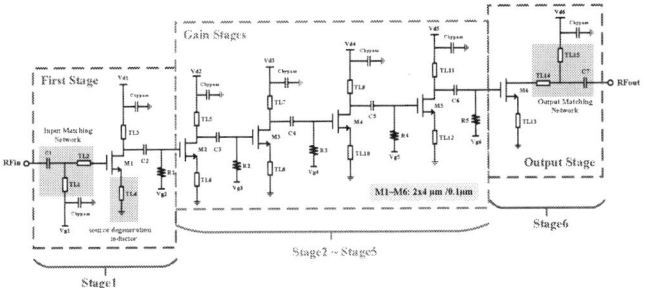

Fig. 3. Complete schematic of the proposed W-band LNA.

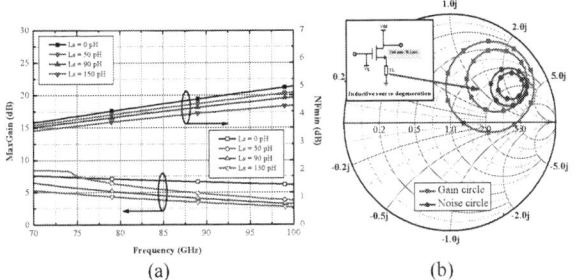

(a)   (b)

Fig. 4. (a) Simulated MAG and NF$_{min}$ under different source degeneration inductor values, and (b) Incorporating the source degeneration inductor (TL4), the gain and the noise circles become closer.

The complete schematic of the proposed W-band LNA is shown in Fig. 3. In order to achieve sufficient gain in the W-band and effectively suppress the noise contribution from the following stages, the proposed LNA consists of six CS stages. The first and second stages are designed using the CS topology to reduce noise figure while maintaining a certain small-signal gain. Similarly, the third, fourth, and fifth stages are also designed using the CS topology to increase small-signal gain by cascading. For the broadband matching, the sixth stage utilizes the CS topology to extend bandwidth and is easier to match 50 ohm at the output, as shown in Fig. 1(a). Besides, the T-type matching networks (C1, TL1, TL2, TL14, TL15, and C8) are utilized at the input of the first stage and the output of the sixth stage to achieve broadband matching, as well as the matching elements (TL3, C2, TL5, C3, TL7, C4, TL9, C5, TL11, and C6) for the interstage networks among the second, third, fourth, and fifth stages. In order to reduce chip area, the source degeneration inductor (TL4) is implemented using transmission lines. Considering the gain, noise, and stability, the value of the source degeneration inductor (TL4) is selected to be 90 pH, as shown in Fig. 5(a). The design of output matching network is depicted in Fig. 5(b), where the matching elements (TL14 and TL15) are also realized using transmission lines to reduce the loss and minimize the chip area. Taking low-power design into consideration, the drain voltage of each stage is set to 1.2 V, and the gate voltage of each stage is set to 0.5 V. The drain current is only 1.16 mA under a gate bias of 0.5 V, which corresponds to 80% of the maximum transconductance (Gm$_{max}$), as shown in Fig. 5(a). Under these bias conditions, the proposed W-band LNA can achieve a decent gain while operating at low power

consumption. Except for the gate bias of first stage, the gate biases of the other stages are fed through large resistors (R1, R2, R3, R4, R5 = 2 kΩ).

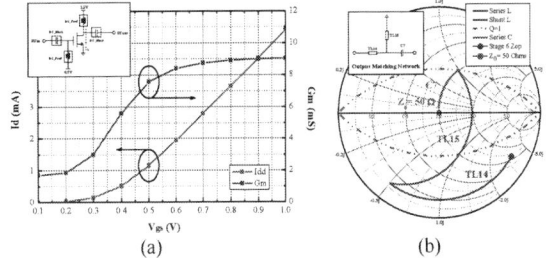

(a)   (b)

Fig. 5. (a) Simulated Gm and Id versus different Vgs at 90 GHz, and (b) the design of output T-type matching network.

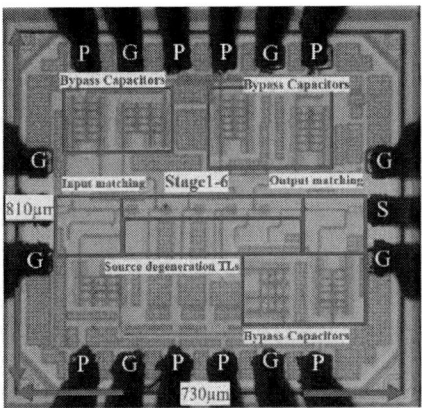

Fig. 6. Chip photo of the proposed W-band LNA, with a dimension of 0.59 mm^2 (0.73 mm × 0.81 mm).

III. MEASREMMENT RESULTS

The chip photo is shown in Fig. 6 and the chip size of this circuit is 0.59 mm^2 (0.73 mm × 0.81 mm), including RF, DC pads, and bypass capacitors. All the measurement results were obtained on-wafer probing, utilizing GSG RF probes and PGPPGP DC probes. The S-parameters are measured with Keysight PNA N5290A. Fig. 7 displays the measured and simulated S-parameters of the proposed LNA. The proposed LNA features a measured maximum small-signal gain of 13 dB with a 3-dB bandwidth from 83.4 to 92.8 GHz, and a DC power consumption of 11.4 mW. The input and output return losses of the proposed LNA are better than 5 dB over the operation bandwidth. Compared with simulation results, the measured S-parameters are in good agreement with the simulated S-parameters considering the process variations in the fabrication process. The measurement of the noise figure is conducted using a Keysight N90x0B X-series signal analyzer with the N5183B microwave signal generator. Furthermore, a Keysight N9029AV10-NSM noise source is employed along with a Keysight N9029AV10-DC9 system downconverter. Fig. 8 reveals the simulated and measured noise figure, where the minimum measured NF is 8 dB at 89 GHz. Compared with the simulated results, the measured noise figure is good agreement with the simulation considering the process variations in the fabrication process, with a difference of approximately 2 dB.

TABLE I.   PERFORMANCE SUMMARY OF THIS WORK AND PUBLISHED MM-WAVE SILICON-BASED LNAs

Reference	This work	[2]	[3]	[4]	[5]	[6]	[7]
Process	90 nm CMOS	28-nm CMOS	90-nm CMOS	90-nm CMOS	28-nm CMOS	130-nm SiGe	65-nm CMOS
Topology	6-stage CS	3-stage CS	2 CR+ Cascode + Current-STR	DTC cell + CS with GB & SRTM	5-stage CS	2-stage CE	4-stage CS
Freq (GHz)	83.4-92.8	53-117	73.7-87.8	50-67	81-91	62-110	54.4-90
Gain (dB)	13	17.7	26.1	16.8	25	13.5	17.7
NF (dB)	8	6	4.8	5.4	6	4.5	4.7
OP$_{1dB}$ (dBm)	-8	1.5	-1	-11.2	-8	0	5
IIP3 (dBm)	-11.2	N/A	N/A	-15.5	N/A	N/A	N/A
P$_{dc}$ (mW)	11.6	38.2	23	5.7	15	5.9	19
Chip Size (mm²)	0.6	0.46	0.52	0.6	0.5	0.08	0.37
FOM*	1.8	31.6	22	5.7	2	125.5	141.7

* CR presents current-reused. * Current-STR presents current-steering.

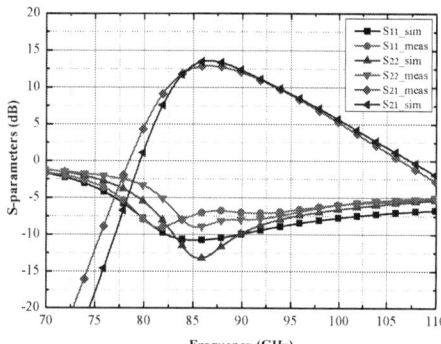

Fig. 7. Simulated and measured *S*-parameters of the proposed *W*-band LNA

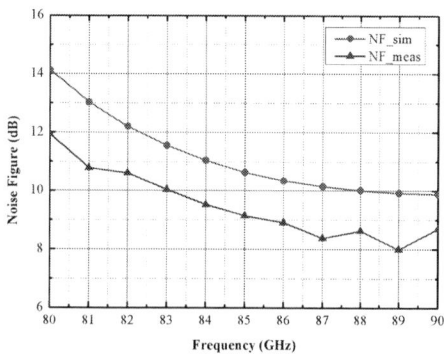

Fig. 8. Simulated and measured noise figure of the proposed *W*-band LNA.

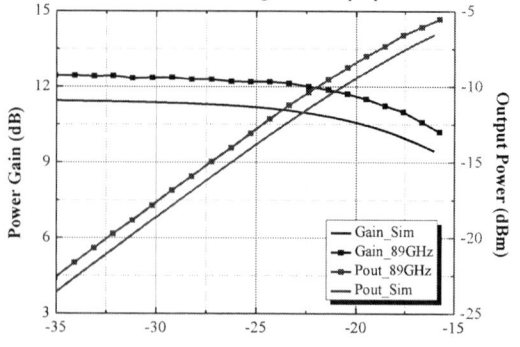

Fig. 9. Simulated and measured input 1-dB compression point (P$_{1dB}$) of the LNA at 89 GHz.

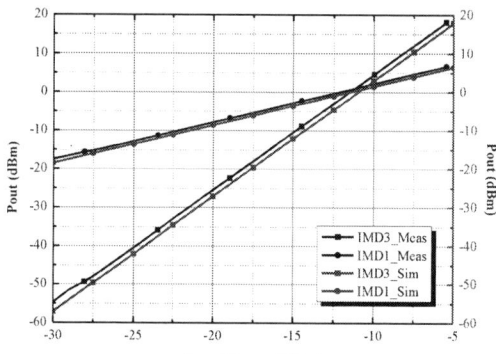

Fig. 10. Simulated and measured input third-order input intercept point (IIP3) of the LNA at 89 GHz.

The simulated and measured large-signal results of the proposed *W*-band LNA are exhibited as follows. Fig. 9 displays the simulated and measured power gain and output power versus input power. The measured power gain of the proposed LNA is higher than 12 dB at 89 GHz, and it achieves an input 1-dB compression point of -18.8 dBm. Futhermore, the proposed LNA also achieves IIP3 of -11.2 dBm with two-tone spacing of 10 MHz at 89 GHz, as depicted in Fig. 10. The performances between the proposed *W*-band LNA and other previously published CMOS LNAs are summarized and compared in Table I. A suitable figure of merit (FOM*) is expressed in (1) to evaluate the performances of the proposed *W*-band LNA.

$$FOM^* = \frac{Gain[dB] \times BW[GHz] \times OP_{1dB}[mW]}{(NF-1)[dB] \times P_{dc}[W] \times f_{max}[GHz]} \quad (1)$$

In (1), NF stands for noise figure in dB. Gain[dB] is the gain in dB. OP$_{1dB}$[mW] is the OP$_{1dB}$ in mW. BW[GHz] is the 3-dB bandwidth in GHz. P$_{dc}$[W] is the power consumption in Watt. f_{max}[GHz] is the maximum oscillation frequency in GHz

IV. CONCLUSION

In this paper, a *W*-band LNA has been proposed and fabricated in TSMC GUTM 90-nm CMOS process. The proposed LNA achieves gain and bandwidth enhancement by cascading six common-source stages, while consuming only

11.6 mW of dc power from a 1.2V supply. This LNA demonstrates low-power performance and is well-suited for application in *W*-band receivers to reduce the noise figure, as well as for further advancements in other *W*-band applications.

ACKNOWLEDGMENT

The chip was fabricated by TSMC in Hsinchu City, Taiwan. This work was supported by the National Science and Technology Council (NSTC), Taiwan, under Grant MOST 110-2221-E-008-029-MY3, and the Taiwan Semiconductor Research Institute (TSRI), Hsinchu City, Taiwan. The dc and RF probes are supported by GGB Inc., FL, USA

REFERENCES

[1] O. Inac, F. Golcuk, T. Kanar and G. M. Rebeiz, "A 90–100-GHz Phased-Array Transmit/Receive Silicon RFIC Module With Built-In Self-Test," *IEEE Trans. Microw. Theory Tech.* vol. 61, no. 10, pp. 3774-3782, Oct. 2013J.

[2] D. Karaca et al., "A 53–117 GHz LNA in 28-nm FDSOI CMOS," *IEEE Microw. Wireless Compon. Lett.* vol. 27, no. 2, pp. 171-173, Feb. 2017

[3] Y. Wang, T. -Y. Chiu, C. -C. Chien, W. -H. Tsai and H. Wang, "An E-Band High-Performance Variable Gain Low Noise Amplifier for Wireless Communications in 90-nm CMOS Process," *IEEE Microw.Wireless Compon. Lett.* vol. 32, no. 9, pp. 1095-1098, Sept. 2022.R. Nicole, "Title of paper with only first word capitalized," J. Name Stand. Abbrev., in press.

[4] M.-H. Li, Y. Wang, and H. Wang, "A 50–67 GHz ultralow-power LNA using double-transformer-coupling technique and self-resonant matching in 90 nm CMOS," *IEEE Microw. Wireless Compon. Lett.* vol. 32, no. 1, pp. 68-71, Jan. 2022.

[5] C.-J. Liang et al., "A 0.6 V VDD W-band neutralized differential low noise amplifier in 28 nm bulk CMOS," *IEEE Microw. Wireless Compon. Lett.* vol. 31, no. 5, pp. 481-484, May. 2021.

[6] V. Eren, P. Sakalas, and S. Michael, "A 5.9 mW E-/W-Band SiGe-HBT LNA with 48 GHz 3-dB bandwidth and 4.5 dB Noise Figure," *IEEE Microw. Wireless Compon. Lett.* vol. 32, no. 12, pp. 1451-1454, Dec. 2022.

[7] Y. Yu, H. Liu, Y. Wu, and K. Kang, "A 54.4–90 GHz low-noise amplifier in 65-nm CMOS," *IEEE J. Solid-State Circuits,* vol. 52, no. 11, pp.2892-2904, Nov. 2017.

A 1.28 mW K-Band Modified Gilbert-Cell Mixer Design in 22nm FDSOI CMOS

Adilet Dossanov
Institute for CMOS Design
Technical University of Braunschweig
Braunschweig, Germany
a.dossanov@tu-braunschweig.de

Vadim Issakov
Institute for CMOS Design
Technical University of Braunschweig
Braunschweig, Germany
v.issakov@tu-braunschweig.de

Abstract—This paper presents a low-power, low-voltage down-conversion mixer fabricated in a 22 nm FDSOI CMOS technology. The proposed mixer design is based on Gilbert-cell architecture and uses a passive transformer instead of an active transconductance stage to overcome voltage headroom limitations in deep-sub-micron CMOS technologies. Measurement results show that the mixer achieves a voltage conversion gain of 7.8 dB, an input-referred 1 dB compression point of -7 dBm, and IIP3 of 4.8 dBm under -8 dBm local oscillator power. The mixer design consumes a low 1.28 mW of power from a single 0.8 V supply voltage, which is a significant improvement compared to the state-of-the-art. Furthermore, the mixer's compact size of 0.54×0.45 mm², including pads, makes it a highly attractive solution for various applications, as e.g. radar.

Index Terms—FDSOI CMOS, down-conversion mixer, low power, low voltage, transformer, linearity.

I. INTRODUCTION

Down conversion mixers play a crucial role in the RF front-ends of wireless systems, including 5G transceivers, short-range automotive radar devices, and many other applications. The modern down-scaled CMOS technologies enable front-end realization at mm-wave frequencies. This is due to improved f_t and $f_{\max}$, which enable the integration of complex systems on the same chip area at a low cost.

On one hand, passive mixers offer such advantages as high linearity and zero DC power consumption, yet on the other hand, they suffer from conversion loss, which degrades overall system performance. Additionally, they require high local oscillator (LO) power, leading to increased DC power dissipation. However, in active mixers like a frequently used Gilbert-cell design [1], high conversion gain (CG), low noise, and good port-to-port isolation performances can be achieved at lower LO power, yet at the cost of lower linearity and DC power consumption. But, with deep sub-micron CMOS technology (e.g., 22 nm FDSOI process with 0.8 V), the Gilbert cell faces limitations due to stacked structures, which cause linearity degradation because of the limited rail-to-rail voltage headroom. To address this issue, several approaches have been proposed, such as employing folded mixer designs with current reuse shunt feedback as an RF stage [2] or using mixer designs with adaptive body effect control [3] to enhance the gain and noise performance.

In our previous work, we introduced a modified Gilbert-cell design that employs transmission lines as an RF stage and LO wave shaping to achieve high linearity and 1/f noise reduction [4], [5]. In this work, we build upon this idea and expand it by realizing the passive gm stage using a transformer as a RF stage to achieve a more compact design at the K frequency band with competitive performance in 22 nm FDSOI CMOS technology.

II. PROPOSED MIXER DESIGN

The schematic diagram of low-voltage down-conversion mixer design is illustrated in Fig. 1. The design uses two transformers to convert single-ended signals to differential and provide an ESD protection during measurement chip handling. Capacitors C_{T1} and C_{T2} are utilized to achieve matching at RF and LO inputs, while C_B acts as a DC block. LO inputs are biased using M_B diode-connected transistor and R_G resistors.

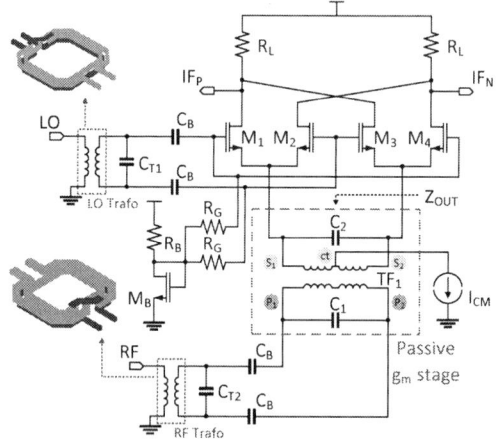

Fig. 1. Proposed low-voltage modified Gilbert-cell mixer design.

The mixer design is realized as a Gilbert cell with a modified RF stage. The transconductance stage of the mixer was removed and replaced with a transformer to mitigate the previously stated limitation in deep sub-micron CMOS technologies. The passive transconductance stage performs the RF voltage to current conversion. In addition, we used a low-voltage cascode current mirror I_{CM} (not shown in the design

979-8-3503-4331-1/24 $31.00 © 2024 IEEE   110   SiRF 2024

schematic) as a tail current source to improve the common mode, power supply rejection ratio, and avoid self-biasing effects.

The EM 3D layout of the passive transconductance stage transformer is illustrated in Fig. 2. The transformer windings are realized in the top thick metals available in the technology to minimize insertion loss. In order to have an accurate design, the transformer was simulated in RFpro ADS Momentum. The optimized EM simulation results show that the primary and secondary inductance at 30 GHz are 200 nH and 300 nH, respectively. Capacitors C_1 and C_2 are designed to achieve resonance at the operating frequency.

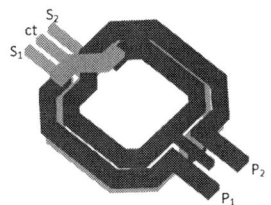

Fig. 2. 3D EM model of the transformer.

Small-signal simulation of the output impedance Z_{OUT} and the transconductance g_m of the passive transconductance stage are shown in Fig. 3. The presented curves shows that the output impedance is relatively high at the operating frequency, and the peak of the passive transconductance stage g_m at this frequency is around 16 mS.

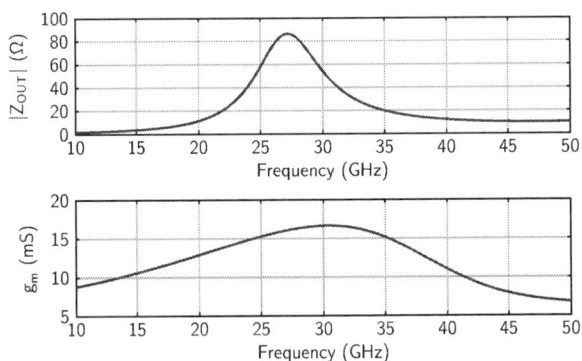

Fig. 3. Simulated output impedance and transconductance of the passive gm-stage over frequency.

The RF current is steered by LO switching quad M_{1-4} and converted to the voltage by the R_L load resistor. The switching quad transistor sizes are optimized to operate near the threshold voltage, which reduces the required driving LO power and noise contribution. For optimal performance, 16 fingers with a 2 μm width and minimum channel length were chosen for M_{1-4}, along with a load resistance of 400 Ω.

III. MEASUREMENT RESULTS

The proposed low-voltage mixer design was fabricated in the 22nm FDSOI CMOS technology from Global Foundries.

The chip microphotograph and its 3D layout are depicted in Fig. 4. The area of the mixer, including pads, occupies 540×450 μm. The mixer core consumes 1.28 mW power from a single 0.8 V supply voltage.

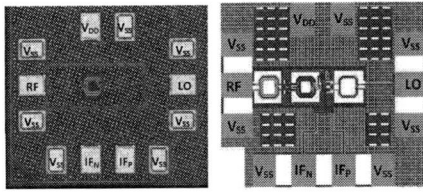

Fig. 4. The mixer chip photo and its 3D layout with pads (540μm × 450μm).

On-wafer S-parameter measurements were performed using MPI TS-200 THz probe station, a low noise Keysight E36312A DC supply, and a PNA Network Analyzer. As a calibration method, the short-open-load-thru (SOLT) technique was used. Fig. 5 shows the measured and simulated RF and LO input matching (S11) and isolation between these ports. The results indicate that the S11 values are lower than -10 dB, and isolation between RF and LO ports is around 50 dB at the operating frequency.

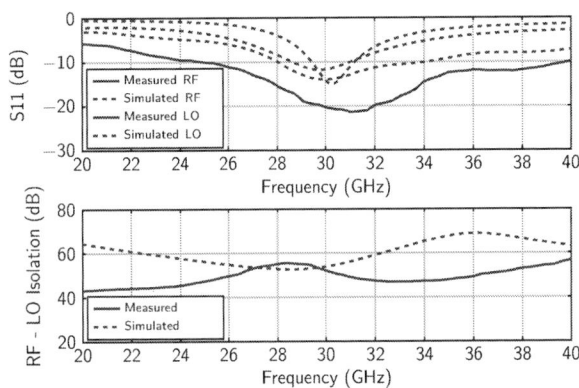

Fig. 5. Measured and simulated RF, LO return losses and isolation.

The mixer parameters were measured using an external buffer that can drive 50 Ω. Input RF and LO signals were fed by Keysight E8257D generators, and the output spectrum was observed by an R&S FSW85 spectrum analyzer.

The measured and simulated results of CG over LO power and IF frequency are shown in Fig. 6. After the de-embedding of cable losses, the measurement results show that the maximum CG can be achieved at -8 dBm LO power. However, the measured CG is slightly lower than the simulations due to losses on adapters and RF probes. Furthermore, the capacitive output of the external buffer leads to a decrease in IF bandwidth, which cannot be de-embedded.

The linearity measurements of the proposed mixer design are presented in Fig. 7. The measurements were performed with an LO power of -8 dBm. The input RF power of the mixer design at 1 dB compression point is -7 dBm, and the corresponding output power and CG are 0.6 dBm and 8.6 dB,

TABLE I
PERFORMANCE SUMMARY AND COMPARISON WITH STATE OF THE ART

Ref.	Technology	Supply	RF	CG	IP1dB	IIP3	PLO	SSB NF	PDC	Area	Topology
		(V)	(GHz)	(dB)	(dBm)	(dBm)	(dBm)	(dB)	(mW)	(mm²)	
This	22nm CMOS	0.8	24-34	7.8	-7	4.8	-8	16† / 12.3‡	1.28	0.243	Gilbert-cell + Passive g_m
[2]	130nm CMOS	1.5	23 - 25	26.1	-17.8	n.a.	-3	7.7	16.8	0.96	Folded Gilbert-cell + IF Buffer
[3]	65nm CMOS	1	26	7.2	-6.1	-2.5	5	12.3	10.3	0.4	Gilbert-cell DB + IF Buffer
[6]	22nm CMOS	1.2	25 - 31	12	-14.8	n.a	0	7-12	25	0.64	Single-balanced + IF Buffer

†Measured NF at 26GHz RF/LO frequency, ‡ Simulated NF at 30 GHz RF/LO frequency.

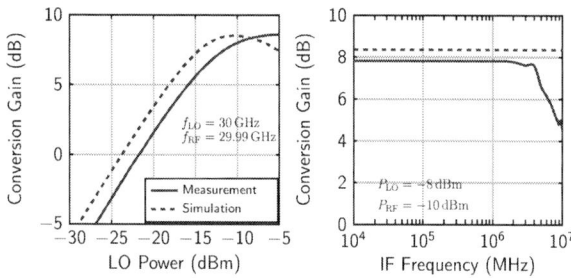

Fig. 6. Measured and simulated voltage CG versus LO power, and IF frequency.

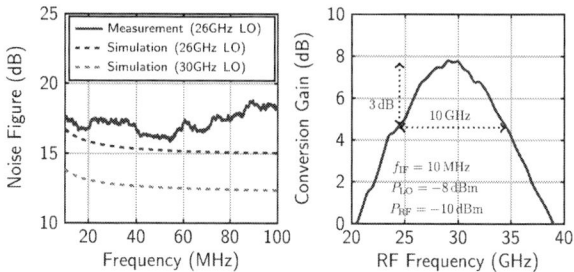

Fig. 8. Noise figure and conversion gain of the mixer.

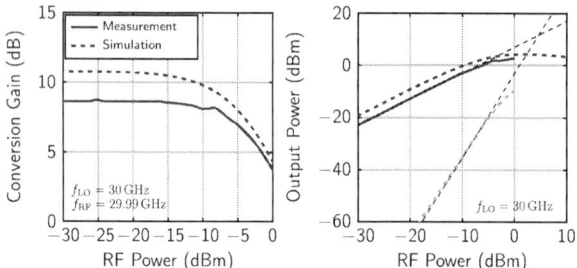

Fig. 7. Measured and simulated IP1dB and IIP3.

respectively. The measured input IP3 is about 4.8 dBm. The linearity of the measurement is higher than the simulated one, due to the lower CG of the fabricated mixer.

The single-sideband (SSB) noise figure has been measured at 26 GHz RF frequency and 26.01 GHz LO frequency by using Agilent 364C (10 MHz - 26.5 GHz) noise source and the spectrum analyzer. The 26 GHz RF frequency was chosen due to equipment limitations. Fig. 8 shows the measured and simulated results of the noise figure, which is 16 dB at an IF of 50 MHz. The mixer demonstrates the 3-dB bandwidth range from 24 to 34 GHz measured at constant IF at 10 MHz.

The measured parameters of the proposed mixer design are summarized and compared with other state-of-the-art mixer designs in Table I. The presented mixer design achieves low power consumption and high linearity at low LO power.

IV. CONCLUSION

The paper presents a modified Gilbert-cell down conversion mixer design, which operates at a low voltage and is fabricated

in the 22 nm FDSOI CMOS technology. Using a passive transformer as an RF stage, the mixer is not only compact in size but also it has a very low power consumption of 1.28 mW, making it a highly suitable option for low-power applications compared to other state-of-the-art mixers.

ACKNOWLEDGMENT

We thank Global Foundries for providing silicon fabrication through the 22FDX university program.

REFERENCES

[1] V. Issakov, H. Knapp, M. Tiebout, A. Thiede, W. Simburger and L. Maurer, "Comparison of 24 GHz low-noise mixers in CMOS and SiGe:C technologies," 2009 European Microwave Integrated Circuits Conference (EuMIC), Rome, Italy, 2009, pp. 184-187.

[2] Y. Peng et al., "A K -Band High-Gain and Low-Noise Folded CMOS Mixer Using Current-Reuse and Cross-Coupled Techniques," in IEEE Access, vol. 7, pp. 133218-133226, 2019, doi: 10.1109/AC-CESS.2019.2941048.

[3] B. Bae and J. Han, "24–40 GHz Gain-Boosted Wideband CMOS Down-Conversion Mixer Employing Body-Effect Control for 5G NR Applications," in IEEE Transactions on Circuits and Systems II: Express Briefs, vol. 69, no. 3, pp. 1034-1038, March 2022, doi: 10.1109/TC-SII.2021.3119995.

[4] R. Ciocoveanu, R. Weigel, A. Hagelauer and V. Issakov, "Bias-switched down-conversion mixer for flicker noise reduction in 28-nm CMOS," 2018 Texas Symposium on Wireless and Microwave Circuits and Systems (WMCS), Waco, TX, USA, 2018, pp. 1-4, doi: 10.1109/WM-CaS.2018.8400634.

[5] R. Ciocoveanu, R. Weigel, A. Hagelauer and V. Issakov, "Modified Gilbert-Cell Mixer With an LO Waveform Shaper and Switched Gate-Biasing for 1/f Noise Reduction in 28-nm CMOS," in IEEE Transactions on Circuits and Systems II: Express Briefs, vol. 66, no. 10, pp. 1688-1692, Oct. 2019, doi: 10.1109/TCSII.2019.2923595.

[6] P. V. Testa, L. Szilagyi, X. Xu, C. Carta and F. Ellinger, "A Low-Power Low-Voltage Down-Conversion Mixer for 5G Applications at 28 GHz in 22-nm FD-SOI CMOS Technology," 2020 IEEE Asia-Pacific Microwave Conference (APMC), Hong Kong, Hong Kong, 2020, pp. 911-913, doi: 10.1109/APMC47863.2020.9331695.

AUTHOR INDEX

A

Aksoyak, İbrahim Kağan . . . . . `Mo3C-1` 29
        `Mo3C-2` 33

B

Bösch, Wolfgang . . . . `IF01-23` 102
Boumaiza, Slim . . . . `Mo2C-2` 17
Boutry, Herve . . . . `Mo4C-3` 48
Breun, Sascha . . . . `Tu1B-4` 67
Bücher, Thomas . . . . `Mo3C-3` 37

C

Cao, Zhibo . . . . `Mo4C-2` 45
Carlson, John . . . . `Mo4C-1` NA
Carta, Corrado . . . . `Mo4C-2` 45
Chandra Prabhu, Arjith . . . . . `Mo3C-3` 37
Chang, Hong-Yeh . . . . `Tu3C-1` 79
        `IF01-25` 106
Chang, Jin-Fa . . . . `Tu1B-1` 56
Chen, Jun-Liang . . . . `Tu3C-1` 79
Chen, Po-Yuan . . . . `Tu3C-1` 79
Chi, Taiyun . . . . `Mo1C-1` NA
Choi, Yongbae . . . . `Mo2C-3` 21
Cuskelly, Lachlan . . . . `Tu1B-3` 63

D

De Filippi, Guglielmo . . . . `Mo2C-4` 25
Dossanov, Adilet . . . . `IF01-30` 110
Dubarry, Christophe . . . . `Mo4C-3` 48
Duriez, Blandine . . . . `Mo4C-3` 48
Dürrwald, Franz Alwin . . . . . . `Mo3C-4` 41

E

Ebrahimi, Najme . . . . `Mo2C-1` 13
Eckel, Selina . . . . `Tu1C-2` 71
Ellinger, Frank . . . . `Mo3C-4` 41
        `Tu3C-2` 83
Engelmann, Andre . . . . `Tu1C-3` 75
        `Tu3C-3` 87

F

Falt, Christopher . . . . `Tu1B-3` 63
Ferrari, Philippe . . . . `Mo4C-4` 52

G

Gaillard, Fred . . . . `Mo4C-3` 48
Ginsburg, Brian . . . . `Tu1C-1` NA
Grzyb, Janusz . . . . `Mo3C-3` 37

H

Haag, Alexander . . . . `Mo1C-2` 1
Hazer Sahlabadi, Mehran . . . . `Mo2C-2` 17
Hesselbarth, Jan . . . . `Tu3C-2` 83
Hetterle, Philip . . . . `Tu3C-3` 87
Hillger, Philipp . . . . `Mo3C-3` 37
Hoyer, Christian . . . . `Mo3C-4` 41

Hsieh, Yi-Ping . . . . `Tu3C-5` 95
Huang, Guo-Wei . . . . `IF01-2` 98
Huang, Rou-Yin . . . . `IF01-25` 106

I

Issakov, Vadim . . . . `IF01-30` 110

J

Jang, Sheng-Lyang . . . . `Tu3C-5` 95
Javadi, Sara . . . . `IF01-23` 102

K

Kaynak, Mehmet . . . . `Mo4C-2` 45
Kim, Junghyun . . . . `Mo2C-3` 21
Kim, Sunghyuk . . . . `Mo2C-3` 21
Knapp, Herbert . . . . `Tu3C-4` 91
Ko, Byunghun . . . . `Mo2C-3` 21
Koch, Manuel . . . . `Tu1B-4` 67
Kraus, Isabel . . . . `Tu3C-4` 91

L

Lai, Wen-Cheng . . . . `Tu3C-5` 95
Lee, Jehwan . . . . `Mo2C-3` 21
Leitner, Michael . . . . `IF01-23` 102
Lin, Chung-Yo . . . . `IF01-2` 98
Lin, Yo-Sheng . . . . `Tu1B-1` 56
Lucci, Luca . . . . `Mo4C-3` 48
Lugo-Alvarez, Jose . . . . `Mo4C-4` 52

M

Mazzanti, Andrea . . . . `Mo2C-4` 25
Medbouhi, Mohammed . . . . . . `Mo4C-4` 52
Meister, Tilo . . . . `Mo3C-4` 41
        `Tu3C-2` 83
Meng, Chinchun . . . . `IF01-2` 98
Möck, Matthias . . . . `Mo3C-1` 29
        `Mo3C-2` 33
Mohan, Ankush . . . . `Mo4C-1` NA
Morvan, Erwan . . . . `Mo4C-4` 52

N

Ng, Yuen-Sum . . . . `Mo1C-3` 5

O

Oliviera, Alexandre . . . . `Mo4C-3` 48

P

Papurcu, Hakan . . . . `Mo1C-4` 9
Pfeiffer, Ullrich . . . . `Mo3C-3` 37
Piotto, Lorenzo . . . . `Mo2C-4` 25
Pirbazari, Mahmoud M. . . . . . `Mo2C-4` 25
Pohl, Nils . . . . `Mo1C-4` 9
        `Tu3C-4` 91
Probst, Florian . . . . `Tu1C-3` 75
        `Tu3C-3` 87
Protze, Florian . . . . `Mo3C-4` 41

R

Rezaee, Behrooz . . . . `IF01-23` 102
Romstadt, Justin . . . . `Mo1C-4` 9
Rücker, Holger . . . . `Mo3C-3` 37

S

Schvan, Peter . . . . `Tu1B-3` 63
Seeholzer, Tina . . . . `Mo4C-1` NA
Seo, Wonwoo . . . . `Mo2C-3` 21
Sharifi, Hasan . . . . `Mo4C-1` NA
Sim, Taejoo . . . . `Mo2C-3` 21
Sodhi, Avantika . . . . `Mo4C-1` NA
Stadler, Manfred . . . . `IF01-23` 102
Stadler, Pascal . . . . `Mo1C-4` 9
Steinweg, Luca . . . . `Mo3C-4` 41
Sterzl, Georg . . . . `Tu3C-2` 83
Su, Yu-Chia . . . . `IF01-25` 106

T

Tsai, Tsung-Ching . . . . `Tu1B-2` 60
Tu, Clayton . . . . `Mo4C-1` NA

U

Ulusoy, Ahmet Çağrı . . . . `Mo1C-2` 1
        `Mo3C-1` 29
        `Mo3C-2` 33
        `Tu1B-2` 60
        `Tu1C-2` 71

V

Valenta, Václav . . . . `Tu1B-2` 60
Valorge, Olivier . . . . `Mo4C-3` 48
Voss, Thomas . . . . `Mo4C-2` 45

W

Wang, Huei . . . . `Mo1C-3` 5
Wang, Yunshan . . . . `Mo1C-3` 5
Weigel, Robert . . . . `Tu1B-4` 67
        `Tu1C-3` 75
        `Tu3C-3` 87
Wietstruck, Matthias . . . . `Mo4C-2` 45

X

Xia, Jingjing . . . . `Mo2C-2` 17

Y

Yu, Hang . . . . `Mo2C-2` 17

Z

Zhu, Yu . . . . `Tu3C-2` 83